Titles in This Series

Volume

Titles in This Series

Volume

35 **Four-manifold theory,** Cameron Gordon and Robion Kirby, Editors

36 **Group actions on manifolds,** Reinhard Schultz, Editor

37 **Conference on algebraic topology in honor of Peter Hilton,** Renzo Piccinini and Denis Sjerve, Editors

38 **Topics in complex analysis,** Dorothy Browne Shaffer, Editor

39 **Errett Bishop: Reflections on him and his research,** Murray Rosenblatt, Editor

40 **Integral bases for affine Lie algebras and their universal enveloping algebras,** David Mitzman

41 **Particle systems, random media and large deviations,** Richard Durrett, Editor

42 **Classical real analysis,** Daniel Waterman, Editor

43 **Group actions on rings,** Susan Montgomery, Editor

44 **Combinatorial methods in topology and algebraic geometry,** John R. Harper and Richard Mandelbaum, Editors

45 **Finite groups–coming of age,** John McKay, Editor

46 **Structure of the standard modules for the affine Lie algebra $A_1^{(1)}$,** James Lepowsky and Mirko Primc

47 **Linear algebra and its role in systems theory,** Richard A. Brualdi, David H. Carlson, Biswa Nath Datta, Charles R. Johnson, and Robert J. Plemmons, Editors

48 **Analytic functions of one complex variable,** Chung-chun Yang and Chi-tai Chuang, Editors

49 **Complex differential geometry and nonlinear differential equations,** Yum-Tong Siu, Editor

50 **Random matrices and their applications,** Joel E. Cohen, Harry Kesten, and Charles M. Newman, Editors

51 **Nonlinear problems in geometry,** Dennis M. DeTurck, Editor

52 **Geometry of normed linear spaces,** R. G. Bartle, N. T. Peck, A. L. Peressini, and J. J. Uhl, Editors

53 **The Selberg trace formula and related topics,** Dennis A. Hejhal, Peter Sarnak, and Audrey Anne Terras, Editors

54 **Differential analysis and infinite dimensional spaces,** Kondagunta Sundaresan and Srinivasa Swaminathan, Editors

55 **Applications of algebraic K-theory to algebraic geometry and number theory,** Spencer J. Bloch, R. Keith Dennis, Eric M. Friedlander, and Michael R. Stein, Editors

56 **Multiparameter bifurcation theory,** Martin Golubitsky and John Guckenheimer, Editors

57 **Combinatorics and ordered sets,** Ivan Rival, Editor

58.I **The Lefschetz centennial conference. Part I: Proceedings on algebraic geometry,** D. Sundararaman, Editor

58.II **The Lefschetz centennial conference. Part II: Proceedings on algebraic topology,** S. Gitler, Editor

58.III **The Lefschetz centennial conference. Part III: Proceedings on differential equations,** A. Verjovsky, Editor

59 **Function estimates,** J. S. Marron, Editor

60 **Nonstrictly hyperbolic conservation laws,** Barbara Lee Keyfitz and Herbert C. Kranzer, Editors

61 **Residues and traces of differential forms via Hochschild homology,** Joseph Lipman

62 **Operator algebras and mathematical physics,** Palle E. T. Jorgensen and Paul S. Muhly, Editors

63 **Integral geometry,** Robert L. Bryant, Victor Guillemin, Sigurdur Helgason, and R. O. Wells, Jr., Editors

64 **The legacy of Sonya Kovalevskaya,** Linda Keen, Editor

65 **Logic and combinatorics,** Stephen G. Simpson, Editor

66 **Free group rings,** Narian Gupta

67 **Current trends in arithmetical algebraic geometry,** Kenneth A. Ribet, Editor

68 **Differential geometry: The interface between pure and applied mathematics,**

Titles in This Series

Volume

Mladen Luksic, Clyde Martin, and William Shadwick, Editors

69 **Methods and applications of mathematical logic,** Walter A. Carnielli and Luiz Paulo de Alcantara, Editors

70 **Index theory of elliptic operators, foliations, and operator algebras,** Jerome Kaminker, Kenneth C. Millett, and Claude Schochet, Editors

71 **Mathematics and general relativity,** James A. Isenberg, Editor

72 **Fixed point theory and its applications,** R. F. Brown, Editor

73 **Geometry of random motion,** Rick Durrett and Mark A. Pinsky, Editors

74 **Geometry of group representations,** William M. Goldman and Andy R. Magid, Editors

75 **The finite calculus associated with Bessel functions,** Frank M. Cholewinski

76 **The structure of finite algebras,** David C. Hobby and Ralph Mckenzie

77 **Number theory and its applications in China,** Wang Yuan, Yang Chung-chun, and Pan Chengbiao, Editors

78 **Braids,** Joan S. Birman and Anatoly Libgober, Editors

79 **Regular differential forms,** Ernst Kunz and Rolf Waldi

80 **Statistical inference from stochastic processes,** N. U. Prabhu, Editor

81 **Hamiltonian dynamical systems,** Kenneth R. Meyer and Donald G. Saari, Editors

82 **Classical groups and related topics,** Alexander J. Hahn, Donald G. James, and Zhe-xian Wan, Editors

83 **Algebraic K-theory and algebraic number theory,** Michael R. Stein and R. Keith Dennis, Editors

84 **Partition problems in topology,** Stevo Todorcevic

85 **Banach space theory,** Bor-Luh Lin, Editor

86 **Representation theory and number theory in connection with the local Langlands conjecture,** J. Ritter, Editor

87 **Abelian group theory,** Laszlo Fuchs, Rüdiger Göbel, and Phillip Schultz, Editors

88 **Invariant theory,** R. Fossum, W. Haboush, M. Hochster, and V. Lakshmibai, Editors

89 **Graphs and algorithms,** R. Bruce Richter, Editor

90 **Singularities,** Richard Randell, Editor

Singularities

CONTEMPORARY MATHEMATICS

Volume 90

Singularities

Proceedings of the IMA Participating Institutions Conference held July 28–August 1, 1986, with support from the Participating Institutions of the Institute for Mathematics and its Applications and the University of Iowa

Richard Randell, Editor

AMERICAN MATHEMATICAL SOCIETY

Providence · Rhode Island

The IMA Participating Institutions Conference on Singularities was held July 28–August 1, 1986 with support from the Participating Institutions of the Institute for Mathematics and Its Applications and the University of Iowa.

1980 *Mathematics Subject Classification* (1985 *Revision*). Primary 14B05, 32B30, 32C40, 57R45.

Library of Congress Cataloging-in-Publication Data

IMA Participating Institutions Conference (1986: University of Iowa)

Singularities; proceedings of the IMA Participating Institutions Conference held July 28–August 1, 1986, with support from the participating institutions of the Institute for Mathematics and Its Applications and the University of Iowa.

(Contemporary mathematics, ISSN 0271-4132; v. 90)

Bibliography: p.

ISBN 0-8218-5096-2 (alk. paper)

1. Singularities (Mathematics)–Congresses. I. Randell, Richard. II. Institute for Mathematics and Its Applications. III. University of Iowa. IV. Title. V. Series: Contemporary mathematics (American Mathematical Society); 90.

QA614.58.I52 1986 516.3′5 89-6662

Printed in the United States of America.
This volume was printed directly from author-prepared copy.
The paper used in this book is acid-free and falls within the guidelines established to ensure permanence and durability. ♾

CONTENTS

LIST OF PARTICIPANTS

Reynir Axelsson
Science Institute
University of Iceland
Raunvisindastofnun Haskolans
Dunhaga 3, Is-107
Reykjavik, Iceland

P. Cartier
Çentre de Mathématiques
École Polytèchnique
91128 Palaiseau Cedex
France

James Damon
Department of Mathematics
University of North Carolina
Chapel Hill, NC 27514

Peter Donelan
Department of Mathematics
Victoria University of Wellington
Private Bag
Wellington, New Zealand

Alan Durfee
Department of Mathematics
Mount Holyoke College
South Hadley, MA 01075

Michael Falk
Department of Mathematics
Northern Arizona University
Flagstaff, AZ 86011

Toshizumi Fukui
Department of Mathematics
Faculty of Science
Tokyo Metropolitan University
2-1-1 Fukazawa, Setagaya-ku
Tokyo, 158 Japan

Yih-nan Gau
Department of Mathematics
SUNY at Geneseo
Geneseo, NY 14454

Herwig Hauser
Department of Mathematics
Universität Innsbruck
Innsbruck A-6020, Austria

Fumio Hidaka
Department of Mathematics
Hokkaido College of Seiishu University
Bibai 079-01, Japan

Per Holm
Mathematik Institut
Universitetet Oslo
Blindern
Oslo 3, Norway

Wu-Chung Hsiang
Department of Mathematics
Princeton University
Princeton, NJ 08540

Shihoko Ishii
Division of Mathematics
Sangyo-Ika University
Yahata-Nishi-ku
Kita-Kyushi City, Japan

Michel Jambu
Department of Mathematics
University of Wuhan
Wuhan, People's Republic of China

Hajime Kaji
1609-304, Shiomidai
Isogoku, Yokohama
Kanagawa 235, Japan

Mark Kannowski
Department of Mathematics
De Pauw University
Greencastle, IN 46135

Mitsuyoshi Kato
1752-121-1-44 Meinohama
Nishi-ku
Fukuoka 814, Japan

Toshitake Kohno
Department of Mathematics
Faculty of Science
Nagoya University
Chikusa-ku
Nagoya 464, Japan

Henry Laufer
Department of Mathematics
SUNY at Stony Brook
Long Island, NY 11794

Dũng Tràng Lê
Centre de Mathématique
de L'École Polytèchnique
91128 Palaiseau, France

Daniel Leborgne
Institut de Mathématiques et d'Informatique
Université de Nantes
2 chemin de la Houssiniere
44072 Nantes-Cedex, France

Joseph Lipman
Department of Mathematics
Purdue University
West Lafayette, IN 47907

Jon Magnusson
Science Institute
University of Iceland
Raunvisindastofnun Haskolans
Dunhaga 3, Is-107
Reykjavik, Iceland

David Mond
Mathematics Institute
University of Warwick
Coventry CV47AL, Great Britain

Donal O'Shea
Department of Mathematics
Mount Holyoke College
South Hadley, MA 01075

Kyungho Oh
1037 Mathematical Sciences
Purdue University
West Lafayette, IN 47907

Mutsuo Oka
Department of Mathematics
Tokyo Institute of Technology
Oh-okayama, Meguro
Tokyo 152, Japan

Peter Orlik
Department of Mathematics
University of Madison
Madison, WI 53706

W. Pardon
Department of Mathematics
Duke University
Durham, NC 27706

George F. Paulik
Department of Mathematics
Purdue University
West Lafayette, IN 47907

G. R. Pellikaan
Department of Mathematics
Vrye Universiteit
De Boelelaan 10011
Amsterdam, The Netherlands

Richard Randell
Department of Mathematics
University of Iowa
Iowa City, IA 52242

Oswald Riemenschneider
Mathematisches Seminar
Universität Hamburg
Bundesstrasse 55
2000 Hamburg 13, West Germany

Dennis Roseman
Department of Mathematics
University of Iowa
Iowa City, IA 52242

Kyoji Saito
Research Institute for
Mathematical Sciences
Kyoto University
Kyoto 606, Japan

Julius Shaneson
Department of Mathematics
Rutgers University
New Brunswick, NJ 08903

Dirk Siersma
Mathematics Institute
Rijksuniversiteit Utrecht
Budapestlaan 6, Box 80.010
3508 TA Utrecht, The
Netherlands

Jonathan Simon
Department of Mathematics
University of Iowa
Iowa City, IA 52242

Andrew Sommese
Department of Mathematics
University of Notre Dame
Notre Dame, IN 46556

Steven Sperber
Institute for Advanced Study
Princeton, NJ 08540

J. Steenbrink
Rijksuniversiteit Leiden
Neils Bohrweg 1
Leiden, The Netherlands

Tatsuo Suwa
Department of Mathematics
Hokkaido University
Sapporo 060, Japan

Susan Szczepanski
Department of Mathematics
Lehigh University
Christmas-Saucon Hall 14
Bethlehem, PA 18015

Hiroaki Terao
International Christian
University
10-2, Osawa 3-Chome
Mitaka-shi
Tokyo 181, Japan

Tadashi Tomaru
726-1 Hassaki Hokkitsu Seta
Gunma, 377 Japan

Günther Trautmann
Department of Mathematics
Universität Kaiserslautern
Postfach 3049
6750 Kaiserslautern, West Germany

Tohsuki Urabe
Department of Mathematics
Faculty of Science
Tokyo Metropolitan University
2-1-1 Fukazawa, Setagaya-ku
Tokyo, 158 Japan

Philip Wagreich
Department of Mathematics
University of Illinois at Chicago
Chicago, IL 60680

Kimio Watanabe
Institute of Mathematics
University of Tsukuba
Sakura-mura Niihari-gun
Ibaraki, 305 Japan

Masayuke Watanabe
1-28-6 Otone Maebashi
Gunma, 371 Japan

Stephen S.T. Yau
Department of Mathematics, Statistics and Computer Sciences
University of Illinois at Chicago
322 Science & Engineering Offices
Box 4348
Chicago, IL 60680

Günter M. Ziegler
Massachusetts Institute of Technology
2-251
Cambridge, MA 02139

PREFACE

The IMA Participating Institutions Conference on Singularities was held on the campus of the University of Iowa in Iowa City during the week of July 28-August 1, 1986, just prior to the International Congress of Mathematicians in Berkeley. The fifty-two participants from eleven countries heard numerous talks and participated in lively informal discussions. Many of the talks are represented by the papers in this volume. All except the problem list are original research papers; I thank the referees for their prompt and careful help.

The financial support of the following is gratefully acknowledged: The Participating Institutions of the Institute for Mathematics and its Applications, Minneapolis; The University of Iowa Mathematics Department; and the University of Iowa Graduate College. I would like to express my appreciation to Joe Lipman, the co-organizer of the conference, and to members of the organizing committee Peter Orlik and Phil Wagreich for their help and support. Thanks are due as well to Michael Falk, Mark Kannowski, Linda Randell, Dennis Roseman, and Jonathan Simon, for their valuable help the week of the meeting.

And finally, special thanks are due to the participants for their delightful combination of scholarly inquiry and cheerful conviviality which makes singularity theory such a pleasant area of research.

Dick Randell
Iowa City

Contemporary Mathematics
Volume **90**, 1989

GENERICALLY AMPLE DIVISORS ON NORMAL GORENSTEIN SURFACES

Marco Andreatta and Andrew John Sommese

Let (S,L) be a pair consisting of a nef and big line bundle, L, on an irreducible normal Gorenstein surface S. In this article we will carry over the structure theory developed by the second author in [So1] to such pairs. For us, the most interesting application of the results is to study surfaces S which are the minimal desingularizations, $\varphi : S \to \varphi(S) \subset \mathbf{P}_\mathbf{C}$, of an irreducible surface in projective space, with L being the pullback of $\mathcal{O}_\mathbf{P}(1)$ to S. Nevertheless no real simplification would occur by considering only such smooth S, since even if we start with such a smooth S, Gorenstein surfaces will occur in both the statements and proofs of the results.

The results in [So1] were for ample line bundles L. If L is merely nef, it is easy to see that the curves C on S that satisfy $L\cdot C = 0$ cause the theorems in [So1] to fail. In §1 we consider the contraction of various classes of smooth rational curves on S that satisfy $L\cdot C = 0$. Let $p : \overline{S} \to S$ denote the desingularization of S which is minimal in the sense that there are no smooth rational curves E on $\overline{S}$ with $E\cdot E = -1$, and $p(E)$ a point. If S is Gorenstein, then $p^*\omega_S \cong \omega_{\overline{S}} + \Delta$ where ω_X denotes the dualizing sheaf of a normal space X, and Δ is a divisor whose support equals exactly the fibres over the nonrational points of S. We say that the pair (S,L) with L nef and big is *a-minimal* if there are no smooth rational

0271-4132/89 $1.00 + $.25 per page

curves E on $\overline{S} - \Delta$ with $E \cdot E = -1$ and $p^*L \cdot E = 0$. We say that (S,L) is *c-minimal* if it is a-minimal and if there are no smooth rational curves E on $\overline{S}$ with $p^*L \cdot E = 0$ and either

a) $\Delta \cdot E = 1$ and $E \cdot E = -1$,

or

b) $\Delta \cdot E = 0$, $E \cdot E = -2$ and $p(E)$ is not a point.

The following result summarizes the key points about the concepts of a-minimality and c-minimality. This result is a straightforward consequence of results of Sakai [Sa3].

REDUCTION THEOREM. *Let (S,L) be a pair consisting of a nef and big line bundle, L, on a normal irreducible Gorenstein surface S. For $i = a$ or c there exists a nef and big line bundle, L_i, on a normal Gorenstein i-minimal surface S_i, and a bimeromorphic holomorphic map φ_i: $S \to S_i$ with $L \cong \varphi_i{}^*L'$ such that the positive dimensional fibres of φ_i consist of rational curves. Further*

a) *if $h^0(\omega_S{}^n \otimes L^n) \neq 0$ for some $n > 0$ then $\omega_{Sa} \otimes L_a$ is nef, $\omega_{Sc} \otimes L_c{}^k$ is ample for $k \geq 2$, and (S_i,L_i) is uniquely determined by (S,L),*

b) *if $h^0(\omega_S{}^n \otimes L^n) = 0$ for all $n > 0$, then (S_a,L_a) is either $(\mathbf{P}_{\mathbf{C}}{}^2, \mathcal{O}_{\mathbf{P}}(e))$ where $e = 1$ or 2, or (S_a,L_a) is $(\mathbf{Q}, \mathcal{O}_{\mathbf{P}^2}(1)|_{\mathbf{Q}})$, a quadric surface in $\mathbf{P}^3$, or the minimal desingularization $\pi : \mathbf{Q} \to \mathbf{Q}$ of the singular quadric in $\mathbf{P}^3$ with $L_a = \pi^* \mathcal{O}_{\mathbf{P}^2}(1)|_{\mathbf{Q}}$, or (S_a,L_a) is geometrically ruled.*

One notable corollary of this is that given a nef and big line bundle L on a normal irreducible c-minimal Gorenstein surface, S, it follows that S is projective. The classification of pairs (S,L) when S is a-minimal proceeds in §2, exactly as in [Sol]. Here is the main result.

(2.7) MAIN THEOREM. *Let (S,L) be an a-minimal generically polarized Gorenstein surface. Assume that $h^0(\omega_S{}^n \otimes L^n) \neq 0$ for some $n > 0$. Then there is an $N > 0$ such that $(\omega_S \otimes L)^N$ is spanned by* *global sections and such that the map* $\varphi : S \to \mathbf{P}_C$ *associated to* $\Gamma((\omega_S \otimes L)^N)$ *has connected fibres and a normal image. The map* $\varphi : S \to S'$ *can be factorized* $\varphi = \varphi' o \varphi_c$, *where* $\varphi_C : S \to S_c$ *is the map in the reduction theorem above,* (S_c, L_c) *is c-minimal and* $\varphi' : S_c \to S'$ *is the map associated to* $\Gamma((\omega_{Sc} \otimes L_c)^N)$. *One of the following holds:*

i) *If* dim $\varphi(S) = 0$ *then* (S_c, L_c) *is a Gorenstein Del Pezzo surface.*

ii) *If* dim $\varphi(S) = 1$ *then* (S_c, L_c) *is a conic bundle and* $\varphi : S \to \mathbf{P}_C$ *is the projection in the definition of conic bundle.*

iii) *If* dim $\varphi(S) = 2$, *then the image* $S' = \varphi(S)$ *is a normal Gorenstein surface.* $L_{S-\varphi^{-1}(F)}$ *extends to a nef and big line bundle* L' *on* S' *such that* $\varphi^*(\omega_{S'} \otimes L') \cong \omega_S \otimes L$ *and* $\omega_{S'} \otimes L'$ *is ample on* S'.

The first author would like to thank Consiglio Nazionale delle Ricerche, and the second author would like to thank the National Science Foundation (DMS 8420 315) for their support. We would like to thank the University of Notre Dame for its support.

§0. BACKGROUND MATERIAL

We consider complex analytic normal surfaces with at most *Gorenstein singularities.* We recall that for a normal surface S, the Gorenstein condition is equivalent to the dualizing sheaf, that we denote by ω_S, being invertible. For the most part we follow the notation of [Sol].

Let $\pi : \bar{S} \to S$ denote the desingularization of S, that is minimal in the sense that the fibres of π contain no smooth rational curves E satisfying $E \cdot E = -1$. If D is a (Weil) divisor we will denote by $\bar{D}$ the proper transform of D. If L is a line bundle we will denote by $\bar{L}$ the inverse image, π^*L except when $L = \omega_S$ since this would lead to a confusion of bars. Note that if $C \in |L|$ it is not necessarily true that $\bar{C} \in |\bar{L}|$. We use without further comment the equality $L \cdot D = \bar{L} \cdot \bar{D}$ where D is a Weil divisor on S. We have the following fundamental facts (see [Sa1]).

(0.1) $\pi^*\omega_S \cong \omega_{\bar{S}} + \Delta$ where Δ is an effective divisor on $\bar{S}$.

(0.2) $\omega_S \cdot C = (\omega_{\bar{S}} + \Delta) \cdot \bar{C} \geqslant \omega_{\bar{S}} \cdot \bar{C}$

(0.3) $\Delta \cdot \bar{C} = 0$ if and only if C meets at worst rational singularities.

(0.4) DEFINITION. A line bundle L on S is said to be *big* if $L \cdot L > 0$ and it is said to be *numerically effective*, or *nef* for short, if $L \cdot C \geqslant 0$ for each irreducible curve C on S.

The proof of the following lemma is straightforward and left to the reader.

(0.4.1) LEMMA. *Let $\mathfrak{L}$ be a nef and big line bundle on a normal compact complex surface, then $h^0(\mathfrak{L}^n) > c \cdot n^2$ for some $c > 0$ and all $n >> 0$. In particular S is Moishezon and has a projective desingularization.*

(0.5) Since S is Cohen-Macaulay and ω_S is invertible, Serre duality for a locally free sheaf E on S takes the usual form $h^i(S,E) = h^{2-i}(S,\omega_S \otimes E^*)$ or $i = 0,1,2$.

The Kodaira-Ramanujam vanishing theorem takes its usual form (see §0 of [So1] for discussion and references).

(0.6) THEOREM. *Let L be a numerically effective and big line bundle on S. Then $h^i(S,\omega_S \otimes L) = 0 = h^{2-i}(S,L^{-1})$ for $i \geqslant 1$. In particular if $C \in |L|$, then C is connected.*

(0.7) We will need some projectivity criteria for surfaces obtained by contracting sets of curves. First we have Nakai's criterion for projectivity of a normal compact complex surface.

(0.7.1) NAKAI CRITERION. *Let L be a line bundle on a normal compact complex surface S. If $L \cdot L > 0$ and $L \cdot C > 0$ for all irreducible curves on S then L is ample, and in particular S is projective.*

PROOF. We follow the usual proof with the minor modifications needed in our case. Replacing L by L^n for a sufficiently large n, we can assume by (0.4.1), that $h^0(L) > 0$. Thus choosing a $C \in |L|$ we have for all $k > 0$ the sequence

$$0 \to L^k \to L^{k+1} \to L_C{}^{k+1} \to 0.$$

Since $L \cdot C' > 0$ for every irreducible component of C it follows that L_C is ample and that $h^1(L_C{}^{k+1}) = 0$ for $k >> 0$. Therefore $h^1(L^k) \geqslant h^1(L^{k+1})$ for all $k >> 0$ implying that $\Gamma(L^{k+1}) \to \Gamma(L_C{}^{k+1}) \to 0$ for all $k >> 0$. Since L_C is ample we conclude from this that L^k is spanned for all large k. From this and the hypothesis we conclude that the map associated to L^k for large k is finite to one, and therefore that L is ample. □

(0.7.2) COROLLARY. *Let S be a normal projective surface and let $\varphi : S \to S'$ be a holomorphic generically one to one surjection of S onto a normal compact complex surface S'. If the images of the positive dimensional fibres of φ are rational double points, then S' is projective.*

PROOF. Let L be a very ample line bundle on S. Let C be a generic element of $|L|$. Since C is generic element of $|L|$, it can be assumed that C is irreducible and doesn't meet the singular set of S. By the rational double point hypothesis it follows that $N \cdot \varphi(C)$ is Cartier for some $N > 0$. By a direct application of the Nakai's criterion (0.8.1) we conclude that $[N \cdot \varphi(C)]$ is ample and S' is projective. □

(0.8) A pair (S,L) consisting of an ample (nef and big) line bundle L on a compact normal surface S is called a *(generically) polarized surface.*

The following classes of special polarized varieties play a key role in this paper.

a) (S,L) is called a *quadric* if S is biholomorphic to a possibly singular quadric $\mathbf{Q} \subset \mathbf{P}^3$ and L is isomorphic to the restriction of the hyperplane bundle, $\mathcal{O}_{\mathbf{P}^3}(1)$, to $\mathbf{Q}$.

b) (S,L) is called a *geometrically ruled surface* if S is a holomorphic $\mathbf{P}^1$ bundle, $p\colon S \to R$, over a nonsingular curve R and the restriction L_f of L to a fibre f of p is $\mathcal{O}_f(1)$.

c) (S,L) is called a *Gorenstein-Del Pezzo surface* if L is ample and $\omega_S^{-1} \cong L$ (see [Br] for a classification of these surfaces).

d) (S,L) is called a *conic bundle* if there is a holomorphic surjection, $p\colon S \to R$, with connected fibres onto a smooth curve R with the properties that L is relatively ample and that $(\omega_S \otimes L)^k \cong p^*\mathfrak{L}$ for some $k > 0$ and some very ample line bundle $\mathfrak{L}$ on R. The reader should note that the general fibre f of the map p : $S \to R$ is a smooth rational curve with $L \cdot f = 2$.

§1. SOME TECHNICAL RESULTS

We assume from now on that L is a *nef* and *big* line bundle on a normal Gorenstein surface S, i.e., that (S,L) is a generically polarized Gorenstein surface. As pointed out in (0.4.1), S is Moishezon.

The following simple example shows how the set

$$\mathcal{U} = \{\text{irreducible curves } C \subset S \text{ such that } L \cdot C = 0\}$$

being nonempty causes the theorems of [So1] on the nefness and ampleness of the line bundles $\omega_S \otimes L$, $\omega_S \otimes L^2$, and $\omega_S \otimes L^3$ to fail.

Let S be the smooth surface obtained by blowing up a point x in $\mathbf{P}^2$, $\tau : S \to \mathbf{P}^2$, and $L = \tau^* \mathcal{O}_{\mathbf{P}}(1)$. L is nef and big and $(S,L) \neq (\mathbf{P}^2, \mathcal{O}(1))$, but $\omega_S \otimes L^i$ is not nef for every i as can be seen by restriction to the exceptional divisor $\tau^{-1}(x)$.

Fortunately the general situation is not much worse than this example. By contracting naturally given finite sets of curves $\subset \mathcal{U}$, we obtain a new Gorenstein surface S' for which the hoped for theorems are true. In this section we will define these sets of curves.

As in §0, let $\pi : \bar{S} \to S$ be the minimal desingularization of S and let $\bar{L} = \pi^*L$. By hypothesis $\bar{S}$ is a smooth projective surface and $\bar{L}$ is nef and big.

(1.1) DEFINITIONS. A smooth rational curve C on $\bar{S}$ is said to be a -1 *curve* if $C \cdot C = -1$; it is said to be a -2 *curve* if $C \cdot C = -2$.

We denote the set of irreducible curves $C \subset \bar{S}$, such that $\bar{L} \cdot C = 0$ by $\bar{\mathcal{U}}$. By a standard argument

(1.2) The sets $\mathcal{U}$ and $\bar{\mathcal{U}}$ are finite.

(1.3) LEMMA. *Consider the set*

$$\mathcal{U}_1 = \{C \in \mathcal{U} \mid \bar{C} \text{ is a -1 curve and } \bar{C} \subset \bar{S} - \Delta\}.$$

The curves in $\mathcal{U}_1$ *are finite, disjoint, and can be contracted to give smooth points on a normal Gorenstein projective surface* S', $\varphi\colon S \to S'$. *Moreover the line bundle* L *descends to a unique nef and big line bundle* L' *on* S' *such that* $\varphi^*L' = L$.

PROOF. This is fairly classical. The first statement follows from a standard fact, Theorem (0.1) of [Li+So], about surface singularities that the second author first learned from H. Laufer. Note that the contractibility hypothesis of that result was only needed to insure negative definiteness of the set of curves $\mathcal{U}_1$, which in our case is an immediate consequence of the Hodge index theorem and the fact that $\bar{L} \cdot E = 0$ for all $E \in \mathcal{U}_1$.

Since the map $\varphi : S \to S'$ resolves only rational singularities, the second statement follows immediately from the fact that $\bar{L} \cdot E = 0$ for all $E \in \mathcal{U}_1$. □

(1.4) LEMMA. *Consider the set*

$$\mathcal{U}_2 = \{C \in \mathcal{U} \mid \bar{C} \subset \bar{S} - \Delta \text{ and } \bar{C} \text{ is a -2 curve}\}.$$

The finite set of curves in $\mathcal{U}_2$ *can be contracted to give (Gorenstein) rational double points in a Gorenstein normal projective surface* S', ψ: $S \to S'$. *The line bundle* L *descends to a unique nef and big line bundle* L' *on* S' *such that* $\psi^*L' = L$ *and* $\psi^*\omega_{S'} = \omega_S$.

PROOF. Let S'' denote $\bar{S}$ with Δ contracted and let p: $S'' \to S$ denote the induced map to S. Denote by $\mathcal{U}_2''$ the proper transforms of elements of $\mathcal{U}_2$ by p. Let denote the union of $\mathcal{U}_2''$ and the curves in the positive dimensional fibres of p. Since $p^*L \cdot C = 0$ for $C \in \mathcal{Y}$, it follows by the Hodge index theorem and Grauert's contraction theorem that $\mathcal{Y}$ can be contracted to give a normal surface S' with induced map $\varphi : S \to S'$. Since the proper transforms of the elements of $\mathcal{U}_2$ in $\bar{S}$ don't meet Δ it follows that all the curves $C \in \mathcal{Y}$ are -2 curves; it is well known that these curves in $\mathcal{Y}$ contract to give rational double points. The projectivity is immediate from (0.8.3). □

(1.5) LEMMA. *The curves in the set*

$$\mathcal{U}_3 = \{C \in \mathcal{U} \mid \bar{C} \cdot \Delta = 1, \bar{C} \text{ is a -1 curve}\}$$

are finite and can be blown down to give a normal Gorenstein surface S', $\eta : S \to S'$.

PROOF. Let $\mathcal{F}$ denote the set of the irreducible curves $C \in \bar{S}$ that are in the fibres of π and let $\bar{\mathcal{U}}_3$ denote the set of proper transforms of irreducible curves $C \in \mathcal{U}_3$.

Since $\bar{L} \cdot C = 0$ for $C \in \bar{\mathcal{U}}$ and since $\bar{L} \cdot \bar{L} > 0$, it follows from the Hodge index theorem that any subset of curves of $\bar{\mathcal{U}}$ has a negative definite intersection matrix. Since $\mathcal{F} \cup \bar{\mathcal{U}}_3 \subset \bar{\mathcal{U}}$ it follows by Grauert's contraction theorem that $\mathcal{F} \cup \bar{\mathcal{U}}_3$ can be contracted, q: $\bar{S} \to S'$, to give a normal surface S'. Since $\mathcal{F}$, the fibres of π, are among the curves contracted, it follows that there is a map, η: $S \to S'$, so that $q = \eta \circ \pi$. It is immediate that the map η is precisely the contraction of the curves in $\mathcal{U}_3$.

It only remains to check that S' is Gorenstein. Let $y \in S'$ be an arbitrary singular point of S' such that η doesn't give a biholomorphism of a neighborhood of y and a neighborhood of $\eta^{-1}(y)$. Since S and S' are normal, it follows from Zariski's main theorem that $\Phi = \eta^{-1}(y)$ is positive dimensional. By the definition of η, we know that there is an irrational singularity, $x \in S$, with $\eta(x) = y$. By Theorem (3.1) of [So1], it follows that there are global holmorphic sections s_k of $\omega_S \otimes L^k$ for $k = 1$ and 2 with $s_k(x) \neq 0$ for $k = 1,2$.

CLAIM. For $k = 1$ and 2, s_k trivializes $\omega_S \otimes L^k$ in a neighborhood of Φ.

PROOF OF THE CLAIM. Since $s_k(x) \neq 0$, it suffices to show $(\omega_S \otimes L^k)\cdot C = 0$ for all of the $C \in \mathcal{U}_3$ such that $x \in C$. This will follow if we show that $[\pi^*(\omega_S \otimes L^k)]\cdot \overline{C} = 0$ where $\overline{C}$ is the proper transform of C. Since $\overline{L}\cdot\overline{C} = L\cdot C = 0$, it suffices to show that $\pi^*\omega_S\cdot\overline{C} = 0$. Since $\pi^*\omega_S = \omega_{\overline{S}} + \Delta$, and since $\Delta\cdot\overline{C} = 1$ by hypothesis, it suffices to show that $\omega_{\overline{S}}\cdot\overline{C} = -1$. But this is an immediate consequence of the hypothesis that $\overline{C}$ is a smooth rational curve with self intersection -1. □

From the claim it follows that $L = (\omega_S \otimes L^2) \otimes (\omega_S \otimes L)^{-1}$ and hence $\omega_S = (\omega_S \otimes L) \otimes L^{-1}$ are trivial in a neighborhood U of Φ. From this it follows that for some open neighborhood V of y, $\omega_{\mathrm{reg}(V)}$ is trivial. From the fact that ω_V is the direct image of $\omega_{\mathrm{reg}(V)}$ under the inclusion of reg(V) in V it will follow that ω_V is free and hence that y is a Gorenstein point of S'. □

(1.5.1) LEMMA. *Let $\eta : S \to S'$ be the contraction of* (1.5). *There is a nef and big bundle L' on S' such that $L \cong \eta^*L'$. Further $\omega_S = \eta^*\omega_{S'}$.*

PROOF. Note the proof of (1.5) showed that ω_S and L are trivial in some neighborhood of any fibre Φ of η. From this it is straightforward to conclude that $L \cong \eta^*L'$ for a nef and big line bundle L' on S' and $\omega_S \cong \eta^*\Omega$ for some line bundle Ω on S'. Since Ω agrees with $\omega_{S'}$ on the Zariski open set $Z \subset S'$ such that η is a biholomorphism from $\eta^{-1}(Z)$ to Z and since $\omega_{S'}$ is the direct image of ω_Z under the inclusion $Z \subset S'$, it follows that $\Omega \cong \omega_{S'}$. □

(1.6) DEFINITION. Let (S,L) a generically polarized Gorenstein surface. (S,L) is

a-minimal if there are no -1 curves C in $\bar{S} - \Delta$ such that $\bar{L}\cdot C = 0$.

b-minimal if it is a-minimal and there are no curves C in S that $L\cdot C = 0$ and its strict transform, $\bar{C}$, in $\bar{S}$ is a -2 curve with $\Delta\cdot\bar{C} = 0$.

c-minimal if it is *b*-minimal and there are no -1 curves C in $\bar{S}$ such that $\bar{L}\cdot C = 0$ and $\Delta\cdot C = 1$.

(1.7) PROPOSITION. *Given any generically polarized Gorenstein surface (S,L) there exist a-minimal, b-minimal, c-minimal generically polarized Gorenstein surfaces (S_i,L_i) and maps φ_i: $S \to S_i$ each obtained by contracting a finite number of curves in $\mathcal{U} = \{C \subset S \mid L\cdot C = 0\}$ such that (S_i,L_i) is i-minimal. There is a commutative diagram of holomorphic maps.*

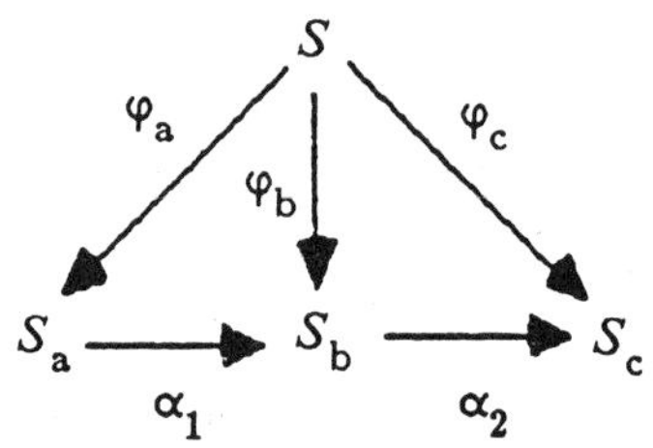

PROOF. It follows by applying Lemma (1.3) a finite number of times, then Lemma (1.4) and finally Lemma (1.5). □

(1.7.1) REMARK. It follows from (0.8.3) that S_a and S_b are projective if S is projective. It will be seen in the next section that the S_c is projective. From the above results it follows that there are nef and big line bundles L_a, L_b, L_c on the spaces S_a, S_b, S_c

respectively, such that $L_a \cong \alpha_1{}^*L_2$, $L_b \cong \alpha_2{}^*L_c$, and $L \cong \varphi_i{}^*L_i$ for all i. Further for all $k \geqslant 0$, the above maps give canonical isomorphisms between $\Gamma(\omega_S \otimes L^k)$, $\Gamma(\omega_a \otimes L_a{}^k)$, $\Gamma(\omega_b \otimes L_b{}^k)$, and $\Gamma(\omega_c \otimes L_c{}^k)$ where ω_i denotes the dualizing sheaf of S_i.

The following result is a slight extension of a result given by [Ko+Oc] in the smooth case and with L ample.

(1.8) LEMMA. *Let (S,L) a generically polarized Gorenstein surface. If $\omega_S \cong L^{-3}$ then $(S,L) \cong (\mathbf{P}^2, \mathcal{O}_{\mathbf{P}^2}(1))$. If $\omega_S \cong L^{-2}$ then either $(S,L) \cong (\mathbf{Q}, \mathcal{O}_{\mathbf{P}^2}(1)|_{\mathbf{Q}})$, a quadratic surface in $\mathbf{P}^3$, or the minimal desingularization $\pi : \mathbf{Q} \to \mathbf{Q}$ of the singular quadric in $\mathbf{P}^3$ with $L = \pi^* \mathcal{O}_{\mathbf{P}^2}(1)|_{\mathbf{Q}}$.*

PROOF. If $\omega_S \cong L^{-3}$ (or L^{-2}) then S has at most rational singularities by proposition (0.2.2) of [So2]. Suppose then the set $\mathcal{U} = \{\text{curve } C \subset S \text{ such that } L\cdot C = 0\}$ is not empty. By our hypothesis, using (0.2) and (0.3) and the fact that there are only rational singularities, we have $\omega_S\cdot C = \omega_{\overline{S}}\cdot\overline{C} = 0$ for $C \in \mathcal{U}$. Moreover, by the Hodge Index theorem, we have that $\overline{C}\cdot\overline{C} < 0$ and thus that the $\overline{C}$'s are -2 curves.

By Lemma (1.4) we can contract all these -2 curves to a generically polarized Gorenstein surface with only rational points, $\varphi: (S,L) \to (S',L')$, such that $\varphi^*L' = L$. It is easy to see, by the Nakai criterion, that L' is ample. Using (0.5), (0.6) and (0.7) with the same argument as works in the case when S' is smooth, L' is ample, it is straightforward to prove that if $\omega_S \cong L^{-3}$ then $\omega_{S'} \cong L'^{-3}$ and $(S',L') \cong (\mathbf{P}^2, \mathcal{O}_{\mathbf{P}^2}(1))$. Since $\mathbf{P}^2$ is smooth the set $\mathcal{U}$ was empty and therefore $(S,L) \cong (\mathbf{P}^2, \mathcal{O}_{\mathbf{P}^2}(1))$.

If instead $\omega_S \cong L^{-2}$ then $\omega_{S'} \cong L'^{-2}$ and $(S',L') \cong (\mathbf{Q}, \mathcal{O}_{\mathbf{Q}}(1))$. This gives the conclusion. □

§2. NUMERICAL EFFECTIVITY AND AMPLENESS OF $\omega_S \otimes L$

In this section, (S,L) will be a generically polarized normal Gorenstein surface in the sense of definition (0.9).

We will recover, under the milder hypothesis we have on L, the same kind of results as in [So1]. For the most part, the proofs proceed as in [So1]. For this reason we will give only the new parts of these proofs, referring to [So1] where they can be read without any change.

(2.1) PROPOSITION. *Let (S,L) be a generically polarized Gorenstein surface. Suppose that (S,L) is a-minimal in the sense of (1.6) and not $\cong$ $(\mathbf{P}^2, \mathcal{O}_{\mathbf{P}^2}(1))$. Then $\omega_S \otimes L^2$ is numerically effective and $h^0(\omega_S \otimes L^2) \neq 0$. In particular, if (S,L) is a-minimal and not $\cong$ $(\mathbf{P}^2, \mathcal{O}_{\mathbf{P}^2}(1))$, then $\omega_S \otimes L^3$ is nef and big.*

PROOF. Suppose there exists an irreducible curve C on S such that $(\omega_S + 2L)\cdot C < 0$; then by (0.2) $(\omega_{\bar{S}} + 2\bar{L})\cdot\bar{C} < 0$. We consider the two following separate cases:

a) $L\cdot C = \bar{L}\cdot\bar{C} = 0$ b) $L\cdot C = \bar{L}\cdot\bar{C} > 0$.

In case a), by the Hodge Index theorem we have $\bar{C}\cdot\bar{C} < 0$ and by our assumption $\omega_{\bar{S}}\cdot\bar{C} < 0$ and $\omega_S\cdot C < 0$. By the adjunction formula $\bar{C}$ is a -1 curve and by (0.2)

$$0 > \omega_S\cdot C \geqslant \omega_{\bar{S}}\cdot\bar{C} = -1.$$

Thus $\omega_S\cdot C = \omega_{\bar{S}}\cdot\bar{C}$ and $\Delta\cdot\bar{C} = 0$, i.e., -1 curve contained in $\bar{S} - \Delta$. By our hypothesis this can not happen.

In the case b), $L\cdot C = \bar{L}\cdot\bar{C} > 0$, the proof goes as in the Lemma (1.1) of [So1], using Lemma (1.8) instead of Lemma (0.7.1) of [So1].

(2.2) PROPOSITION. *Let (S,L) be a generically polarized Gorenstein surface and suppose that it is c-minimal. If there exists an irreducible curve C on S such that $(\omega_S + 2L)\cdot C = 0$ then either (S,L) is a quadric or it is geometrically ruled. In the latter case C is a smooth* *fibre of the map $p: S \to R$ giving the ruling and $\omega_S \otimes L^2 = p^*\mathfrak{L}$ for some ample line bundle $\mathfrak{L}$ on R.*

PROOF. By the hypothesis we have that $\pi^*(\omega_S + 2L)\cdot C = 0$ and therefore by (0.2) that $(\omega_{\bar{S}} + 2\bar{L})\cdot\bar{C} \leq 0$.

Suppose first $L\cdot C = \bar{L}\cdot\bar{C} = 0$ and thus $\omega_S\cdot C = 0$, $\omega_{\bar{S}}\cdot\bar{C} \leq 0$ and $\bar{C}\cdot\bar{C} < 0$. By the adjunction formula the two following cases are possible.

i) $\omega_S\cdot C = \omega_{\bar{S}}\cdot\bar{C} = \Delta\cdot\bar{C} = 0$ and $\bar{C}\cdot\bar{C} = -2$.
ii) $\omega_{\bar{S}}\cdot\bar{C} = \bar{C}\cdot\bar{C} = -\Delta\cdot\bar{C} = 1$.

Since by hypothesis $\bar{S}$ does not contain such curves, we can thus assume $L\cdot C = \bar{L}\cdot\bar{C} \geq 1$.

Using the proof of Theorem (1.2) of [So1], with our lemma (1.1) instead of his Lemma (0.7.1), we conclude that either (S,L) is a quadric or that $L\cdot C = 1$ and there is a map $p: S \to R$ from S to a curve R such that p is a $\mathbf{P}^1$ bundle near C with C a smooth fibre and $(\omega_S \otimes L^2)^N = p^*\mathfrak{L}$ for some $N > 0$ and some ample line bundle $\mathfrak{L}$ on R.

Suppose there exists a singular fibre $f = \Sigma_{j\in J}\alpha_j F_j$. By the equalities

$$0 = f\cdot F_i = F_i\cdot\Sigma\,\alpha_j F_j = \alpha_i F_i^2 + \Sigma_{j\neq i}\alpha_j F_j\cdot F_i$$

we get that $F_i\cdot F_i < 0$ for each i and by $1 = L\cdot C = L\cdot f = L\cdot\Sigma\,\alpha_j F_j$ that $L\cdot F_k \geq 1$ for at least one $k \in J$.

Moreover by $(\omega_S \otimes L^2)^N = p^*\mathfrak{L}$ we have $0 > (\omega_S + 2L)\cdot f = (\omega_S + 2L)\cdot\Sigma\,\alpha_j F_j$. Thus, by the nefness of $(\omega_S + 2L)$, $(\omega_S + 2L)\cdot F_j = 0$ for each i. This leads to the absurdity

$$\omega_S\cdot F_k + F_k\cdot F_k = -2\,L\cdot F_k + F_k\cdot F_k < -2.$$

This proves that S is geometrically ruled. □

(2.3) COROLLARY. *Let (S,L) be a generically polarized Gorenstein surface and suppose it is c-minimal. Then either:*

a) $(\omega_S \otimes L^2)$ *is ample,*
b) (S,L) *is* $(\mathbf{P}^2, \mathcal{O}_{\mathbf{P}^2}(1))$ *or a quadric,*
c) (S,L) *is geometrically ruled and* $\omega_S \otimes L^2 = p^*\mathfrak{L}$ *for some ample line bundle* $\mathfrak{L}$ *on* R *where* $p: S \to R$ *gives the ruling.*

PROOF. This follows immediately from (2.1) and (2.2), using Nakai criterion (0.8.1). □

(2.4) COROLLARY. *Let* (S,L) *be a generically polarized Gorenstein surface and suppose it is c-minimal. Then* S *is projective.*

PROOF. In case a) of Corollary (2.3), $\omega_S \otimes L^2$ ample and S is therefore projective by the Nakai criterion (0.8.3). In case b) or c), S is trivially projective. □

(2.5) THEOREM. *Let* (S,L) *be a generically polarized Gorenstein surface and suppose it is a-minimal. The following are equivalent:*

a) (S,L) *is neither geometrically ruled nor a quadric nor the minimal resolution of a quadric, nor equal to* $(\mathbf{P}^2, \mathcal{O}_{\mathbf{P}^2}(e))$ *for* $e = 1$ *or* 2,
b) $(\omega_S \otimes L)$ *is numerically effective.*
c) $h^0((\omega_S \otimes L)^N) \neq 0$ *for some* $N > 0$.

PROOF. To prove the equivalence a) $\Leftrightarrow$ b) note that we can suppose (S,L) to be c-minimal. In fact, let $\varphi_c : S \to S_c$ be the map given in (1.7) where (S_c,L_c) is c-minimal. Then $\omega_S \otimes L$ is nef if and only if $\omega_{S_c} \otimes L_{S_c}$ is nef and (S_c,L_c) is as in a) if and only if (S,L) is.

By the above observation we can apply Corollary (2.3); with this the proof is the same as that of Theorem (1.3) in [So1]. □

(2.6) DEFINITION. Let (S,L) be a generically polarized surface and A denotes the set of positive integer n such that $h^0((\omega_S \otimes L)^n) > 0$. For each n in A let $\varphi_n : S \to \mathbf{P}_\mathbf{C}$ denote the meromorphic map associated to $\Gamma((\omega_S \otimes L)^n)$. The *logarithmic Kodaira dimension* of the pair (S,L) denoted by $\ln(S,L)$, is defined as follows

$\ln(S,L) = -\infty$ if A is empty and

$\ln(S,L) = \max_{n\in A}\dim \varphi_n(S)$ otherwise.

The generically polarized Gorenstein surfaces with $\ln(S,L) = -\infty$ were classified in Theorem (2.5).

(2.7) MAIN THEOREM. *Let (S,L) be an a-minimal generically polarized Gorenstein surface. Assume that $h^0(\omega_S{}^n \otimes L^n) \neq 0$ for some $n > 0$. Then there is an $N > 0$ such that $(\omega_S \otimes L)^N$ is spanned by global sections and such that the map $\varphi : S \to \mathbf{P}_C$ associated to $\Gamma((\omega_S \otimes L)^N)$ has connected fibres and a normal image. The map $\varphi : S \to S'$ can be factorized $\varphi = \varphi' o \varphi_c$, where $\varphi_c : S \to S_c$ is the map in the reduction theorem above, (S_c,L_c) is c-minimal and $\varphi': S_c \to S'$ is the map associated to $\Gamma((\omega_{Sc} \otimes L_c)^N)$. One of the following holds:*

i) *If* dim $\varphi(S) = 0$ *then (S_c,L_c) is a Gorenstein Del Pezzo surface.*

ii) *If* dim $\varphi(S) = 1$ *then (S_c,L_c) is a conic bundle and $\varphi : S \to \mathbf{P}_c$ is the projection in the definition of conic bundle.*

iii) *If* dim $\varphi(S) = 2$, *then the image $S' = \varphi(S)$ is a normal Gorenstein surface. $L_{S-\varphi^{-1}(F)}$ extends to a nef and big line bundle L' on S' such that $\varphi^*(\omega_{S'} \otimes L') \cong \omega_S \otimes L$ and $\omega_{S'} \otimes L'$ is ample on S'.*

PROOF. Except for iii), the proof is exactly as the one of the main theorem in [So1], the nefness of $\omega_S \otimes L$ being insured by our hypothesis of *a*-minimality and our Theorem (2.3), or by Theorem (7.4) of [Sa3].

To prove that point we first observe that each fibre of the map $\varphi_c : S \to S_c$ is contained in a fibre of the map $\varphi : S \to S'$ and that $\varphi^*(\omega_{Sc} \otimes L_c') \cong \omega_S \otimes L$. This is clear since, as we have seen in Chapter 1, every irreducible component of a fibre of the map φ_c: $S \to S_c$ is a curve C either in the set $\mathcal{U}_2$ or in the set $\mathcal{U}_3$ (see Lemmas (1.4), (1.5) and in both cases it is easy to compute that $(\omega_S \otimes L)\cdot C = 0$.

We can thus suppose that S is *c*-minimal. We will prove that each positive dimensional fibre of φ: $S \to S'$ is an irreducible curve C biholomorphic to $\mathbf{P}^1$ such that $L\cdot C = 1$ and such that $\varphi(C)$ is a smooth point of S'. Since L is algebraic L_{S-C} extends over the smooth puncture on S' to give a nef and big line bundle; the rest of the theorem is now trivial.

Let now C be an irreducible curve in S such that $(\omega_S \otimes L)\cdot C = 0$. If $L\cdot C = \bar{L}\cdot\bar{C} = 0$ then $\omega_{\bar{S}}\cdot\bar{C} \leqslant \omega_S\cdot C \leqslant 0$ and $\bar{C}\cdot\bar{C} < 0$ by the Hodge index theorem.

That implies, by the adjunction formula and (0.1), (0.2), that either $\bar{C}$ is a -2 curve such that $\Delta\cdot\bar{C} = 0$ or $\bar{C}$ is a -1 curve such that $\Delta\cdot\bar{C} = 1$; i.e., C is either in the set $\mathcal{U}_2$ or in the set $\mathcal{U}_3$; that is impossible since S is c-minimal.

Therefore we can suppose $L\cdot C = \bar{L}\cdot\bar{C} > 0$. In this case $\omega_{\bar{S}}\cdot\bar{C} \leqslant \omega_S\cdot C \leqslant 1$.

The hypothesis dim $\varphi(S) = 2$ implies

$$(\omega_S \otimes L)\cdot(\omega_S \otimes L) = \pi^*(\omega_S \otimes L)\cdot\pi^*(\omega_S \otimes L) > 0.$$

Since $\pi^*(\omega_S \otimes L)\cdot\bar{C} = 0$, it follows by the index theorem and what is above that:

$$\bar{C}\cdot\bar{C} = -1 = \omega_{\bar{S}}\cdot\bar{C} = \omega_S\cdot C \quad\text{and}\quad L\cdot C = \bar{L}\cdot\bar{C} = 1.$$

This in particular implies that $\bar{C}$ is a (-1) curve contained in $\bar{S} - \Delta$. The conclusion follows from Lemma (1.3) using the nef and big line bundle $(\omega_S \otimes L)$ instead of L. □

(2.8) THEOREM. *Let (S,L) be a generically polarized Gorenstein surface. Assume that L is spanned by global sections and the genus, $g(L)$ of a smooth $C \in |L|$ equals $h^{1,0}(S)$. Then* $\ln(S,L) = -\infty$ *and the associated a-minimal pair of (S,L) is classified in Theorem* (2.5a).

PROOF. Consider the sequence

$$0 \to \omega_S \to \omega_S \otimes L \to \omega_C \to 0.$$

Since $h^1(\omega_S \otimes L) = 0$ by the Kodaira-Ramanujam vanishing theorem (0.6), we conclude that

$$\Gamma(\omega_C) \to H^1(\omega_S) \to 0.$$

By the hypothesis that $g(L) = h^{1,0}(S)$ we conclude that $\Gamma(\omega_C) \cong H^1(\omega_S)$. Since C can be chosen to be a general element of $|L|$, we conclude that the map $\Gamma(\omega_S \otimes L) \to \Gamma(\omega_C)$ can be the 0 map only if $\Gamma(\omega_S \otimes L) = 0$. Therefore we conclude that S has only rational singularities by Corollary (0.2.2) of [So2]. From this it is easy to see that

$$p^*: \Gamma((\omega_S \otimes L)^n) \to \Gamma((\omega_{\bar{S}} \otimes \bar{L})^n)$$

is an isomorphism where $p : \bar{S} \to S$ is a desingularization of S. Therefore we can assume that S is smooth without loss of generality.

If $g(L) = 0$, then $(\omega_S + L)\cdot L < 0$ and therefore $h^0((\omega_S \otimes L)^n) = 0$ for all $n > 0$ and hence $\ln(S,L) = -\infty$.

Therefore we can assume that $h^{1,0}(S) > 0$. We know that the Albanese mapping $\alpha : S \to R$ has a one dimensional image since $h^2(\omega_S) < h^0(\omega_S \otimes L) = 0$. The general fibre, f, if α is smooth and connected. Being a general fibre of a map between smooth varieties we also know that $\omega_{S,f} \cong \omega_f$. If we can show that the $L\cdot f = 1$ then we will be done. This will follow because L_f being spanned and $\deg(L_f) = 1$ for a irreducible curve f, imply that f is a smooth $\mathbf{P}_{\mathbf{C}}^1$. From this we conclude that

$$\deg(\omega_S \otimes L)_f = \deg(\omega_S) + \deg(L_f) = \deg(\omega_f) + \deg(L_f) = -2 + 1 < 0.$$

To see that $L\cdot f = 1$ it suffices to show that α_C gives a biholomorphism between C and R. Since the genus of C equals $h^{1,0}(S)$, the genus of R, this is obvious by the Hurwitz formula if $h^{1,0}(S) > 1$. If $h^{1,0}(S) = 1$, then the Hurwitz formula tells us that α_C is an unbranched cover. We also know from the vanishing theorem (0.6) that elements of $|L|$ are connected. Using these two facts the argument of the third paragraph of the proof of Theorem (3.6) of [So2] applies without change to our situation here. □

Finally we would like to call attention to [Sa4] where the stable results in [So1] are generalized to pairs (S,L) where L is an ample line bundle on a Q-Gorenstein surface.

REFERENCES

[Br] L. Brenton, On singular complex surfaces with negative canonical bundle, with applications to singular compactifications of C^2 and to three dimensional rational singularities, Math. Ann. **248** (1980), 117-124.

[Ko+Oc] S. Kobayashi and T. Ochiai, Characterization of the complex projective space and hyperquadrics, J. Math. Kyoto Univ. **13** (1972), 31-47.

[Li+So] J. Lipman and A.J. Sommese, On Blowing Down Projective Spaces in Singular Varieties, Jour. für die reine und ange. Math. **362** (1985), 52-62.

[Sa1] F. Sakai, Curves with trivial dualizing sheaf on algebraic surfaces, Amer. J. Math. **104** (1982), 1217-1231.

[Sa2] F. Sakai, Enriques classification of normal Gorenstein surfaces, Amer. J. Math. **104** (1982), 1233-1241.

[Sa3] F. Sakai, Weil divisors on normal surfaces, Duke Math. J. **51** (1984), 877-887.

[Sa4] F. Sakai, Ample Cartier divisors on normal surfaces, J. für reine und angew. Math. **366** (1986), 121-128.

[So1] A.J. Sommese, Ample divisors on normal Gorenstein surfaces, Abh. Math. Sem. Univ. Hamburg **55** (1985), 151-170.

[So2] Andrew J. Sommese, On the adjunction theoretic structure of projective varieties, *Proceedings of the Complex Analysis and Algebraic Geometry Conference* (ed. by H. Grauert), Göttingen, 1985, Springer Lecture Notes, **1194** (1986), 175-213.

Marco Andreatta
DIPARTIMENTO DI MATEMATICA "F. ENRIQUES"
DELL'UNIVERSITA MILAN
VIA C. SALDINI, 50
1-20133 MILANO, ITALIA

Andrew J. Sommese
DEPARTMENT OF MATHEMATICS
UNIVERSITY OF NOTRE DAME
NOTRE DAME, INDIANA 46556
U.S.A.

Contemporary Mathematics
Volume **90**, 1989

TOPOLOGICAL EQUIVALENCE FOR NONISOLATED SINGULARITIES AND GLOBAL AFFINE HYPERSURFACES

James Damon

In this paper, we shall derive sufficient conditions that germs defining nonisolated singularities $f_0 : k^n,0 \to k,0$ or affine varieties $f_0^{-1}(0)$ defined by global polynomial mappings $f_0 : k^n \to k$ are topologically equivalent (here the germs may be real analytic if $k = \mathbb{R}$ or holomorphic if $k = \mathbb{C}$). The conditions will be given in terms of the Newton polyhedra and Newton filtration associated to the germs or polynomials. These results will follow from several basic results (Theorems 1, 2, and 3 of §3) which give sufficient conditions that a deformation of a germ defining a nonisolated singularity is topologically trivial. They generalize results in [DG] for isolated singularities.

When $k = \mathbb{C}$, we will conclude that germs defining nonisolated singularities which are non-degenerate (in a sense to be defined) are determined up to topological type by their Newton polyhedrons. This extends a result of Kouchnirenko [K] to the nonisolated case. This notion of non-degenerate places a restriction on the Newton polyhedron; however, without any restriction on the Newton polyhedron we also obtain the above results under the weak notion of "topological equivalence off of S", where S is the generic

*Partially supported by a Grant from the National Science Foundation.

0271-4132/89 $1.00 + $.25 per page

singular set for the Newton polyhedron. As a corollary, this implies in the complex case that the Milnor fibration is a topological invariant (this was obtained by Oka [O] using a different method). Likewise, we deduce in the complex case that if a polynomial f_0 has a global Newton polyhedron satisfying a certain property (see §6) then f_0 nondegenerate implies the global topological type of $f_0^{-1}(0)$ is determined by its Newton polyhedron.

Originally the result on the topological type of f_0 was announced [D2] without any restriction on the Newton polyhedron. While it still appears that this is true, our method of proof forces us to either restrict the Newton polyhedron or the form of topological equivalence in the conclusion. Still, the restrictions are sufficiently mild that the results are widely applicable. They apply to germs which are still of infinite codimension in the sense of Siersma [S] or Pellikaan [P] and polynomials which are not tame in the sense of Broughton [B].

1. NEWTON POLYHEDRON AND FILTRATION FOR NONISOLATED SINGULARITIES

Let $f_0 : k^n,0 \to k,0$ be a holomorphic germ if $k = \mathbb{C}$ or an analytic germ if $k = \mathbb{R}$. If f_0 has a Taylor expansion $\Sigma c_\alpha x^\alpha$ relative to a fixed set of coordinates $\mathbf{x} = (x_1, \ldots, x_n)$, we define, as in [K], $\Gamma_+(f_0)$ to be the convex hull of the set $\cup\{\alpha + \mathbb{R}^n_+ : c_\alpha \neq 0\}$. Here $\mathbb{R}^n_+ = \{(x_1, \ldots, x_n) \in \mathbb{R}^n : x_i \geq 0, \text{ all } i\}$. We let $\Gamma(f_0)$ denote the boundary of $\Gamma_+(f_0)$ minus those non-compact faces lying in a coordinate hyperplane. We also denote the union of the compact faces of $\Gamma(f_0)$ by $\Gamma_0(f_0)$. Then, $\Gamma(f_0)$ is called the Newton polyhedron of f_0. For example, for $f_0 = x^2y^2 + x^2z^2 + y^2z^2$, $\Gamma(f_0)$ has four 2-dimensional faces as shown in (1.1) and $\Gamma_0(f_0)$ consists of the triangle (and its vertices and edges).

(1.1)

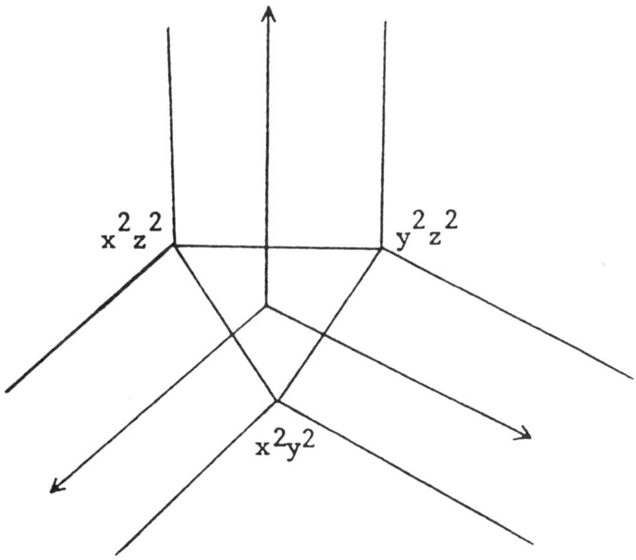

For an open or closed face $\Delta \subset \Gamma(f_0)$, we let $C(\Delta)$ denote the cone formed from half-rays from the origin passing through points of Δ. Likewise, we let $\bar{\Delta}$ denote the closure of an open face Δ; and $\overline{C(\Delta)}$, the closure of $C(\Delta)$ in $\mathbb{R}^n_+$. The Newton filtration associated to $\Gamma(f_0)$ is defined via a piecewise linear map $\phi : \mathbb{R}^n_+ \to \mathbb{R}$ satisfying:

i) ϕ is linear on each $\overline{C(\Delta)}$.

ii) ϕ takes non-negative integer values on the lattice points of $\mathbb{R}^n_+\backslash\{0\}$.

iii) $\phi|\Gamma(f_0) \equiv m_0$ (for some positive integer m_0).

For a polynomial ring $k[\mathbf{x}_n] = k[x_1, \ldots, x_n]$, formal power series ring $k[[\mathbf{x}_n]]$, or convergent power series ring $\mathcal{C}_{\mathbf{x}} = k\{x_n\}$, we define $\mathrm{fil}(x^{\boldsymbol{\alpha}}) = \phi(\boldsymbol{\alpha})$ and

$$\mathrm{fil}(\Sigma a_{\boldsymbol{\alpha}} x^{\boldsymbol{\alpha}}) = \min\{\mathrm{fil}(x^{\boldsymbol{\alpha}}) : a_{\boldsymbol{\alpha}} \neq 0\}.$$

For $\mathcal{C}_{\mathbf{x},\mathbf{u}} = k\{x_1, \ldots, x_n, u_1, \ldots, u_q\}$ we extend the filtration so that $\mathrm{fil}(x^{\boldsymbol{\alpha}} u^{\boldsymbol{\beta}}) = \mathrm{fil}(x^{\boldsymbol{\alpha}})$. Then, we denote the set of g with $\mathrm{fil}(g) \geq \ell$ in $k[\mathbf{x}_n]$, $k[[\mathbf{x}_n]]$, $\mathcal{C}_{\mathbf{x}}$, or $\mathcal{C}_{\mathbf{x},\mathbf{u}}$ by A_ℓ, $\mathcal{A}_\ell$, $\mathcal{A}_{\ell,\mathbf{x}}$ or $\mathcal{A}_{\ell,\mathbf{x},\mathbf{u}}$ (respectively). These define Newton filtrations on the respective rings having properties analogous to those previously defined when f_0 has an isolated singularity ([K] or [DG]). The one key difference is that now the Newton filtration will in general have infinite codimension.

Likewise, given a face $\Delta \subset \Gamma(f_0)$, we let $k[\bar{\Delta}]$, $k[[\bar{\Delta}]]$, $\mathcal{C}_{\bar{\Delta}} = k\{\bar{\Delta}\}$, etc. denote the ring of polynomials, formal power series, etc. whose monomials x^{α} belong to $\overline{C(\Delta)}$ (i.e. $\alpha \in \overline{C(\Delta)}$). Note that for non-compact faces Δ of $\Gamma(f)$, $\overline{C(\Delta)} \neq C(\bar{\Delta})$. The initial part of a germ g, in(g), is defined to be the sum of terms of the Taylor expansion of g of filtration = fil(g). Unlike the situation for isolated singularities, in(g) may contain an infinite number of terms. We also call A_{ℓ} a *fit level* of the filtration if the vertices of $\phi^{-1}(\ell)$ are lattice points of $\mathbb{R}^n_+$.

In addition to the global Newton filtration defined on $k[\mathbf{x}_n]$, etc., we also define a local filtration fil_v near a vertex v of $\Gamma(f_0)$. The formulation given here is equivalent to that given in [DG]. By star(v) we mean the union of open faces $\Delta \subset \Gamma(f_0)$ such that $v \in \bar{\Delta}$. for each maximal dimension face $\Delta \subset \Gamma(f_0)$, there exists a unique vector $w_{\Delta} \in \mathbb{R}^n_+$ so that $\phi(\alpha) = (w_{\Delta}, \alpha)$ for all $\alpha \in \overline{C(\Delta)}$, where (,) denotes inner product in $\mathbb{R}^n$. We define

$$\mathrm{fil}_v(x^{\alpha}) = \min_{\Delta \subset \mathrm{star}(v)} (w_{\Delta}, \alpha).$$

It is with fil_v that we shall measure the degeneracy of functions in §3.

One final notion must be introduced for the Newton polyhedron of nonisolated singularities.

For a compact face $\Delta \subset \Gamma_0(f_0)$, we say that x_j is an *unbounded direction for* Δ *in* $\Gamma(f_0)$ if $\mathbb{R}_+\epsilon_j + \Delta \subset \Gamma(f_0)$ ($\epsilon_j = (0, \ldots, 0,1,0, \ldots, 0)$ with 1 in the jth position). For example, in (1.1) the edge of the triangle joining x^2y^2 and x^2z^2 is unbounded in the x-direction; however, the vertices of the triangle do *not* have any unbounded directions in $\Gamma(f_0)$, for the lines parallel to the coordinate axes and beginning at these vertices do *not* belong to $\Gamma(f_0)$.

§2 CRITERION FOR TOPOLOGICAL TRIVIALITY

In [DG], a sufficient condition for topological triviality of deformations was given for the case of isolated singularities. Here we consider a version which handles the extra complications arising with nonisolated singularities.

Let $f : k^{n+q},0 \to k,0$ be a deformation of f_0 (in the appropriate category), and let u denote local coordinates for k^q. By f being a topologically trivial deformation of f_0 we mean it in the usual sense of topological right equivalence. Furthermore, let $0 \in S \subset k^n$ satisfy $f(S \times k^q) = 0$.

DEFINITION 2.1. We say that f is *trivial off of S* if there are neighborhoods U and V of 0 in k^{n+q} and a diffeomorphism ϕ : $U\backslash(S \times k^q) \to V\backslash(S \times k^q)$ such that:

i) $f \circ \bar{\phi}(x,u) = f(x,0)$ where $\phi(x,u) = (\bar{\phi}(x,u),u)$

ii) Given any neighborhood V' of $\{0\} \times k^q$ in V, there is a neighborhood U' of $\{0\} \times k^q$ in U so that $\phi(U'\backslash S) \subset V'\backslash S$.

REMARK. Even if f is only trivial off of S, since $S \times k^q \subset f^{-1}(0)$, it follows that the Milnor fibrations of $f_u(x) = f(x,u)$ (for fixed u) are equivalent. Hence, all information contained in the Milnor fibration is constant under such deformations.

To state the conditions for triviality we define a filtration on the module of vector fields along $\pi : k^{n+q} \to k^n$ by

$$(2.2)\quad V_{\ell,x,u} = \left\{\zeta \in A_{\ell+1,x,u}\left\{\frac{\partial}{\partial x_1}, \ldots, \frac{\partial}{\partial x_n}\right\} : \zeta(A_{k,x}) \subset A_{k+\ell,x} \text{ all } k \geq 0\right\}$$

Note:

$$A_{\ell,x,u}\left\{x_1\frac{\partial}{\partial x_1}, \ldots, x_n\frac{\partial}{\partial x_n}\right\} \cap A_{\ell+1,x,u}\left\{\frac{\partial}{\partial x_1}, \ldots, \frac{\partial}{\partial x_n}\right\} \subset V_{\ell,x,u}$$

We also let

$$V_{\ell,x,u}(f) = \{\zeta(f) : \zeta \in V_{\ell,x,u}\}$$

and

$$S(f_0) = \{x \in k^n : x^{\alpha} = 0 \text{ for all vertices } \alpha \in \Gamma(f_0)\}.$$

PROPOSITION 2.3. *Let* A_ℓ *be a fit level.* 1) *If*

$$\text{ver}(A_\ell)\cdot \frac{\partial f}{\partial u_i} \subset A_{\ell,x,u}\left\{x_1\frac{\partial f}{\partial x_1}, \ldots, x_n\frac{\partial f}{\partial x_n}\right\} \quad 1 \leqslant i \leqslant q \tag{2.3a}$$

then f *is a trivial deformation of* f_0 *off of* $S = S(f_0)$.

2) *If*

$$\text{ver}(A_\ell)\cdot \frac{\partial f}{\partial u_i} \subset V_{\ell,x,u}(f) \quad 1 \leqslant i \leqslant q \tag{2.3b}$$

then f *is a topologically trivial deformation of* f_0.

PROOF. From 2.3a) or b), we may solve for each i and each vertex $x^{\alpha} \in \text{ver}(A_\ell)$

$$x^{\alpha}\cdot \frac{\partial f}{\partial u_i} = \xi_{\alpha}^{(i)}(f) \tag{2.4}$$

with $\xi_{\alpha}^{(i)} \in A_{\ell,x,u}\{x_1(\partial/\partial x_1), \ldots, x_n(\partial/\partial x_n)\}$ or $V_{\ell,x,u}$. Multiplying (2.4) by $-\bar{x}^{\alpha}$ (= -complex conjugate of x^{α} if $k = \mathbb{C}$, or $-x^{\alpha}$ if $k = \mathbb{R}$) and summing over $\text{ver}(A_\ell)$, we obtain

$$-\rho\frac{\partial f}{\partial u_i} = \xi_i(f) \quad \text{where} \quad \rho = \sum_{\alpha\in\text{ver}(A_\ell)} |x^{\alpha}|^2.$$

The convexity lemma (lemma 3.5) from [DG] is valid as well for filtrations of infinite codimension so we have

LEMMA 2.5. *If* $g \in A_{m,x,u}$, *then there is a neighborhood* U *of* 0 *and a constant* $C > 0$ *so that*

$$|g| \leqslant C\rho^{(m/2\ell)} \quad \text{on} \quad U.$$

Thus, we have by Lemma 2.5

$$|\xi_i(\rho)| \leqslant \Sigma\,|\bar{x}^{\alpha}|\;|\xi_{\alpha}^{(i)}(\rho)| \tag{2.6}$$

where $\xi_i = \Sigma_{\alpha}\,\bar{x}^{\alpha}\xi_{\alpha}^{(i)}$. In the case of 2.1a),

$$\xi^{(i)}_{\alpha} = \Sigma\, g^{(i)}_{\alpha,j}\, x_j \frac{\partial\rho}{\partial x_j} \quad \text{with} \quad g^{(i)}_{\alpha,j} \in A_{\ell,x,u}$$

so by lemma 2.5,

$$(2.7) \qquad |\xi^{(i)}_{\alpha}(\rho)| \leq \sum_j |g^{(i)}_{\alpha,j}| \cdot \left| x_j \frac{\partial\rho}{\partial x_j} \right| \leq C' \cdot \rho^{(\ell/2\ell)} \cdot \rho^{(2\ell/2\ell)} .$$

(since $x_j(\partial\rho/\partial x_j)$ again consists of monomials $\overline{x}^{\alpha}x^{\alpha}$ with $x^{\alpha} \in$ ver(A_ℓ)). Thus, from (2.6) and (2.7)

$$(2.8) \qquad |\xi_i(\rho)| \leq C \cdot \rho^{2\ell}.$$

We let $\xi_i' = (\partial/\partial u_i) + \rho^{-1}\xi_i$. Then, from (2.8)

$$(2.9) \qquad |\xi_i'(\rho)| \leq C \cdot \rho.$$

We use three facts about ξ_i':

1) By an argument in [L] (or see [D1]), (2.9) implies that if $\gamma(t)$ is an integral curve of ξ_i' with $\gamma(0) \notin S$ then

$$\rho(\gamma(0))e^{-C\cdot t} \leq \rho(\gamma(t)) \leq \rho(\gamma(0))e^{Ct}.$$

Then, $\gamma(t)$ remains away from S.
2) $\|\xi_i'\|$ is bounded in a neighborhood of 0, for by lemma 2.5,

$$\|\xi_i\| \leq \Sigma\, |\overline{x}^{\alpha} g^{(i)}_{\alpha,j}\, x_j| \leq \sum_j C_j\rho \cdot |x_j| \leq C'\rho.$$

3) if $\rho' = \Sigma|x_i|^2$, then

$$|\xi_i'(\rho')| \leq C \cdot \rho' \text{ in a neighborhood of } 0.$$

This follows because

$$|\xi_i'(\rho')| = |\rho^{-1}\xi_i(\rho')| \leq \rho^{-1}\left(\Sigma\, |\overline{x}^{\alpha} g^{(i)}_{\alpha,j}|\, |x_j|^2\right) \leq C \cdot \rho'$$

where $C = m \cdot \max\{C_j\}$, with C_j as in step 2 and m = number of vertices. We argue as follows:

It is sufficient to prove it for one ξ_i' using the same reasoning of [D1,§6]. We carry it out for ξ_1'. Pick $\epsilon > 0$ so that ξ_1' is defined on $B_\epsilon(0)\backslash(S \times k^q)$ and $\|\xi_1'\| \leq C$ on $B_\epsilon(0)$. Let $\epsilon_1 < (\epsilon/4)\cdot\min\{1,C\}$. Then, an integral curve of ξ_1' beginning in $B_{\epsilon/2}(0)\backslash(S \times k^q)$ remains in $B_\epsilon(0)\backslash(S \times k^q)$ for time ϵ_1 by fact 1) and the choice of ϵ_1. Hence, there is defined a smooth flow

$$\alpha : (B_{\epsilon/2}(0)\backslash(S \times k^q) \times (-\epsilon_1,\epsilon_1) \to k^{n+q}\backslash(S \times k^q).$$

If $|t_0| < \epsilon_1$ and $x_0 \in B_{\epsilon/4}(0)$, then there is an $x' \in B_{\epsilon/2}(0)$ so that $\alpha(x',t_0) = x_0$. In fact, if we start from x_0 following the integral curve of $-\xi_1'$, then after time $-t_0$, we will still be in $B_{\epsilon/2}(0)$ and we will exactly reach the desired x'. Hence, Image(α) contains $(B_{\epsilon/4}(0)\backslash S) \times (-\epsilon_1,\epsilon_1)$. Lastly, fact 3) implies that integral curves beginning in a neighborhood $\{(x,u) : \rho'(x) < \delta\}$ will remain within the neighborhood $\{(x,u) : \rho'(x) < \delta e^{C\epsilon_1}\}$. Thus, for each i, $\xi_i' \mid k^n \times (\{0\} \times k^{q-i+1})$ defines a flow α_t and we define

$$\phi_i : (k^n\backslash S) \times (\{0\} \times k^{q-i+1}) \to (k^n\backslash S) \times (\{0\} \times k^{q-i+1})$$

by

$$\phi_i(\mathbf{x},(0, \ldots, 0, u_i, \ldots, u_q)) = \alpha_{u_i}(\mathbf{x},(0, \ldots, 0,u_{i+1}, \ldots, u_q)).$$

Then, we apply standard arguments (see e.g. [D1,§6]) to compose the ϕ_i to obtain the desired trivialization.

For the case of 2.3b), we first observe that

$$\begin{aligned} |\xi_i(\rho)| &\leqslant \Sigma|\bar{x}^{\alpha}| \cdot |\xi_\alpha^{(i)}(\rho)| \leqslant \Sigma|\bar{x}^{\alpha}|\,|\bar{x}^{\beta}| \cdot |\xi_\alpha^{(i)}(x^{\beta})| \\ &\leqslant C\cdot\rho^{(\ell/2\ell)} \cdot \rho^{(\ell/2\ell)} \cdot \rho^{(2\ell/2\ell)} = C \cdot \rho^2 \end{aligned}$$

so again (2.9) holds. This time we claim that $\rho^{-1}\xi_i$ extends continuously to be 0 on S. In fact

$$\|\rho^{-1}\xi_i\| \leqslant \rho^{-1}\,\Sigma|\bar{x}^{\alpha}| \cdot \|\xi_\alpha^{(i)}\|.$$

Then, $\xi_\alpha^{(i)} \in A_{\ell+1,x,u}\{(\partial/\partial x_1), \ldots, (\partial/\partial x_n)\}$, with say $\xi_\alpha^{(i)} = \Sigma h_{\alpha,j}^{(i)}(\partial/\partial x_j)$. By lemma 2.5, there is a $C > 0$, so that on a neighborhood of 0,

$$|h_{\alpha,j}^{(i)}| \leqslant C \cdot \rho^{(\ell+1/2\ell)}.$$

Thus,

$$\|\rho^{-1}\xi_i\| \leqslant \rho^{-1}(C' \cdot \rho^{(\ell/2\ell)} \cdot \rho^{(\ell+1/2\ell)}) = C' \cdot \rho^{(1/2\ell)}.$$

Thus, we define

$$\xi_i' = \begin{cases} \dfrac{\partial}{\partial u_i} & \text{on } S \\ \dfrac{\partial}{\partial u_i} + \rho^{-1}\xi_i & \text{off of } S \end{cases}$$

To see that ξ_i' is a locally integrable vector field we can repeat verbatum the arguments given in [D1,§9] replacing the linear subspace there by S which is a union of coordinate planes. Again the argument is completed as in [D1,§6].

§3. STATEMENTS OF THE LOCAL RESULTS

As in [DG], the main results are stated in terms of non-degeneracy and local patching data.

DEFINITION 3.1. A set of *local patching data* for a germ f_0 consists of: for each *compact* face $\Delta \subset \Gamma_0(f_0)$, germs of polynomial vector fields $\zeta_{\Delta,i} = \Sigma_j g_{\Delta,ij} x_j(\partial/\partial x_j)$ such that $\{\text{in}(\zeta_{\Delta,i}(f_0))|_{C(\bar{\Delta})}\}$ generate an ideal of finite codimension in $k[[\bar{\Delta}]]$. We say in addition that $\{\zeta_{\Delta,i}\}$ is a set of *strong local patching data* if for each compact Δ, $\text{fil}(g_{\Delta,ij} \cdot x_j) > \text{fil}(g_{\Delta,ij})$ (or $g_{\Delta,ij} = 0$) for any direction x_j in which Δ is unbounded (as defined in §1).

DEFINITION 3.2. f_0 is *non-degenerate* if $\{x_i(\partial/\partial x_i)\}_{i=1}^n$ is a set of local patching data for each compact face. We also say f_0 is *strongly non-degenerate* if it contains a set of strong local patching data (for each compact face).

REMARK. The additional "strong" condition is really a statement about the Newton polyhedron of f_0.

Example (1.1) is strongly non-degenerate. This is seen from the following lemma (whose proof we postpone until §4). For a cone $C(\Delta)$, we let $TC(\Delta)$ denote the linear subspace spanned by $C(\Delta)$ (i.e. the vectors in $C(\Delta)$).

LEMMA 3.3. *Suppose that for each compact face* $\Delta \subset \Gamma_0(f_0)$, $TC(\Delta)$ *is linearly independent from the subspace spanned by the unbounded directions for* Δ *in* $\Gamma(f_0)$ *(see* §1*). Then, if* f_0 *is non-degenerate, it is strongly non-degenerate.*

REMARK. For a compact face $\Delta \subset \Gamma(f_0)$, $k\{\bar{\Delta}\}$ is a ring of dimension $= \dim_{\mathbb{R}} TC(\Delta)$; hence, for f_0 to be strongly non-degenerate, $\Gamma(f_0)$ must satisfy the condition of lemma 3.3. Examples for which lemma 3.3 fails have "degenerate" Newton polyhedrons. For example, if $f_0 = x^2y^2z^2 + x^8y^8 + x^8z^8 + y^8z^8$ then $\Gamma(f_0)$ is shown in figure 3.4. The compact part $\Gamma_0(f_0)$ consists of just the 3 edges emanating from the central vertex; each of the edges fail to satisfy the conditions of lemma 3.3. Thus, although f_0 is non-degenerate it fails to be strongly nondegenerate.

In the degenerate case, just as in [DG] the degeneracy is measured near each vertex in terms of jump conditions,

For $\zeta = \Sigma g_j x_j(\partial/\partial x_j)$, we define

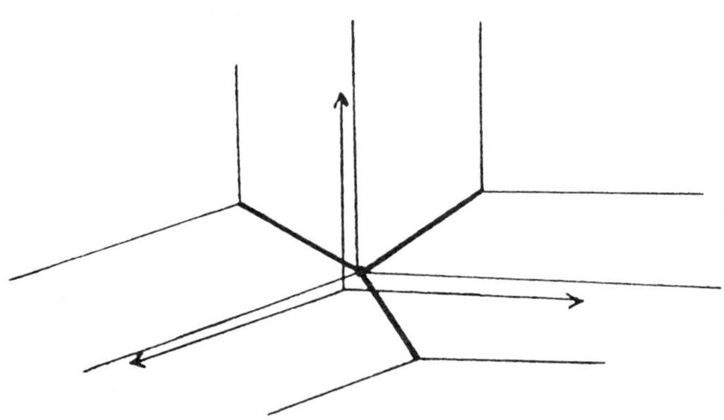

Figure (3.4)

$$\text{jump}(\zeta) = \text{fil}(\zeta(f_0)) - \min_i \text{fil}\left[g_i x_i \frac{\partial f_0}{\partial x_i}\right].$$

For a compact face Δ, we let

$$\text{jump}(\Delta) = \max_i \ \text{jump}(\zeta_{\Delta,i});$$

and for a vertex $v \in \Gamma_0(f_0)$, we define

$$\text{jump}(v) = \max\{\text{jump}(\Delta) : \Delta \subset \text{star}(v)\}$$

In terms of the local filtration fil_v defined in §1, with f a deformation of f_0, we say that ϕ satisfies a *simple jump condition* if

(3.5) $\quad \text{fil}_v(\phi) \geqslant \text{fil}(f_0) + \text{jump}(v) \quad$ for all vertices $\quad v \in \Gamma(f_0)$.

or a *general jump condition* if for all vertices $v \in \Gamma(f_0)$ there is a $\xi \in V_{\ell,x,u}$, with A_ℓ fit, and $x^\alpha \in \text{ver}(A_\ell) \cap C(v)$ so that

(3.6) $\quad \text{fil}_v(x^\alpha \cdot \phi - \xi(f)) \geqslant \text{fil}(f_0) + \ell + \text{jump}(v)$.

We shall say that ϕ satisfies a *jump condition* if it satisfies (3.5) or (3.6), or a *strict jump condition* if the inequalities are strict.

Then, we can state the principal results

THEOREM 1. *Let f_0 have strong local patching data $\{\zeta_{\Delta,i}\}$. If f is a deformation of non-decreasing Newton filtration which satisfies*

i) $\text{fil}(\zeta_{\Delta,i}(f)) = \text{fil}(\zeta_{\Delta,i}(f_0))$ *all* i,Δ.

ii) $\partial f/\partial u_i$ *satisfies a jump condition for* $1 \leqslant i \leqslant q$ *then* f *is a topologically trivial deformation of* f_0.

As an immediate corollary, we have

COROLLARY 2. *If f_0 is strongly non-degenerate then any deformation of non-decreasing Newton filtration is topologically trivial.*

Since by Kouchnirenko [K], even in the nonisolated case the set of non-degenerate germs with a given Newton polyhedron is still Zariski open, we can conclude.

COROLLARY 3. *If* f_0 *is strongly non-degenerate, and* $k = \mathbb{C}$ *then the topological type of* f_0 *only depends on* $\Gamma(f_0)$.

COROLLARY 4. *If* f_0 *is strongly non-degenerate and* $k = \mathbb{R}$ *then the topological type of* f_0 *only depends on* $\Gamma(f_0)$ *and its component among all non-degenerate germs with the same Newton polyhedron.*

Lastly, the order of topological determinancy is given by

COROLLARY 5. *Let* $L = \max\{\text{jump}(v) : v \in \Gamma(f_0)\}$. *If* $g \equiv f_0 \bmod \mathcal{A}_{L+m_0,x}$ *(where* $m_0 = \text{fil}(f_0)$*), then* g *is topologically equivalent to* f_0.

We obtain slightly weaker results if the "strong" condition is removed.

THEOREM 2. *Suppose that* f_0 *has local patching data* $\{\zeta_{\Delta,i}\}$ *or is non-degenerate, then the preceding theorems and corollaries remain valid provided we replace topological equivalence by equivalence off of* $S(f_0)$ $(= \{x \in k^n: x^{\alpha} = 0$ *for all vertices* $\alpha \in \Gamma(f_0)\})$ or we require the deformations to satisfy the strict jump condition.

Since $S(f_0) \subset f_1^{-1}(0)$ for any f_1 with $\Gamma(f_1) = \Gamma(f_0)$ we obtain

COROLLARY 6. *If* f_0 *has local patching data* $\{\zeta_{\Delta,i}\}$ *or is non-degenerate, then the preceding theorem and corollaries remain valid if we replace topological equivalence by topological equivalence of Milnor fibrations.*

The last result we give shows that corollaries 2 and 3 can be slightly extended when f_0 has the form $f_0 = x^{\beta} \cdot g_0$. Then, $\Gamma_+(f_0) = \Gamma_+(g_0) + \beta$; however, $\Gamma(g_0)$ will, in general, have fewer non- compact faces then $\Gamma(f_0)$. In particular, if Δ is a compact face of $\Gamma(g_0)$ then it may have fewer unbounded directions in $\Gamma(g_0)$ then $\Delta + \beta$ will have in

$\Gamma(f_0)$. Thus, the following condition is a weaker restriction than that given in lemma 3.3.

(3.7) For every compact face $\Delta \subset \Gamma(f_0)$, $TC(\Delta)$ is linearly independent from the subspace spanned by the unbounded directions for $\Delta - \boldsymbol{\beta}$ in $\Gamma(g_0)$.

THEOREM 3. *If $f_0 = x^{\boldsymbol{\beta}} \cdot g_0$ is non-degenerate and $\Gamma(f_0)$ satisfies (3.7), then any deformation of f_0 of non-decreasing Newton filtration is topologically trivial. If $k = \mathbb{C}$, then the topological type of f_0 is determined by $\Gamma(f_0)$; and if $k = \mathbb{R}$, it is determined by its connected component among germs with the same Newton polyhedron.*

§4. SEVERAL LEMMAS

The main problem in proving Theorems 1, 2 or 3 is to overcome the algebraic difficulties in working with an infinite codimension filtration. Then, we shall be able to derive the algebraic conditions 2.3a) and b) which we use to deduce the topological results. In this section we give the lemmas needed to overcome these problems.

We first begin with the proof of lemma 3.3. Let Δ be a compact face of $\Gamma(f_0)$. Suppose after reordering the coordinates that $\{x_1, \ldots, x_r\}$ are unbounded directions for Δ in $\Gamma(f_0)$. By assumption

$$\langle \frac{\partial}{\partial x_1}, \ldots, \frac{\partial}{\partial x_r} \rangle \cap TC(\Delta) = (0)$$

Thus,

$$\langle \frac{\partial}{\partial x_{r+1}}, \ldots, \frac{\partial}{\partial x_n} \rangle + TC(\Delta)^{\perp} = \mathbb{R}^n \tag{4.1}$$

If $v = (v_1, \ldots, v_n)$, let $\eta_v = \Sigma v_i x_i(\partial/\partial x_i)$. Then, if $v \in TC(\Delta)^{\perp}$, $\eta_v(g) = 0$ for $g \in k\{\overline{\Delta}\}$. Thus, if $v \in TC(\Delta)^{\perp}$, $\eta_v(f)|_{\overline{C(\Delta)}} = 0$. If $\{\epsilon_1, \ldots, \epsilon_n\}$ denote the standard basis vector for $\mathbb{R}^n$, then, by (4.1) we may write $\epsilon_j = w_j + v_j$ with $w_j \in \langle \epsilon_{r+1}, \ldots, \epsilon_n \rangle$ and $v_j \in TC(\Delta)^{\perp}$ for $j = 1, \ldots, r$. Then,

$$(4.2)\qquad \mathrm{in}(\eta_{\epsilon_j}(f_0))\,|\,\overline{C(\Delta)} = \mathrm{in}(\eta_{w_j}(f_0))\,|\,\overline{C(\Delta)} + \mathrm{in}(\eta_{v_j}(f_0))\,|\,\overline{C(\Delta)}\,.$$

The last term is zero; thus, (4.2) implies that for $1 \leqslant i \leqslant n$,

$$\mathrm{in}\left[x_i \frac{\partial f_0}{\partial x_i}\right]\Big|\,\overline{C(\Delta)} \in \langle \mathrm{in}\left[x_j \frac{\partial f_0}{\partial x_j}\right]\Big|\,\overline{C(\Delta)}\,, \quad j = r+1, \ldots, n\rangle.$$

Hence, $\{\mathrm{in}(x_j(\partial f_0/\partial x_j))\,|\,\overline{C(\Delta)}\}_{j=r+1}^{n}$ generates the same ideal as the corresponding set for $i = 1, \ldots, n$. Thus, we may use $\{x_j(\partial/\partial x_j)\}_{j=r+1}^{n}$ as local patching data for Δ.

The first algebraic lemma we need is a modified version of the contraction lemma of [DG].

LEMMA 4.3. *Suppose the Newton filtration* $\{A_{\ell,x}\}$ *of* f_0 *is fit at level* ℓ. *For each* $s > 0$, *there exists an* $L > 0$ *such that for any vertex* $x^{\alpha} \in A_{\ell,x}$ *and* $m > L$, *if* $\psi \in C_{x,u}$ *with* $\mathrm{fil}(\psi) = q$ *then there is a* $\psi_1 \in A_{m\ell+s+q,x,u}$ *so that*

$$\mathrm{supp}((x^{\alpha})^m\psi - \psi_1) \subset \mathrm{star}(\alpha).$$

Note: for $g = \Sigma\, a_{\alpha}x^{\alpha}$, $\mathrm{supp}(g) = \{\alpha : a_{\alpha} \neq 0\}$.

PROOF. For each maximal dimensional face $\Delta \not\subset \mathrm{star}(\alpha)$, let ϕ_{Δ} denote the linear extension of $\phi\,|\,\overline{C(\Delta)}$.

By the strict convexity of ϕ, $\phi_{\Delta}(\alpha) > \mathrm{fil}(x^{\alpha})$. Since there are only finitely many such Δ, there is a $\delta > 0$ so that $\phi_{\Delta}(\alpha) > \mathrm{fil}(x^{\alpha}) + \delta$ for all maximal $\Delta \not\subset \mathrm{star}(\alpha)$. We choose L so that $L\cdot\delta > s$. For any monomial x^{γ} and $m > L$, if $(x^{\alpha})^m \cdot x^{\gamma} \notin \mathrm{star}(\alpha)$, then $m\alpha + \gamma \in \overline{C(\Delta)}$ for some maximal dimensional face $\Delta \not\subset \mathrm{star}(\alpha)$. Thus,

$$\begin{aligned}\mathrm{fil}((x^{\alpha})^m \cdot x^{\gamma}) = \phi_{\Delta}(m\cdot\alpha + \gamma) &= m\phi_{\Delta}(\alpha) + \phi_{\Delta}(\gamma)\\ &> m(\ell + \delta) + \mathrm{fil}(x^{\gamma})\\ &> m\ell + s + \mathrm{fil}(x^{\gamma}).\end{aligned}$$

From this follows the result for a general ψ. $\square$

We observe that as a consequence of this, for v a vertex of $\Gamma(f_0)$ and $\alpha \in \mathrm{ver}(A_{\ell}) \cap C(v)$ we have

(4.4) $$\mathrm{fil}_v(\psi) = \mathrm{fil}((x^\alpha)^m \cdot \psi) - \mathrm{fil}((x^\alpha)^m) \qquad m >> 0$$

By the lemma,

$$(x^\alpha)^m \cdot \psi = \psi_1 + \psi_2 \quad \text{with} \quad \mathrm{supp}(\psi_2) \subset \mathrm{star}(\alpha)$$

and $\psi_1 \in A_{m\ell+q+s,x,u}$ with s to be chosen. If x^γ is any non-zero monomial of ψ, then $(x^\alpha)^{m_1} \cdot x^\gamma$ belongs to star(α) for some m_1 sufficiently large. Let $s > \mathrm{fil}((x^\alpha)^{m_1} x^\gamma)$. Then, $\mathrm{supp}(\psi_2) \neq \phi$; thus, from the definition of fil_v (in §1), since $\alpha \in C(v)$

$$\mathrm{fil}_v(x^\alpha \psi) = \mathrm{fil}(x^\alpha) + f_v(\psi)$$

Hence,

(4.5) $$\begin{aligned} \mathrm{fil}_v(\psi) &= \mathrm{fil}_v((x^\alpha)^m \cdot \psi) - \mathrm{fil}((x^\alpha)^m) \\ &= \mathrm{fil}_v(\psi_1 + \psi_2) - \mathrm{fil}((x^\alpha)^m) \end{aligned}$$

Then, $\mathrm{fil}(\psi_1) \geq m \cdot \ell + \mathrm{fil}(\psi) + s$, and $s > \mathrm{fil}((x^\alpha)^m \cdot x^\gamma)$ so $(x^\alpha)^m \cdot x^\gamma$ appears in ψ_2. Thus, $\mathrm{fil}_v(\psi_1 + \psi_2) = \mathrm{fil}_v(\psi_2) = \mathrm{fil}(\psi_2) = \mathrm{fil}((x^\alpha)^m \cdot \psi)$. Together with (4.5) this gives (4.4).

Next, we consider a non-compact face $\Delta \subset \Gamma(f_0)$. Let $C = \overline{C(\Delta)}$. Then $\phi|_C$ is linear and ≥ 0. It has the minimum value 0 on a positive dimensional face C' of C. Likewise $\phi|_C > 0$ on $C''\backslash\{0\}$ for some maximal dimensional face C'' of C. Then, $C'' = \overline{C(\Delta')}$, and Δ' is a compact face in Δ. Also, C' represents the directions (in C) along which Δ' is infinite in $\Gamma(f_0)$. We call Δ' the compact part of Δ. We also let A denote the algebra of convergent power series with monomials x^α having $\alpha \in C$.

Then, the basic lemma which allows us to overcome the infinite codimension problem is the following.

LEMMA 4.6. *Let $\Delta \subset \Gamma(f_0)$ have Δ' for its compact part. Suppose for $\{g_i\}_{i=1}^s \subset A$, $\{\mathrm{in}\{g_i\}\,|\,\overline{C(\Delta')}\}_{i=1}^s$ generates an ideal of finite codimension in $k\{\Delta'\}$ (or equivalently $k[[\Delta']]$). Then, there exists $N > 0$ so that $\{\mathrm{in}(g_i)\,|\,\overline{C(\Delta)}\}_{i=1}^s$ generates an ideal I which contains $A_{N,x} \cap A$.*

PROOF. 1) Since $\{\mathrm{in}(g_i)\,|\,\overline{C(\Delta')}\}$ generates an ideal I' of finite codimension in $k\{\overline{\Delta}'\}$, there exists a level ℓ such that $A_{\ell,x} \cap k\{\overline{\Delta}'\} \subset I'$.

Pick monomials $\{x^{\alpha_i}\}$ belonging to the edges of C'' and $\{x^{\beta_j}\}$ belonging to the edges of C', such that $\mathrm{fil}(x^{\alpha_i}) > \ell$. Then, $\{x^{\alpha_i}, x^{\beta_j}\}$ generate an ideal of finite codimension in A by Gordon's lemma (see[Da, 1.3]). Furthermore, the proof shows that all but a finite number of monomials, say $x^{\gamma_1}, \ldots, x^{\gamma_r}$, are divisible by one of these monomials. Pick $N > \max\{\mathrm{fil}(x^{\gamma_i})\}$.

2) We may pick a finite set of monomial generators for $A_{N,x} \cap A$ because A is noetherian and hence $A_{N,x} \cap A$ is finitely generated. Let x^{γ} be a monomial generator with $\mathrm{fil}(x^{\gamma}) = m \geq N$. Thus, $x^{\gamma} = x^{\gamma'}x^{\alpha_i}$ or $x^{\gamma'}x^{\beta_j}$ for some i or j. In the second case

$$\mathrm{fil}(x^{\gamma'} \cdot x^{\beta_j}) = \mathrm{fil}(x^{\gamma'}) + \mathrm{fil}(x^{\beta_j}) = \mathrm{fil}(x^{\gamma'}).$$

Thus, again $\mathrm{fil}(x^{\gamma'}) \geq N$. This may be repeated for $x^{\gamma'}$ and eventually we must have the first case. Thus, $A_{N,x} \cap A$ is contained in the ideal generated by $\{x^{\alpha_i}\}$.

3) Next, let σ_0 be a linear functional on $\mathbb{R}^n$ whose kernel is a supporting hyperplane for the face C'' of C; which takes on integral values on lattice points of C; and for which $\sigma_0|C \geq 0$. Define $\sigma(x^{\alpha}) = \sigma_0(\alpha)$; and for $g = \Sigma a_{\alpha}x^{\alpha} \in A$, define

$$\mathrm{fil}'(\Sigma\, a_{\alpha}x^{\alpha}) = \min\{\sigma(x^{\alpha}) : a_{\alpha} \neq 0\}.$$

Then fil' is a second filtration on A. Let $p = \max\{\sigma(x^{\gamma_i})\}$. Also let J_m be the ideal in A generated by monomials x^{α} with $\sigma(x^{\alpha}) \geq m$ (and $J_0 = A$). We claim that for $m = 0,1, \ldots, p$.

(4.7m) $$J_m \cap \mathcal{A}_{N,x} \subset I + m_A(\mathcal{A}_{N,x} \cap A) + J_{m+1} \cap \mathcal{A}_{N,x}.$$

where m_A denotes the maximal ideal of A (and I denotes the ideal generated by $\{\mathrm{in}(g_i)|\overline{C(\Delta)}\}$). If this is true, then (4.7m) $m = 0,1, \ldots, p$ together imply

(4.8) $$\mathcal{A}_{N,x} \cap A \subset I + m_A \cdot (\mathcal{A}_{N,x} \cap A) + J_{p+1} \cap \mathcal{A}_{N,x}$$

4) We claim that the result follows from (4.8).

We observe that for any x^α, we may repeatedly divide by some x^{α_i} or x^{β_j} until we obtain some x^{γ_i}. Thus, we may write $x^\alpha = m_1 \cdot m_2 \cdot x^{\gamma_i}$ with m_1 a product of terms from $\{x^{\alpha_i}\}$ and m_2 a product of terms from $\{x^{\beta_j}\}$. Then, $\mathrm{fil}(x^\alpha) = \mathrm{fil}(m_1 \cdot x^{\gamma_i})$ since $\mathrm{fil}(x^{\beta_i}) = 0$ for all i. Also, $\mathrm{fil}'(x^\alpha) = \mathrm{fil}(m_2 \cdot x^{\gamma_i})$ since $m_1 \in k\{\bar{\Delta}'\}$. Thus if $x^\alpha \in J_{p+1} \cap \mathcal{A}_{N,x}$, then $\mathrm{fil}(m_1 \cdot x^{\gamma_i}) = \mathrm{fil}(x^\alpha) \geq N$ and $\mathrm{fil}'(m_2 \cdot x^{\gamma_i}) = \mathrm{fil}'(x^\alpha) > p$. Hence, $m_2 \neq 1$; we have $J_{p+1} \cap \mathcal{A}_{N,x} \subset m_A \cdot \mathcal{A}_{N,x}$. Thus, (4.8) implies

$$\mathcal{A}_{N,x} \cap A \subset I + m_A \cdot (\mathcal{A}_{N,x} \cap A)$$

which via Nakayama's lemma gives the desired result.

5) It remains to establish (4.7m). Again, we may find a set of monomial generators for $J_m \cap \mathcal{A}_{N,x}$. Let x^γ be such a generator. As $x^\gamma \in \mathcal{A}_{N,x}$ by step 2, $x^\gamma = x^{\alpha_i} \cdot x^{\gamma'}$. Also, $\mathrm{fil}'(x^\gamma) = \mathrm{fil}'(x^{\gamma'})$. By step 1, we may write

(4.9) $$x^{\alpha_i} = \sum_j h_j \,\mathrm{in}(g_j)|\overline{C(\Delta')}$$

with $\mathrm{fil}(h_j) = \mathrm{fil}(x^{\alpha_i}) - \mathrm{fil}(g_j)$ and $h_j \in k\{\bar{\Delta}'\}$. Let

$$\psi = x^{\alpha_i} - \sum_j h_j \,\mathrm{in}(g_j)|\overline{C(\Delta)}\,.$$

Then, (4.9) implies $\mathrm{fil}(\psi) \geq \mathrm{fil}(x^{\alpha_i})$ and $\mathrm{fil}'(\psi) > 0$. Thus, $\mathrm{fil}(x^\gamma \cdot \psi)$

$\geqslant N$ and $\mathrm{fil}'(x^{\gamma'} \cdot \psi) > \mathrm{fil}(x^{\gamma'})$. Thus

$$x^{\gamma} = \Sigma\, x^{\gamma'} \cdot h_i\ \mathrm{in}(g_i)\,|\,\overline{C(\Delta)} + x^{\gamma'} \cdot \psi.$$

with

$$x^{\gamma'} \cdot \psi \in J_{m+1} \cap A_{N,x} \quad \text{as desired.} \qquad \square$$

The third lemma we need relates the Newton filtration with the filtration by powers of the maximal ideal.

LEMMA 4.10. *Given ℓ, there is an ℓ' so that*

$$m_x \cdot A_{\ell,x} \supset A_{\ell',x}.$$

Hence, given ℓ there is a k so that $m_x^{\ell} \supset A_{k,x}$.

PROOF. Let Δ be a face with compact part Δ'. Then, $A_{\ell,x} \cap k\{\overline{\Delta'}\}$ has finite codimension; hence so does $m_{\overline{\Delta}'}(A_{\ell,x} \cap k\{\overline{\Delta'}\})$, where $m_{\overline{\Delta}'}$ denotes the maximal ideal of $k\{\overline{\Delta'}\}$. Thus, by lemma 4.6, its generators generate an ideal in A containing $A_{N,x} \cap A$ for some $N > 0$. Thus $m_x \cdot A_{\ell,x} \supset A_{N,x} \cap A$. Let ℓ' be the maximum of N for each face Δ. Then, $m_x \cdot A_{\ell,x} \supset A_{\ell',x}$. Hence, there exist ℓ_i so that

$$m_x \supset A_{\ell_1,x},\quad m_x \cdot A_{\ell_1,x} \supset A_{\ell_2,x}, \ldots, \quad m_x \cdot A_{\ell_{k-1},x} \supset A_{\ell_k,x}$$

or

$$m_x^k \supset m_x^{k-1} \cdot A_{\ell_1,x} \supset m_x^{k-2} \cdot A_{\ell_2,x} \supset \cdots \supset A_{\ell_k,x}. \qquad \square$$

§5. PROOFS OF THE LOCAL THEOREMS

To prove theorems 1 and 2, it is only necessary to deduce 2.3a) and b). To do this, we use the method of [DG], with lemmas of §4 allowing us to carry through arguments for the infinite codimension case.

We first show that under the assumptions of theorem 2, we may solve 2.3a). Rather than repeating the arguments of [DG], modifying them as needed, we refer the reader to §8 and 9 [DG] (denoted by §DG8, §DG9) and indicate where the argument must be modified and how we do it.

Before describing the modifications, we make a general remark about the use of the modified contraction lemma. The contraction lemma was used in [DG] to pull terms into star(v) to increase the filtration of the terms. This is necessary to overcome the effects of jumps in filtration for the $\zeta_{\Delta,i}(f)$. This would be unnecessary for terms of filtration greater than $N_0 = m_0 + \max\{\text{jump}(v)$: v vertex of $\Gamma(f_0)\}$. Thus, if we replace the contraction lemma by the modified contraction lemma 4.6 with $s > N_0$, then we do not have to worry about the terms not pulled into star(v) as they will have high enough filtration.

Then, we first consider §DG8. There, lemma 8.1, called the push-off lemma, is proven and used to establish, via a filtered version of Nakayama's lemma, a proposition which allows us to represent all germs of sufficiently high filtration in terms of $\{\zeta_{\Delta,i}(f)\}$.

Modification of the proof of push-off lemma: Although the proof uses formal power series, it works as well with convergent power series. Then, for a non-compact face Δ with compact part Δ', it is only necessary to replace $A_{r\ell} \cap k[[\bar{\Delta}]]$ by $A_{r\ell,x} \cap A$ and $A_{r\ell} \cap C_u[[\bar{\Delta}]]$ by $A_{r\ell,x,u} \cap A_u$ (where A and A_u denote the algebras of convergent power series with monomials x^{α} or $x^{\alpha}u^{\beta}$ where $\alpha \in \overline{C(\Delta)}$). By lemma 4.6, for r large enough, $\{\text{in}(\zeta_{\Delta',i}(f_0))\,|\,\overline{C(\Delta)}\}$ generate an ideal in A containing $A_{r\ell,x} \cap A$. Then, step 1 proceeds in the same way. In step 2, the contraction lemma is replaced by the modified contraction lemma (4.3) as described above. Also, $h_i^{(\alpha)}$ need not now be a polynomial; however this does not affect the proof. Now proposition 8.3 of [DG] will follow in the same way provided the filtered version of Nakayama's lemma [DG, lemma 1.1] can be used. However, its proof only depends on $m_x \cdot A_{\ell,x} \supset A_{\ell',x}$ for any ℓ (and some ℓ' depending on ℓ). This follows by lemma 4.10.

Modification of the proofs in §DG9. Both the stable push-off lemma ([DG, 9.1]) and its corollary 9.2 only depend on the push-off lemma and the contraction lemma. Both of these lemmas are only used modulo $A_{m,x,u}$ for m sufficiently large. Hence, we may replace the polynomial germ ψ in them by any germ and use the modified contraction lemma. However, conclusion ii) of (DG, 9.1) becomes

$$\mathrm{supp}(((x^{\alpha})^r \cdot \psi - \psi_1)_{\mathrm{fil}(\psi)+r\ell+s}) \subset \mathrm{star}(v)$$

(with s from the modified contraction lemma). Then, the remaining steps of the proof in §DG9 follow exactly the same way. Thus, we can solve 2.3a).

Next, we show that under the additional assumption that $\{\zeta_{\Delta,i}\}$ is a set of strong local patching data. We may solve 2.3b). The preceding argument allows us to solve

$$(x^{\alpha})^r \cdot \frac{\partial f}{\partial u_i} = \sum_{\Delta,i,j} h_{\Delta,i} \cdot g_{\Delta,i,j} x_j \frac{\partial f}{\partial x_j} \tag{5.1}$$

for $\alpha \in C(v)$ with v any vertex of $\Gamma(f_0)$. Here $\mathrm{fil}(h_{\Delta,i}) \geqslant r \cdot \mathrm{fil}(x^{\alpha}) - \mathrm{fil}(g_{\Delta,ij})$. We wish to show that either $g_{\Delta,ij} = 0$ or

$$\mathrm{fil}(x_j \cdot h_{\Delta,i} g_{\Delta,i,j}) > r \cdot \mathrm{fil}(x^{\alpha}). \tag{5.2}$$

Then, in fact, (5.1) solves 2.3b). Again referring to §DG8 together with our lemma 4.6, we see that modulo sufficiently high filtration,

$$\mathrm{supp}(h_{\Delta,i}) \subset \bigcup_k C(\Delta_k)$$

where Δ_k are faces with Δ as compact part or a face. By assumption, if x_j is an unbounded direction for Δ in $\Gamma(f_0)$, then either $g_{\Delta,ij} = 0$ or $\mathrm{fil}(x_j \cdot g_{\Delta,ij}) > \mathrm{fil}(g_{\Delta,ij})$. Thus, if $g_{\Delta,ij} \neq 0$ then

$$\mathrm{fil}(x_j \cdot h_{\Delta,i} \cdot g_{\Delta,ij}) > \mathrm{fil}(h_{\Delta,i}) + \mathrm{fil}(g_{\Delta,ij}) \geqslant r \cdot \mathrm{fil}(x^{\alpha})$$

By the proof of §DG9, (5.2) follows. □

For the proof of theorem 3, we will now solve 2.3b) in the form

$$(5.3) \qquad (x^{\alpha})^{r} \cdot \frac{\partial f}{\partial u_i} = \Sigma h_j \cdot x_j \frac{\partial f}{\partial x_j}$$

where now we use the Newton filtration associated to g_0 and $\alpha \in \mathrm{ver}(A_{\ell,x})$ for g_0. In (5.3), we will find h_j with $\mathrm{fil}(h_j) \geq r \cdot \mathrm{fil}(x^{\alpha})$ and $\mathrm{fil}(h_j \cdot x_j) > r \cdot \mathrm{fil}(x^{\alpha})$. Replacing $f = x^{\beta} \cdot g$ and substituting into (5.3) we see that to solve (5.3) it is sufficient to solve

$$(5.4) \qquad (x^{\alpha})^{r} \cdot \frac{\partial g}{\partial u_i} = \Sigma_j h_j \left[\beta_j g + x_j \frac{\partial g}{\partial x_j}\right]$$

where $\beta = (\beta_1, \ldots, \beta_n)$, and then multiply by x^{β}. Now, we want to be able to replace in the proof of theorem 1, $\{\zeta_{\Delta,i}(f)\}$ and $\{\zeta_{\Delta,i}(f_0)\}$ by $\{\beta_j g + x_j(\partial g/\partial x_j)\}$ and $\{\beta_j g_0 + x_j(\partial g_0/\partial x_j)\}$. For this, let Δ be a compact face of $\Gamma(g_0)$ for which $\{x_{j_1}, \ldots, x_{j_r}\}$ are not unbounded directions in $\Gamma(g_0)$. Then, it is sufficient that the set of elements

$$\left\{\mathrm{in}\left[\beta_{j_i} g_0 + x_{j_i} \frac{\partial g_0}{\partial x_{j_i}}\right|_{\overline{C(\Delta)}}\right\}_{i=1}^{r}$$

generates an ideal of finite codimension in $k\{\bar{\Delta}\}$. By the criterion of Kouchnirenko [K, Thm. 6.2] it is enough to know that for any fact $\Delta' \subset \bar{\Delta}$,

$$(5.5) \qquad \mathrm{in}\left[\beta_{j_i} g_0 + x_{j_i} \frac{\partial g_0}{\partial x_{j_i}}\right]_{\overline{C(\Delta')}} = 0 \ \text{ in } \mathbb{C}^n, \quad i = 1, \ldots, r$$

only when $\prod_{i=1}^{n} x_i = 0$. However

$$x^{\beta} \cdot \left[\mathrm{in}\left[\beta_{j_i} g_0 + x_{j_i} \frac{\partial g_0}{\partial x_{j_i}}\right]\Big|_{\bar{\Delta}'}\right] = \mathrm{in}'\left[x_{j_i} \frac{\partial f_0}{\partial x_{j_i}}\right]\Big|_{\overline{\Delta_1'}}$$

where $\Delta_1' = \Delta' + \beta$ and in' denotes the initial terms relative to the Newton filtration of f_0. Hence, (5.5) holds exactly if it holds for

$\left\{\mathrm{in}\left[x_{j_i}\frac{\partial f_0}{\partial x_{j_i}}\right]\Big|_{\overline{\Delta_1'}}\right\}$. However, this is true by Kouchnirenko's criterion since, by assumption, $\left\{\mathrm{in}\left[x_{j_i}\frac{\partial f_0}{\partial x_{j_i}}\right]\Big|_{\overline{\Delta_1'}}\right\}_{i=1}^{r}$ generates an ideal of finite codimension in $k\{\overline{\Delta_1'}\}$. Thus, the proof given for theorem 2 can be applied to allow us to solve (5.3). Then, the same reasoning used to prove theorem 1 shows that we may choose h_i so that $\mathrm{fil}(h_i x_i) > r \cdot \mathrm{fil}(x^{\alpha})$. Thus, we actually solve 2.3a), yielding the result. □

§6. GLOBAL TOPOLOGICAL TYPE OF AFFINE HYPERSURFACES

In this section we shall prove a result about the global topological type of $f^{-1}(0)$ for a polynomial f which is non-degenerate. In the complex case, if S denotes the common zero set of the non-zero monomials in f, then we shall prove that the global topological type of $f^{-1}(0)$ off of S is determined by its Newton polyhedron. From this it will follow that for non-degenerate "commode" polynomials the global topological type of $f^{-1}(0)$ is determined by its Newton polyhedron, and that the global Milnor fibration is locally trivial off of the set where degeneracy occurs. Alternately if its Newton polyhedron is "well-situated" in a certain sense and f is non-degenerate then we shall also prove that the global topological type of $f^{-1}(0)$ is determined by its Newton polyhedron. Let $f_0 : k^n \to k$ be a polynomial of degree m. If we write $f_0(x) = \Sigma\, c_{\alpha} x^{\alpha}$ with $0 \leq |\alpha| \leq m$, then the global Newton polyhedron of f, denoted by $\breve{\Gamma}_+(f_0)$, is the convex hull of the set of α such that $c_{\alpha} \neq 0$. It has a boundary $\breve{\Gamma}(f_0)$. Following [V, §7], we say that f_0 is *globally non-degenerate* if there are no simultaneous solutions in $\mathbb{C}^n\backslash\{\Pi_{i=1}^{n} x_i = 0\}$ to $f_0 = 0$, $x_i(\partial f_0/\partial x_i) = 0$, $i = 1, \ldots, n$ nor $f_0|_{\Delta} = 0$, $x_i \frac{\partial f_0}{\partial x_i}\Big|_{\Delta} = 0$ for any face $\Delta \subset \breve{\Gamma}(f_0)$. We also say f_0 is

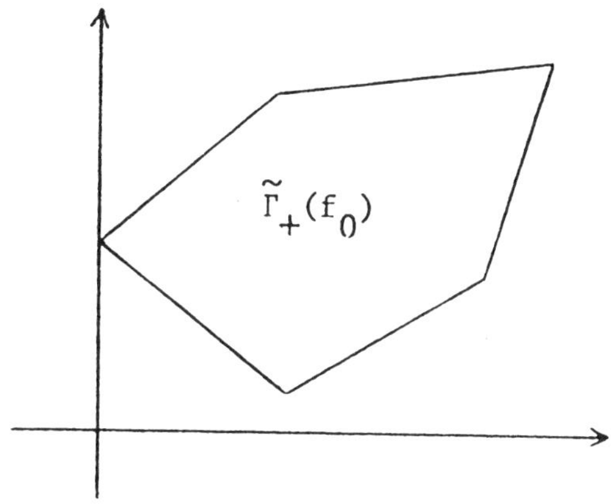

"*commode*" if either $f_0(0) \neq 0$ or for each i, some monomial $x_i^{k_i}$ appears with non-zero coefficients.

EXAMPLE 6.1. $f_0(x,y) = x^5 + y^5 + z^5 + x^{10}y^{10}z^{10}$ is globally non-degenerate and commode with Newton polyhedron shown in (6.2)

(6.2)

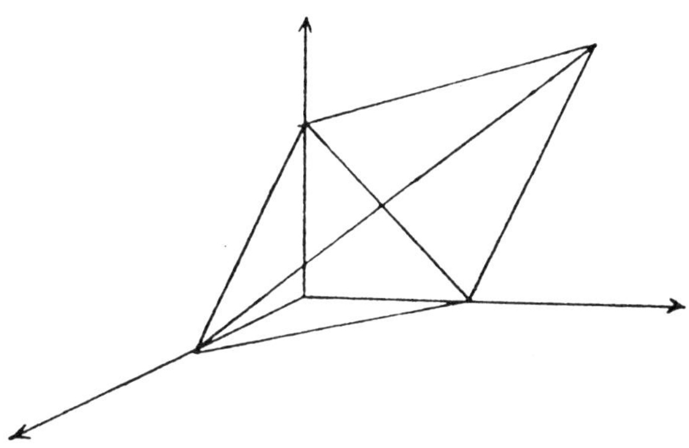

Next, let F_0 be the homogenization of f_0, i.e.

$$F_0(x_0,x_1, \ldots, x_n) = x_0^m(f_0((x_1/x_0)), \ldots, (x_n/x_0))$$

We shall deduce the global results for $f_0^{-1}(0)$ from local results for

F_0. First, $\Gamma_0(F_0)$ has exactly one n-dimensional face which projects along the x_0-axis to $\widetilde{\Gamma}_+(f_0)$, and maps the compact faces of $\Gamma(F_0)$ bijectively to the faces of $\Gamma_+(f_0)$.

The first relation between f_0 and F_0 is

LEMMA 6.3. *If f_0 is globally non-degenerate then F_0 is non-degenerate.*

We give the proof later.

Next, we define

DEFINITION 6.4. $\widetilde{\Gamma}(f_0)$ is *well-situated* if $\Gamma(F_0)$ satisfies condition (3.7) (for theorem 3).

We make several observations:

(6.5) In the case of polynomial $f_0(x,y)$, it is easy to see that if $\widetilde{\Gamma}_+(f_0)$ is 2-dimensional, then $\widetilde{\Gamma}(f_0)$ is well-situated.

(6.6) If $\widetilde{\Gamma}(f_0)$ is well-situated, then so it $\widetilde{\Gamma}(h_0)$ for $h_0 = x^{\beta} \cdot f_0$ for any $\beta \in \mathbf{Z}^n_+$.

EXAMPLE (6.7). For $f_0 = z^3x^4y + z^3xy^4 + x^2y^2 + yz^2 + xz$, $\widetilde{\Gamma}(f_0)$ is not well-situated, for the 2-dimensional face in $\Gamma(F_0)$ corresponding to that in $\widetilde{\Gamma}(f_0)$ with vertices for z^3x^4y, z^3xy^4, x^2y^2 has both x and y as unbounded directions.

The main global result is given by the next theorem.

THEOREM 4 1) *If $f_0 : k^n \to k$ is a polynomial which is non-degenerate with $\widetilde{\Gamma}(f_0)$ well-situated, then for any deformation f of f_0 with the same Newton polyhedron, $f^{-1}(0)$ is topologically trivial. Hence, when $k = \mathbb{C}$, the global topological type of $f_0^{-1}(0)$ is determined by $\widetilde{\Gamma}(f_0)$, and when $k = \mathbb{R}$ it is determined by its connected component in the set of non-degenerate polynomials with the same Newton polyhedron.*

2) *In the case that f_0 is non-degenerate with no condition on $\widetilde{\Gamma}(f_0)$, then the above results hold provided we replace topological equivalence by topological equivalence off of $S = \{x \in k^n \backslash \{0\} : x^{\alpha} = 0$ for all vertices α of $\widetilde{\Gamma}_+(f_0)\}$.*

If f_0 is commode then $S = \phi$, hence we have as a first corollary.

COROLLARY 7. *If f_0 is non-degenerate and commode then the topological type $f_0^{-1}(0)$ is determined by $\tilde{\Gamma}(f_0)$.*

As a corollary, consider a complex polynomial f_0 such that $f_0(0) = 0$. Suppose for some $a_0 \neq 0$, $f_0(x) - a_0$ is non-degenerate.

COROLLARY 8. *If A denotes the set of $a \in \mathbb{C}^*$ such that $f_0(x) - a$ is non-degenerate, then the global Milnor fibration $\mathbb{C}^n \backslash f^{-1}(A) \to \mathbb{C}^* \backslash A$ is a locally trivial fibration.*

NOTE. As pointed out by the referee, this is similar to lemma 5.9 of [O].

PROOF (of the corollary). If $a' \in \mathbb{C}^* \backslash A$, then $f_0(x) - a'$ is non-degenerate and commode. Thus, allowing a to vary near a' gives a deformation of $f_0(x) = a'$ with the same global Newton polyhedron. We conclude by theorem 4, that $(f_0(x) - a)^{-1}(0)$ is a topologically trivial fibration for a near a'. □

PROOF (of theorem 4). The outline of the proof is as follows. The homogenization of f with respect to $\mathbf{x}$ gives a germ homogeneous of degree m in $(x_0, x_1, \ldots, x_n)$ which is a deformation of f_0 of constant degree. By lemma 6.3 (which must still be proven), F_0 is non-degenerate; and in the case of part 1, $\tilde{\Gamma}(f_0)$ well-situated means that $\Gamma(F_0)$ satisfies condition (3.7). Then, we see that the proof of theorem 3 not only proves that F is a topologically trivial deformation of F_0 as a germ; but, in fact, the family of homeomorphisms extends to a k^*-equivariant family on k^{n+1} which preserves the hyperplane $x_0 = 0$. This is used to induce a homeomorphism of the $x_0 = 1$ affine plane. This homeomorphism only preserves $F^{-1}(0)$; however, this will yield the result.

The proof of theorem 3 in the case of 1) or theorem 1 in the case of 2) allows us to solve

$$(x^{\alpha})^r \cdot \frac{\partial F}{\partial u_i} = \Sigma\, h^{(i)}_{\alpha,j}\, x_j \frac{\partial F}{\partial x_j}$$

Since $(\partial F/\partial u_i)$ and $x_j(\partial F/\partial x_j)$ have the same degree, taking terms of $h^{(i)}_{\alpha,j}$ of degree $= r \cdot |\alpha|$, we must have equality. Thus, we may assume $\deg(h^{(i)}_{\alpha,j}) = r \cdot |\alpha|$. Also, the proof allows us to assume $\mathrm{fil}(h^{(i)}_{\alpha,j} \cdot x_j) > r \cdot |\alpha|$. Multiplying by $-(\overline{x}^{\alpha})^r$ and summing over $\alpha \in \mathrm{ver}(A_{\ell,x})$, with A_ℓ a fit level for F_0, we obtain

$$-\rho \frac{\partial F}{\partial u_i} = \xi_i(F) = \Sigma\, \xi^{(i)}_j\, x_j \frac{\partial F}{\partial x_j}$$

where

$$\xi^{(i)}_j = \Sigma - (\overline{x}^{\alpha})^r \cdot h^{(i)}_{\alpha,j}$$

Then, for $\lambda \in k^*$

$$\rho(\lambda \cdot x) = |\lambda|^{2r\ell}\rho(x) \quad \text{and} \quad \xi^{(i)}(\lambda \cdot x,u) = |\lambda|^{2r\ell}\xi^{(i)}(x,u)$$

Thus, letting k^* act trivially on k^q, we conclude that

$$\xi_i' = \begin{cases} \dfrac{\partial}{\partial u_i} & \text{on } S \text{ (where appropriate)} \\[2ex] \dfrac{\partial}{\partial u_i} + \rho^{-1}\xi_i' & \text{off } S \end{cases}$$

is k^*-equivariant. Thus, by the k^*-action, we can define it on $k^{n+1} \times U$, for U a neighborhood of 0 in k^q (or $(k^{n+1}\backslash S) \times U$ as appropriate). Thus, the integral curves of ξ_i' are transformed to integral curves by the k^*-action. Also, when defined where $x_0 = 0$, ξ_i' is tangent to this plane. Since ξ_i' is locally integrable, the k^*-equivariance implies it is integrable on $k^{n+1} \times U_1$ (or $(k^{n+1}\backslash S) \times U_1$). For example, for $i = 1$ and $\widehat{\Gamma}(f_0)$ well-situated, we obtain a flow α_t and define ϕ as in §2, $\phi(x,u) = \alpha_{u_1}(x,0,u_2, \ldots, u_q)$. Actually, ϕ is not defined at 0. However, this corresponds to the x_0-axis in

k^{n+1}. By the same argument used in §2, "fact 3", the integral curves near the x_0-axis remain close to it. Hence, ϕ maps arbitrarily small neighborhoods of 0 into neighborhoods of 0 and so extends continuously to 0. Also, if $F_1 = F|k^{n+1} \times k^{q-1}$, then

$$(6.8) \qquad (F_1 \times id) \circ \phi = F.$$

Now, we define ϕ': $k^n \times U_1 \to k^n \times U_1$. Let ϕ_1 be the x_0-coordinate of ϕ and let ϕ_2 denote the composition with projection onto the remaining x-coordinates. Then $\phi_1(x,u) \neq 0$ if $x_0 \neq 0$, because ϕ preserves $\{x_0 = 0\}$. Thus, we define

$$(6.9) \qquad \phi(x_1, ..., x_n, \mathbf{u}) = ((\phi_1(1,x_2, ..., x_n,\mathbf{u}))^{-1} \cdot \phi_2(1,x_2, ..., x_n,\mathbf{u}),\mathbf{u}).$$

Then, 6.8 and the k^*-equivariance of ϕ imply

$$(6.10) \qquad (f_1 \times id) \circ \phi = \lambda(x,u) \cdot f$$

where $f_1 = f|k^n \times k^{q-1}$ and $\lambda(x,u) \in \mathbb{C}^*$. If ϕ is a homeomorphism, (6.10) gives topological contact equivalence between f and $f_1 \times id$. That ϕ is a homeomorphism follows because its inverse is obtained by applying the same argument to $-\xi_1'$. This implies the ϕ is a homeomorphism. We obtain the result by applying this argument together with standard arguments as in [D1, §6] to obtain global topological triviality of the family. □

To prove lemma 6.3, for a polynomial of degree k, $g(x)$, we let homog(g) denote the homogenization of g with respect to x_0. Suppose Δ is a compact face of $\Gamma(F_0)$ which projects along the x_0-axis to $\Delta' \subset \check{\Gamma}_+(f_0)$. Then

$$x_i \frac{\partial F_0}{\partial x_i}\Big|_{\bar{\Delta}} = x_0^{m-r_i} \operatorname{homog}\left(x_i \frac{\partial f_0}{\partial x_i}\Big|_{\bar{\Delta}'}\right)$$

where $r_i = \deg\left(x_i \frac{\partial f_0}{\partial x_i}\Big|_{\bar{\Delta}'}\right)$. We also have

$$x_0 \frac{\partial F_0}{\partial x_0}\Big|_{\bar{\Delta}} = \text{homog}\left(\left[\sum_{\alpha}(n - |\alpha|)a_{\alpha}x^{\alpha}\right]\Big|_{\bar{\Delta}'}\right)$$

where $f_0 = \Sigma\, a_{\alpha}x^{\alpha}$. Thus,

$$x_0 \frac{\partial F_0}{\partial x_0}\Big|_{\bar{\Delta}} = \text{homog}\left(n \cdot f_0|_{\bar{\Delta}'} - \sum_{i=1}^{n} x_i \frac{\partial f_0}{\partial x_i}\Big|_{\bar{\Delta}'}\right)$$

$$= n \cdot F_0|_{\bar{\Delta}} - \sum_{i=1}^{n} x_i \frac{\partial F_0}{\partial x_i}\Big|_{\bar{\Delta}'}.$$

Hence, if $x_i \frac{\partial F_0}{\partial x_i}\Big|_{\bar{\Delta}} = 0$, $i = 0, \ldots, n$ at $(x_0, x_1, \ldots, x_n)$ where $x_0 \neq 0$,

then $f_0|_{\bar{\Delta}'} = 0$ and all $x_i \frac{\partial f_0}{\partial x_i}\Big|_{\bar{\Delta}'} = 0$ at $((x_1/x_0), \ldots, (x_n/x_0))$.

However, either $\bar{\Delta}'$ is a face of $\tilde{\Gamma}(f_0)$ or $= \tilde{\Gamma}_+(f_0)$; thus, some $x_j/x_0 = 0$ by global non-degeneracy. Hence, F_0 is non-degenerate. □

§7. EXAMPLES

Local Examples

1. $f_0 = x^2y^2 + x^2z^2 + y^2z^2$. As was pointed out in §3, f_0 is strongly nondegenerate; hence in the complex case its topological type is determined by its Newton polyhedron. In particular, if $\text{supp}(h) \subset \Gamma_+(f_0)\backslash\Gamma_0(f_0)$, then $f_0 + h$ is topologically equivalent to f_0.

2. $f_0 = x^8y^8 + x^8z^8 + z^8y^8 + x^2y^2z^2$. Then, f_0 is non-degenerate, so its topological type off of S = union of the coordinate axes is determined by its Newton polyhedron if $k = \mathbb{C}$.

3. $f_0 = x^2y^2z^2 + x^3y^7 + y^3x^7 = x^2y^2(z^2 + xy^5 + yx^5) = x^2y^2 \cdot g_0(x,y)$. Then, f_0 is non-degenerate with $\Gamma(f_0)$ as shown in (7.1).

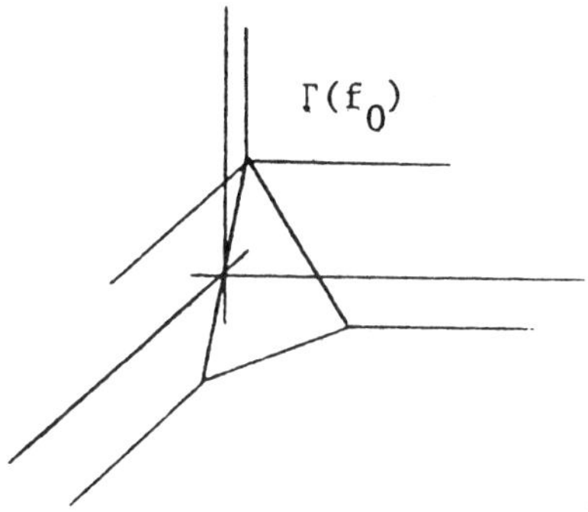

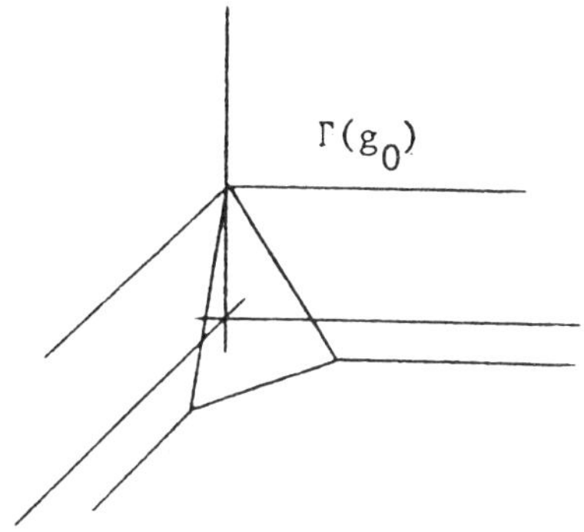

Since $\Gamma(g_0)$, as shown, has at most one unbounded direction for an edge or 2 unbounded directions for z^2, f_0 satisfies condition (3.7) so again the topological type of f_0 is determined by $\Gamma(f_0)$.

4) *transversal A_1-singularities* (see [S] or [P])

$$f_0 = \Sigma\, c_i x^{T_i} y_i^2 \,(\mathrm{mod}\ (y_1, \ldots, y_n)^3), \quad c_i \neq 0.$$

If f_0 is non-degenerate (and say commode for $(y_1, \ldots, y_n)$) then, the only unbounded direction is x and f_0 is strongly non-degenerate.

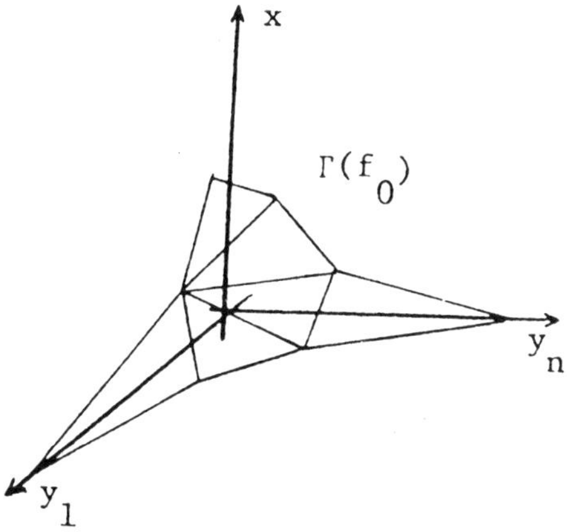

5) $f_0 = (z^2 + xy^3)^2 + x^2y^7 + x^4y^2 + x^8y$. $\Gamma(f_0)$ is shown in Fig. (7.2). It is non-degenerate (and hence strongly non-degenerate) on

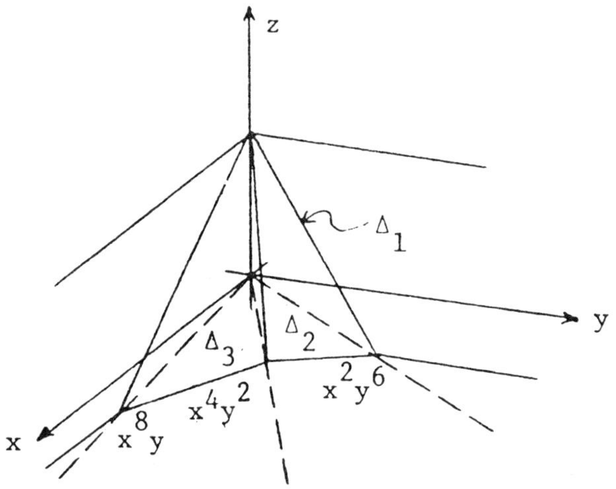

Figure 7.2

all faces except Δ_1 and Δ_2. We may use weights $w_{\Delta_2} = (12,6,15)$ and $w_{\Delta_3} = (5,20,15)$ with $m_0 = 60$. For Δ_1 we have

$$x \frac{\partial f_0}{\partial x}\Big|_{\Delta_1}, \quad y \frac{\partial f_0}{\partial y}\Big|_{\Delta_1} \quad \text{and} \quad z \frac{\partial f_0}{\partial z}\Big|_{\Delta_1}$$

divisible by $z^2 + xy^3$. Then,

$$2z^2x \frac{\partial f_0}{\partial x} - xy^3z \frac{\partial f_0}{\partial z} = 2z^2(2x^2y^7 + 4x^4y^2 + 8x^8y)$$

and

$$y \frac{\partial f_0}{\partial y} - 3x \frac{\partial f_0}{\partial x} = x^2y^7 - 10x^4y^2 - 23x^8y.$$

Thus, we let

$$\begin{aligned} \zeta &= 5xy^2\left[2z^2x \frac{\partial}{\partial x} - xy^3 \cdot z \frac{\partial}{\partial z}\right] + 4z^2xy^2\left[y \frac{\partial}{\partial y} - 3x \frac{\partial}{\partial x}\right] \\ &= -2z^2xy^2 \cdot x \frac{\partial}{\partial x} - 5x^2y^5z \frac{\partial}{\partial z} + 4z^2xy^2 \cdot y \frac{\partial}{\partial y} \end{aligned}$$

Then, $\text{in}(\zeta(f_0)) = 24z^2x^3y^9$. Thus, $\{x(\partial/\partial x),\ z(\partial/\partial z),\ \zeta\}$ is a set of local patching data for both faces on which f_0 is degenerate, namely Δ_1 and Δ_2. Since $\mathrm{fil}(z^2xy^2 \cdot y) > \mathrm{fil}(z^2xy^2)$ for the unbounded direction y for Δ_1, this is a set of strong local patching data. Then, jump(ζ) = 120 − 114 = 6; hence jump(Δ_2) = jump(Δ_1) = 6. Thus, a deformation of f_0 by terms of filtration $\geqslant 66$ satisfies a simple jump condition and hence is topologically trivial. Moreover, the terms x^3yz^2, x^5y^2, x^8y and x^9y are terms of filtration < 66 which satisfy a general jump condition. For example, for x^3yz^2,

$$z^4 \cdot x^3yz^2 \equiv x^3yz^2 \cdot z \frac{\partial f_0}{\partial z} \bmod A_{126,x}$$

$$x^8y \cdot x^3yz^2 \equiv (1/6)x^3yz^2x \frac{\partial f_0}{\partial x} - (1/3)x^3yz^2 \cdot y \frac{\partial f_0}{\partial y} \bmod A_{126,x}$$

$$x^4y^2 \cdot x^3yz^2 \equiv (-1/12)x^3yz^2x \frac{\partial f_0}{\partial x} + (2/3)x^3yz^2 \cdot y \frac{\partial f_0}{\partial y} \bmod A_{126,x}$$

and

$$x^2y^6 \cdot x^3yz^2 \in A_{126,x}\,.$$

Global Examples

6) $f_0 = x^5y + x^2y^6 + x^9y^2 + x^{10}y^8$. A computation shows that f_0 is non-degenerate. $\breve{\Gamma}(f_0)$ is as shown below,

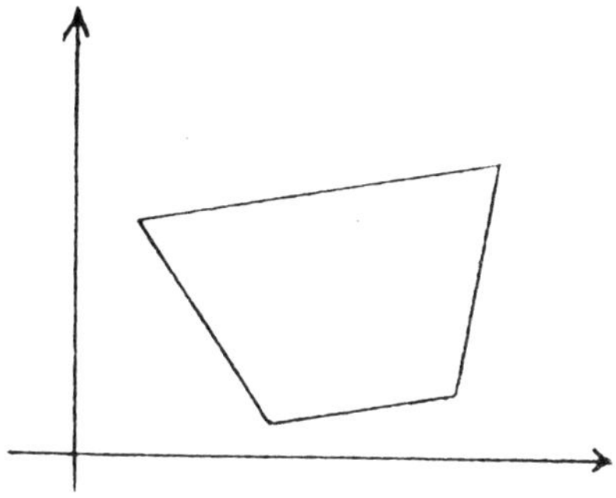

By a remark in §6, $\breve{\Gamma}(f_0)$ is well-situated. Hence, for any complex polynomial g which is non-degenerate with $\breve{\Gamma}(g) = \breve{\Gamma}(f_0)$, $g^{-1}(0)$ is topologically equivalent to $f_0^{-1}(0)$.

7) $f_0 = x^5 + y^5 + z^5 + x^{10}y^{10}z^{10}$ is both non-degenerate and commode. Hence, in the complex case, the topological type of $f_0^{-1}(0)$ is determined by $\tilde{\Gamma}(f_0)$. Moreover, we see that if $a \neq$ critical value of f_0 then $h_0 = f_0 - a$ is non-degenerate. For a face of $\tilde{\Gamma}(h_0)$, the non-degeneracy follows from that of f_0 and $a \neq 0$. For $\tilde{\Gamma}_+(h_0)$ itself, the non-degeneracy requires that $x_i(\partial h_0/\partial x_i) = x_i(\partial f_0/\partial x_i) = 0$ and $h_0 = 0$ i.e., $f_0 = a$ then $\Pi_{i=1}^{n} x_i = 0$. If all $x_i \neq 0$, this implies a is a critical value of f_0. Hence, by Corollary 8, the global Milnor fibration is a locally trivial fibration off the set of critical values.

8) $f_0 = x^8y^8 + 2x^5y^5z + x^2y^2z^2 + x^8z^8 + z^8y^8$ is a degenerate version of example 2. Although it is degenerate, if $a \neq 0$, $h_0 = f_0 - a$ is non-degenerate with $\tilde{\Gamma}(h_0)$ as shown below provided a is not a critical value. Hence, the global Milnor fibration is locally trivial off of the set of critical values.

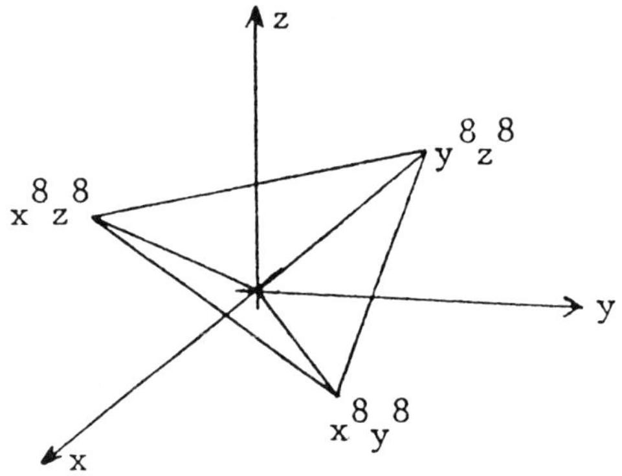

BIBLIOGRAPHY

B Broughton, S. A. *On the Topology of Polynomial Hypersurfaces*, Proc. Sym. Pure Math. 40 (1983) pp. 167-178.

D1 Damon, J. *Finite Determinacy and Topological Triviality I*, Invent. Math. 62 (1980), pp. 299-324.

D2 Damon, J. *Newton Filtrations, Monomial Algebras, and Nonisolated and Equivariant Singularities*, Proc. Sym. Pure Math. 40 (1983) pp. 267-278.

DG Damon, J. and Gaffney, T. *Topological Triviality and Deformations of Functions and Newton Filtrations*, Invent. Math. 72 (1983) pp. 335-358.

Da Danilov, V. I. *The Geometry of Toric Varieties*, Russian Math. Surveys 33 (1978) pp. 97-154.

K Kouchnirenko, A. G. *Polyedres de Newton et Nombres de Milnor*, Invent. Math. 32 (1976) pp. 1-31.

L Looijenga, E. *Semi-universal Deformation of a Simple Elliptic Hypersurface Singularity I: Unimodularity*, Topology 16 (1977) pp. 257-262.

O Oka, M. *On the Topology of the Newton Boundary III*, J. Math. Soc. Japan 34 (1982) pp. 541-549.

P Pellikaan, G. R. *Hypersurface Singularities and Resolutions of Jacobi Modules*, Thesis, Univ. of Utrecht, 1985.

S Siersma, D. *Isolated Line Singularities*, Proc. Sym. Pure Math. 40 (1983) pt II, pp. 485-496.

V Varchenko, A. N. *Zeta Function of Monodromy and Newton's Diagram*, Invent. Math. 37 (1976) pp. 253-262.

DEPARTMENT OF MATHEMATICS
UNIVERSITY OF NORTH CAROLINA
CHAPEL HILL, NORTH CAROLINA 27599

Contemporary Mathematics
Volume 90, 1989

THE COHOMOLOGY AND FUNDAMENTAL GROUP OF A HYPERPLANE COMPLEMENT

Michael Falk

§1. INTRODUCTION

Let $\mathbf{A} = H_1, ..., H_n$ be an arrangement of hyperplanes through 0 in $\mathbf{C}^\ell$, with complement $M = C^\ell - \cup_{i=1}^n H_i$. The cohomology algebra $H^*(M)$ is generated by degree one elements $\omega_1, ..., \omega_n$ dual to the hyperplanes. The relations among the ω_i have been described explicitly in terms of the intersection lattice of the arrangement [13], and this structure has been studied extensively—recent examples include [16,§3] and several papers in these Proceedings. In this article we will describe a relatively new technique for the analysis of these algebras, the Sullivan one-minimal model. The theory of minimal models is well suited algebraically to the study of arrangements, because M is a formal space, in the sense of [15], and also because the presentation of $H^*(M)$ is so nice. The topological consequences are somewhat limited, however, because $\pi_1(M)$ is so complicated. One does obtain information on the factors of the lower central series of $\pi_1(M)$—these results have been described at length in other papers [4,5,11]. In this paper we will view the minimal model mainly as a tool for distinguishing the cohomology algebras associated with combinatorially distinct arrangements.

It is known [1] that $H^p(M;\mathbf{Z})$ is free abelian, and the rank of $H^p(M)$ may be computed using Mobius functions [13]. So it is easy to determine whether two arrangements have cohomology rings that are isomorphic as graded modules. It is much more difficult to discern the existence of ring isomorphisms in this setting. This fact is not surprising, in that the standard description of $H^*(M)$ is in terms of generators and relators. On the other hand, this presentation of $H^*(M)$ is precisely what facilitates the construction of the one-minimal model S. In many cases, certain numerical invariants $\{\phi_n\}$ of S will suffice to distinguish the algebras $H^*(M;\mathbf{R})$ for arrangements with identical sets of betti numbers.

According to Sullivan's theory [15], these numerical invariants have homotopy-theoretic significance in terms of the lower central series of $\pi_1(M)$. The main result of [6] says that, for a pair of *fiber-type* arrangements with the same betti numbers, the invariants ϕ_n will necessarily agree. It is still an open question whether the cohomology algebras are isomorphic in this case. We have found a second sequence $\{\psi_n\}$ of *combinatorial* invariants in the one-minimal model which may serve to classify the cohomology algebras of these fiber-type arrangements. At this stage we can only conjecture that the ψ_n are actually algebraic invariants.

Here is a brief outline of this paper. In §2 we construct the one-minimal model S from the presentation of $H^*(M)$ given in [13]. We then derive the main results of [4] relating to the lower central series of $\pi_1(M)$. In particular we reproduce the main result of [6] concerning the LCS formula, without using the commutator calculus in $\pi_1(M)$. Rigorous proofs are not given, as the reader may consult [4] for all the details.

In §3 we develop some other consequences of the construction. These include geometric criteria which have some bearing on the LCS formula. Also, we derive some bounds on the ϕ_n. The first such result (3.4) shows that arrangements satisfying the LCS formula have in some sense the "smallest" fundamental group (modulo the fourth commutator subgroup) among arrangements with a given set of betti numbers. The more interesting result is the second (3.8), which provides a lower bound ψ_n on ϕ_n for each n. This value for $n = 3$ distinguishes local and global "second-order" relations among the generators ω_i of $H^*(M)$. In §4 we illustrate these ideas with two examples. Here we state the conjecture that the ψ_n are algebraic invariants. We also conjecture that, together with ϕ_3 and the betti numbers of M, ψ_3 suffices to determine the isomorphism type of the algebra $H^*(M)$, for arrangements in 3 dimensions.

Many aspects of this work parallel the work of T. Kohno on the holonomy Lie algebra of M [8,9,11], which is dual to the one-minimal model. Kohno's construction holds in a more general setting, but its development requires a considerable measure of sophisticated machinery. We have benefited from a lively correspondence with Kohno over the last two years, for which we are grateful. We also thank the editor of this volume for our many enjoyable discussions concerning arrangements of hyperplanes.

§2. CONSTRUCTION AND BASIC RESULTS

In this section we describe the one-minimal model S and its relation to the cohomology algebra $H^*(M)$. We then show that the LCS formula of [6] will hold under the hypothesis that $H^*(S) \cong H^*(M)$. This condition is satisfied by fiber-type arrangements, so the main result of [6] follow from our cohomological approach. The proof of this result in [6] relies on Witt's formula [12] and some delicate commutator calculations in $\pi_1(M)$. By contrast, the argument here does not involve $\pi_1(M)$ at all, and Witt's formula becomes a corollary. In addition, the proof does not depend on the existence of iterated fiberings, and may apply in situations where the Poincare polynomial does not even factor. The problem of finding such examples will be addressed in §3.

A thorough treatment of this material appears in [4], to which we refer the reader for proofs.

Let $\mathbf{A} = \{H_1, ..., H_n\}$ be an arrangement of hyperplanes in $\mathbf{C}^\ell$, with complement $M = \mathbf{C}^\ell - \cup_{i=1}^n H_i$. Orlik and Solomon [13] provide us with a presentation of $H^*(M)$ as a quotient of the exterior algebra $E = \Lambda(e_1, \ldots, e_n)$ by a certain homogeneous ideal I. (We work over the reals throughout this paper.) This algebra $A = E/I$ is discussed in several papers in this volume. We postpone further discussion of I until we need it. For now we merely note that $I^1 = 0$. Let us describe the differential graded algebra S. Set $S(1) = E$, and ρ: $S(1) \to A$ the natural projection. The differentials on $S(1)$ and on A are identically zero. For $n \geqslant 2$, assume we have constructed $S(n-1)$ and a map ρ: $S(n-1) \to A$ satisfying $\rho \cdot d = 0$. Set $V_n = \ker(\rho^*: H^2(S(n-1)) \to A^2)$ and define $S(n) = S(n-1) \otimes \Lambda(V_n)$. The differential is extended from $S(n-1)$ to $S(n)$ by taking d: $V_n \to S(n-1)^2$ to be a linear choice of representatives, and extending via Leibniz' rule. The map ρ is extended to $S(n)$ by setting $\rho|V_n = 0$. Clearly these extensions satisfy $\rho \cdot d = 0$. Let $S = \cup_{n=1}^{\infty} S(n)$. We have a d.g. algebra map ρ: $S \to A$ which induces isomorphisms $H^p(S) \to A^p$ for $p = 1,2$. That S is the one-minimal model of M follows from the fact [1] that A is isomorphic to a d.g. subalgebra of the DeRham complex of M which carries the cohomology of M (i.e. M is a formal space).

Let $S(0) = R$, and $V_1 = E^1$, so that $S(1) = S(0) \otimes \Lambda(V_1)$. The following result will be useful in §3.

PROPOSITION 2.1. *The differential* d: $S^1 \to S^2$ *is zero on* V_1 *and injective on* $\oplus_{n=2}^{\infty} V_n$.

Now, by definition of the relation ideal I, the degree-two part I^2 is spanned by the elements $e_{pq} - e_{pr} + e_{qr}$, where $H_p, H_q, H_r \in \mathbf{A}$ satisfy $\operatorname{codim}(H_p \cap H_q \cap H_r) = 2$. (Here e_{ij} stands for $e_i \wedge e_j$.) The next result follows from the definition of $S(2)$.

PROPOSITION 2.2. *The differential* d: $V_2 \to S(1)^2 = E^2$ *is an isomorphism onto* I^2. *The differential* d: $S(2)^p \to S(2)^{p+1}$, *restricted to* $E^{p-1} \wedge V_2$, *is equivalent to the multiplication map* $E^{p-1} \otimes I^2 \to E^{p+1}$.

Next we define a graded subspace U_k of S consisting of the elements of "index sum" k. Precisely, $U_k^p = \bigoplus_I V_{i_1} \wedge \cdots \wedge V_{i_p}$, summing over all $I = \{i_1, \ldots, i_p\}$ satisfying $1 \leqslant i_1 \leqslant \cdots \leqslant i_p$ and $i_1 + \cdots + i_p = k$. By convention $U_0 = S(0)$.

PROPOSITION 2.3.

(i) $S = \bigoplus_{k \geqslant 0} U_k$

(ii) $U_k^p = 0$ *if* $p > k$

(iii) $U_k^1 = V_k$

(iv) $U_k^k = E^k$

(v) $U_k^{k-1} = E^{k-2} \wedge V_2$

(vi) $\rho|_{U_k^p} = 0$ *for* $p < k$.

The following technical result is the basis for all our calculations.

PROPOSITION 2.4. *The differential* $d\colon S^p \to S^{p+1}$ *preserves index sum.*

Thus U_k forms a subcomplex of S, finite in length according to (2.3), and $H^*(S) = \oplus_{k \geqslant 0} H^*(U_k)$. The proof of the next result is an adaptation of one of Kohno's arguments [11] to our setting.

PROPOSITION 2.5. $\prod_{n \geqslant 1} (1 - t^n)^{\dim(V_n)} = \sum_{k \geqslant 0} \chi(U_k) t^k.$

Here the $\chi(U_k)$ denotes the Euler characteristic of U_k.

Now, according to Sullivan, $\dim(V_n)$ is equal to the rank ϕ_n of the nth factor in the lower central series of $\pi_1(M)$. The LCS formula of [6] relates these ranks to the betti numbers of M. It reads

$$(*) \qquad \prod_{n \geqslant 1} (1 - t^n)^{\phi_n} = P_M(-t),$$

where $P_M(t) = \sum_{p \geqslant 0} \dim(H^p(M)) t^p$. Immediately we obtain the following.

PROPOSITION 2.6. *The* LCS *formula* (*) *holds if and only if* $\chi(U_p) = (-1)^p\dim(H^p(M))$ *for all* p.

Now, suppose $\rho^*\colon H^p(S) \to A^p$ is an isomorphism for all p. Then, by (2.3) (vi), we must have $H^p(U_k) = 0$ for $p < k$. Then (2.3) (ii) implies $H^p(S) = H^p(U_p)$. Then the standard Euler characteristic argument shows that $\chi(U_p) = (-1)^p\dim H^p(U_p) = (-1)^p\dim H^p(S) = (-1)^p\dim A^p$. Since $A \cong H^*(M)$ we have the following result.

THEOREM 2.7. *If* $\rho^* = H^p(S) \to A^p$ *is an isomorphism for all* p, *then the* LCS *formula* (*) *holds.*

This result has also been obtained by Kohno [8] using different methods.

Arrangements satisfying the hypothesis of (2.7) are called *rational* $K(\pi,1)$. The terminology is justified by rational homotopy theory. Under this assumption, S is the minimal model of M, and the subspace of indecomposable elements of S of degree p is dual to the pth homotopy group (tensored with R) of $\mathbf{Q}_\infty M$, the rational nilpotent completion of M [2,3]. Thus $\mathbf{Q}_\infty M$ is aspherical precisely when $\mathbf{A}$ is a rational $K(\pi,1)$.

This observation can be used to show that fiber-type arrangements are rational $K(\pi,1)$. The linear fibrations in the definition of fiber-type are nilpotent fibrations (in fact, the action is trivial), so they are preserved upon passing to the completions. One then uses induction in the exact homotopy sequences, together with the fact [3] that $\mathbf{Q}_\infty F$ is aspherical when F is aspherical with $\pi_1(F)$ free of finite rank.

Thus (2.7) implies that fiber-type arrangements satisfy the LCS formula, ostensibly generalizing the main result of [6]. In actuality, the existence of rational $K(\pi,1)$ arrangements which are not fiber-type has not been clearly established (cf. (3.2) and the remark following (2.8)).

Now, from the discussion preceding (2.7), we see that a rational $K(\pi,1)$ arrangement will necessarily have $\rho^*\colon H^p(U_p) \to A^p$ an isomorphism for all p. Using (2.3), we see that $H^p(U_p)$ is the cokernel of the map $d\colon E^{p-2} \wedge V_2 \to E^p$. By (2.2), the image of this map is J^p, where J is the ideal generated by I^2, the degree-two part of the relation ideal. Thus $d\colon H^p(U_p) \to A^p$ is an isomorphism for all p if and only if $J = I$. In this case we say the arrangement is *2-determined*.

PROPOSITION 2.8. *If* **A** *is rational* $K(\pi,1)$, *then* **A** *is 2-determined.*

The converse holds under the additional assumption that $H^k(U_p) = 0$ for $k < p$. It is not known whether this condition can be dropped.

We close this section with a result which is useful for the computation of dim V_n.

PROPOSITION 2.9. *The dimension of* V_n *is equal to the kernel rank of the map* $d\colon U_n^2 \to U_n^3$.

PROOF. We know that $\rho^*\colon H^p(S) \to A^p$ is an isomorphism for $p = 1,2$. Thus, by (2.3), $H^1(U_n) = H^2(U_n) = 0$, so the sequence $0 \to U_n^1 \to U_n^2 \to U_n^3$ is exact. Since $U_n^1 = V_n$, we are done.

§3. PERIPHERAL RESULTS

We start this section with some observations concerning 2-determined arrangements. The pictures in §3 and §4 depict the projective image of real arrangements in $\mathbf{R}^3$.

PROPOSITION 3.1. *A 3-arrangement* **A** *is 2-determined if and only if* $J^3 = I^3$.

PROOF. We need to show $J^3 = I^3$ implies $J = I$. We will show $J^p = I^p$ by induction. Let $p \geqslant 4$. Since we are working in $\mathbf{C}^3$, $A^p = 0$, so $I^p = E^p$. Also, any set of p hyperplanes is dependent, so for any $e = e_{i_1} \cdots e_{i_p} \in E^p$, $\partial e \in I^{p-1}$. (We use the notation and terminology of [13].) Thus $\partial e \in J^{p-1}$ by the inductive hypothesis. The observation that $e = e_{i_1}(\partial e)$ finishes the proof.

A 3-arrangement **A** is *parallel* if for any four hyperplanes H_p, H_q, H_r, $H_s \in$ **A** in general position, there is a fifth hyperplane $H_t \in$ **A** satisfying $H_q \cap H_r \subseteq H_t$ and $H_p \cap H_t \subseteq H_s$. The terminology comes from the picture:

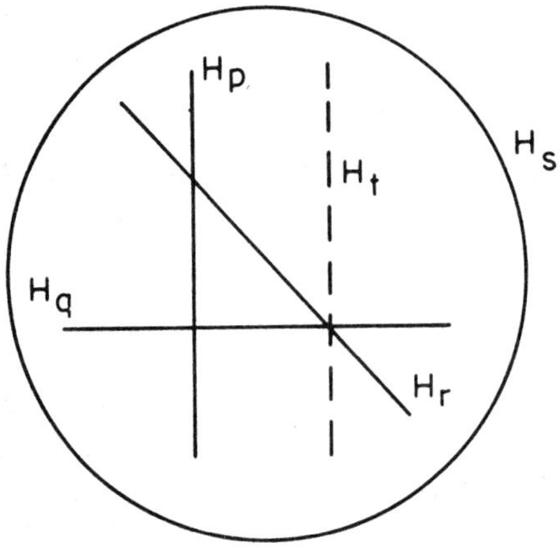

PROPOSITION 3.2. *If* **A** *is a parallel* 3*-arrangement, then* **A** *is* 2*-determined.*

PROOF. By (3.1), we must show $I^3 \subseteq J^3$. For this it suffices to take a dependent set $\{H_p, H_q, H_r, H_s\}$ of four hyperplanes in **A** and show that $\partial e_{pqrs} \in J^3$. If $H_p \cap H_q \cap H_r$ has codimension two, we have $\partial e_{pqrs} = (e_s - e_p)\partial e_{pqr} \in J^3$. If the four hyperplanes are in general position, then, by assumption, there is an $H_t \in$ **A** as in the picture above. Note that ∂e_{qrt} and ∂e_{pst} are in $I^2 \subseteq J$. The result then follows from the identity $(e_r - e_q)\partial e_{pst} + (e_s - e_p)\partial e_{qrt} = \partial e_{pqrs}$.

The notion of parallel was introduced by Kohno in [10], where he proves a version of (3.2). He asserts that parallel 3-arrangements are rational $K(\pi,1)$ (see remarks following (2.8)), but the details are sketchy and have not been independently verified. (This work remains unpublished.)

There are 2-determined 3-arrangements which are not parallel. For example,

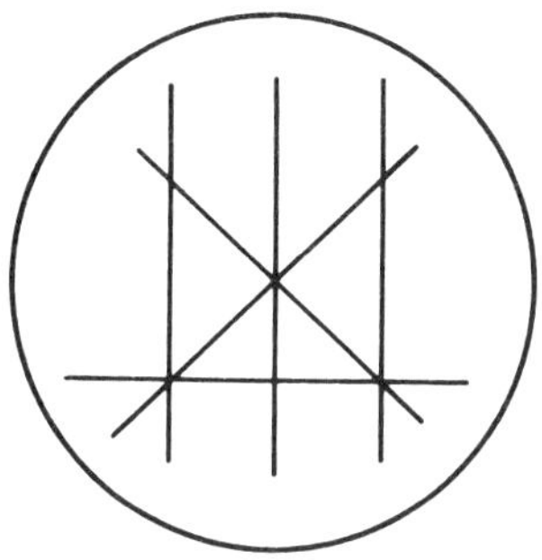

is fiber-type, hence 2-determined, but not parallel. At the other extreme we have the following result, proof of which appears in [4].

PROPOSITION 3.3. *Suppose* **A** *is an* ℓ*-arrangement containing* $(\ell+1)$ *hyperplanes in general position whose codimension two intersections have multiplicity two in* **A**. *Then* **A** *is not* 2*-determined.*

Let us return to the LCS formula of (2.5) and (2.6). Let $\pi = \pi_1(M)$, with lower central series $\pi = \pi^1 \supseteq \pi^2 \supseteq \pi^3 \supseteq \pi^4 \supseteq \cdots$. The rank of π^n/π^{n+1} is ϕ_n. The next result says that the quotient π/π^4 is "smallest" among arrangements with given betti numbers when the LCS formula is satisfied.

PROPOSITION 3.4. *Suppose* M *has betti numbers* $b_1, \ldots, b_\ell$. *Then* $\phi_1 = b_1$, $\phi_2 = \binom{b_1}{2} - b_2$, and $\phi_3 \geq b_3 + \phi_1\phi_2 - \binom{\phi_1}{3}$, *with equality when the* LCS *formula is satisfied to degree* 3.

PROOF. The definition of V_1 gives the first assertion; the second follows from (2.2), since $\dim(E^2/I^2) = b_2$. To establish the third, we examine U_3:

$$0 \to V_3 \to V_1 \wedge V_2 \to E^3 \to 0.$$

Now $H^1(U_3) = H^2(U_3) = 0$ (as in the proof of 2.9), and (2.2) implies $H^3(U_3) = E^3/J^3$. Since $J^3 \subseteq I^3$, and $\dim(E^3/I^3) = b_3$, we have

$\dim(E^3/J^3) \geq b_3$, and $\chi(U_3) = -\phi_3 + \phi_1\phi_2 - \binom{\phi_1}{3} = -\dim(E^3/J^3) \leq -b_3$. The result follows.

Set $\gamma_3 = b_3 + \phi_1\phi_2 - \binom{b_1}{3}$, so that $\phi_3 = \gamma_3$ when the LCS formula holds to degree 3.

PROPOSITION 3.5. *If* **A** *is a 2-determined arrangement, then* $\phi_3 = \gamma_3$.

PROOF. The assumption gives us $J^3 = I^3$, and the result follows from the proof of (3.4).

The converse holds for 3-arrangements. Indeed, the proof of (3.4) shows that $\phi_3 = \gamma_3$ implies $J^3 = I^3$. Then (3.1) tells us that **A** is 2-determined.

PROPOSITION 3.6. *If* **A** *is a 3-arrangement, then* **A** *is 2-determined if and only if* $\phi_3 = \gamma_3$.

Next we establish a lower bound on ϕ_n, for each n, which arises from "local versus global" considerations. Let L be the intersection lattice of the arrangement, $L = \{\cap_{H\in \mathbf{B}} H \mid \mathbf{B} \subseteq \mathbf{A}\}$, ordered by reverse inclusion. The *rank* of $X \in L$ is $r(X) = \operatorname{codim}(X)$. For $X \in L$, let $\mathbf{A}_X$ be the arrangement $\{H \in \mathbf{A} \mid H \supseteq X\}$ with complement M_X, and cohomology algebra $A_X = E_X/I_X$. Note that $E_X \subseteq E$ and $I_X \subseteq I$. Let $S(X)$ be the one-minimal model of A_X as constructed in §2, with canonical series $V_n(X)$. Set $\phi_n(X) = \dim(V_n(X))$. The *multiplicity* of X is $m(X) = \#\mathbf{A}_X$.

LEMMA 3.7. *For* $X \in L$ *with* $r(X) = 2$, *we have* $\phi_n(X) = N_n^{m(x)-1}$, *for all* $n \geq 2$.

Here N_n^r denotes the Witt number [12]; namely ϕ_n for the free group of rank r. The lemma holds because $\mathbf{A}_X$ is essentially a 2-arrangement, hence is fiber-type, with exponents $d_1 = 1$ and $d_2 = m(X) - 1$. By [6], this yields $\phi_n(X) = N_n^1 + N_n^{m(x)-1}$ for all n, and

$N_n^1 = 0$ for $n \geq 2$.

The Witt numbers are given by the formula $N_n^r = \frac{1}{n}\Sigma_{d|n}\mu(d)r^{n/d}$, where μ is the number-theoretic Möbius function. For example, with $n = 3$ we obtain $N_3^1 = 0$, $N_3^2 = 2$, $N_3^3 = 8$, $N_3^4 = 20$, and $N_3^5 = 40$.

We aim to prove the following.

PROPOSITION 3.8. *For all $n \geq 2$*

$$\phi_n \geq \sum_{\substack{X\in L \\ r(X)=2}} N_n^{m(X)-1}, \quad \textit{or, equivalently,}$$

$$\phi_n \geq \sum_{r\geq 3} c(r)N_n^{r-1}, \quad \textit{where } c(r) \textit{ is the number of rank 2 lattice elements of multiplicity } r.$$

The inductive proof will be accomplished by a sequence of lemmas. First of all, the inclusions $M \to M_X$ induce homomorphisms $A_X \to A$ and $S(X) \to S$. The latter restrict to maps i_X: $V_n(X) \to V_n$ for each n. In this way we obtain a homomorphism i_n: $\oplus_X V_n(X) \to V_n$ for each n. Here $\oplus_X$ denotes the sum over $X \in L$ with $r(X) = 2$. By (3.7) we need to show that i_n is injective for $n \geq 2$.

LEMMA 3.9. *The map i_2 is injective.*

PROOF. We are just asserting that $\oplus_X I_X^2 \to I^2$ is injective. This follows from the definition of I: I^2 is spanned by ∂e_{pqr}, where $H_p \cap H_q \cap H_r = X$ has codimension two.

Now recall $U_m^p = \oplus_I V_{i_1} \wedge \cdots \wedge V_{i_p}$, summing over all $I = \{i_1, ..., i_p\}$ with $1 \leq i_1 \leq \cdots \leq i_p$ and $i_1 + \cdots + i_p = m$. Set

$$U_m^p(X) = \bigoplus_I V_{i_1}(X) \wedge \cdots \wedge V_{i_p}(X).$$

LEMMA 3.10. *Suppose i_n: $\oplus_X V_n(X) \to V_n$ is injective for $2 \leq n < m$. Then the induced map $\hat{i}_m$: $\oplus_X U_m^p(X) \to U_m^p$ is injective for $2 \leq p < m$.*

PROOF. The key here is that, because of the restriction on p, each term $V_{i_1}(X) \wedge \cdots \wedge V_{i_p}(X) \subseteq U^p_m(X)$ must have $2 \leqslant i_p < m$. Thus, by assumption, $\bigoplus_X V_{i_p}(X) \to V_{i_p}$ is injective. It follows that $\bigoplus_X V_{i_1}(X) \wedge \cdots \wedge V_{i_p}(X) \to V_{i_1} \wedge \cdots \wedge V_{i_p}$ is injective for each index set $1 \leqslant i_1 \leqslant \cdots \leqslant i_p$ with $i_1 + \cdots + i_p = m$. Hence $\bigoplus_X U^p_m(X) \to U^p_m$ is injective.

LEMMA 3.11. *Suppose* $i_n\colon \bigoplus_X V_n(X) \to V_n$ *is injective for* $2 \leqslant n < m$. *Then* $i_m = \bigoplus_X V_m(X) \to V_m$ *is injective.*

PROOF. Consider the commutative diagram

$$\begin{array}{ccc} \bigoplus_X V_m(X) & \xrightarrow{\oplus d_X} & \bigoplus_X U^2_m(X) \\ \big\downarrow{\scriptstyle i_m} & & \big\downarrow{\scriptstyle \hat{i}_m} \\ V_m & \xrightarrow{\ d\ } & U^2_m\,. \end{array}$$

By (3.10), $\hat{i}_m$ is injective. By (2.1), each d_X is injective, so $\oplus_X d_X$ is injective. Thus i_m is injective.

The main result (3.8) now follows inductively from (3.9), (3.11), and (3.7).

A real 3-arrangement **A** is a *product* arrangement if, for some choice of coordinates, **A** is a disjoint union $\mathbf{A}_1 \cup \mathbf{A}_2 \cup \mathbf{A}_3$ where $\mathbf{A}_1$ consists of the single hyperplane $x_1 = 0$ and for $i = 2$ or 3, the hyperplanes of $\mathbf{A}_i$ have defining equations which are independent of x_i. The picture is

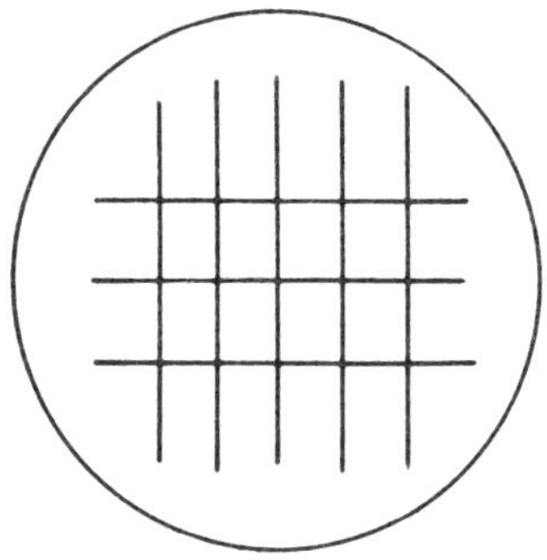

$X_1 = 0$

Product arrangements for fiber-type with exponents 1, d_2 = #$\mathbf{A}_2$, and d_3 = #$\mathbf{A}_3$.

PROPOSITION 3.12. *If* **A** *is a product* 3-*arrangement, then equality holds in* (3.8) *for all* $n \geq 2$.

PROOF. Using the notation of (3.8), we have $c(r) = 1$ for $r = d_2 + 1$ or $d_3 + 1$, and $c(r) = 0$ for all other $r \geq 3$. On the other hand, the main result of [6] shows that, for a fiber-type arrangement with exponents 1, d_2, d_3, one has $\phi_n = N_n^{d_2 - 1} + N_n^{d_3 - 1}$ for $n \geq 2$. This completes the proof.

By (2.9) and (2.2) we have that V_3 is isomorphic to the kernel of the map $E^1 \otimes I^2 \to E^3$. That is, V_3 consists of relations among the (degree 2) relations in the natural generators of $H^*(M)$. The subspace $\oplus_X V_3(X)$ consists of those relations in V_3 which are "supported" *locally* near the codimension two intersections. Thus the gap in the inequality of (3.8) for $n = 3$ is measuring the number of independent *global* relations in V_3. A further discussion (with examples) of this phenomenon will be reserved for the next section.

We close this section with some remarks on reflection arrangements and the LCS formula. As we reported in [5], the LCS formula is known to hold for Coxeter arrangements of types A_ℓ, B_ℓ, D_ℓ, G_2, and $I_2(p)$, and is known to fail for the arrangement of type H_3 and the arrangement corresponding to the Shephard group G_{26}. The minimal model approach affords a possible explanation for this behavior, as follows. The action on the reflection group G on $\mathbf{C}^{\ell}$ induces a G-module structure on $V_1 \cong H^1(M)$. Using the injectivity (2.1) of $d\colon V_n \to U_n^2$, and exterior and tensor products of modules, one obtains inductively a G-module structure on V_n for all n. Now the LCS formula relates the betti numbers of M with the dimensions ϕ_n of V_n, $n \geqslant 1$. By [14] and the preceding remarks, we see that *both* sides of the LCS formula are connected with the representation theory of G. Thus the LCS formula may have a representation-theoretic interpretation.

§4. EXAMPLES AND CONJECTURES

We now discuss the problem of classifying the algebras which arise as cohomology rings of hyerplane complements, using the results of the preceding sections. Suppose $\mathbf{A}_1$ and $\mathbf{A}_2$ are arrangements in $\mathbf{C}^{\ell}$, with complements M_1 and M_2. Let $A_i = H^*(M_i)$ for $i = 1,2$. As we noted in §1, the betti numbers $\{b_n^{(i)}\}_{n=0}^{\ell}$ of A_i are easily computed from the intersection lattice of $\mathbf{A}_i$, and if $b_n^{(1)} = b_n^{(2)}$ for all n, then A_1 and A_2 are isomorphic as graded modules, but not necessarily as rings.

EXAMPLE 4. Let A_1 be the arrangement associated with the divisor $Q_1 = xyz(x+z)(x-z)(y+z)(y-z)$; the hyperplanes in $\mathbf{A}_1$ are the irreducible components of Q_1. Similarly, let $\mathbf{A}_2$ be defined by $Q_2 = (x+y)(x-y)(x+z)(x-z)(y+z)(y-z)z$. The pictures are

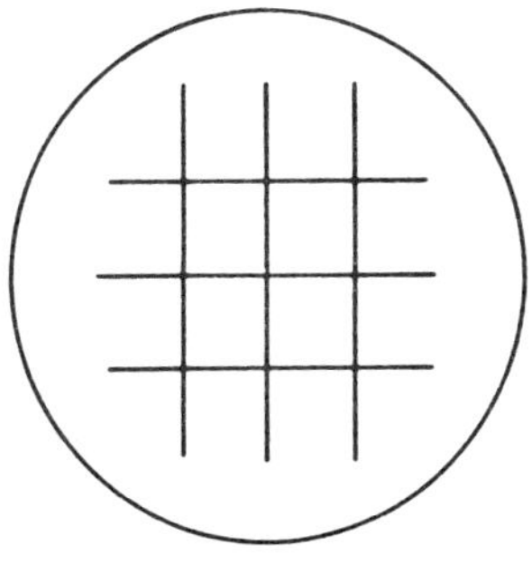

$\mathbf{A}_1$

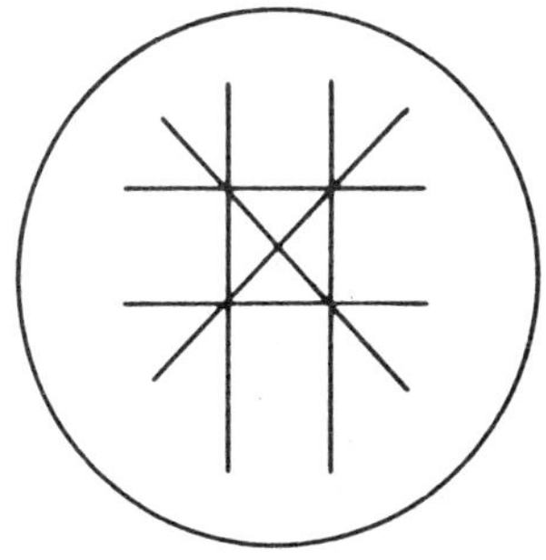

$\mathbf{A}_2$

Then the betti numbers of A_1 and A_2 agree. Specifically, $b_0 = 1$, $b_1 = 7$, $b_2 = 15$, and $b_3 = 9$. Let us compare the one-minimal models. The arrangement $\mathbf{A}_1$ is fiber-type (in fact a product) with exponents $d_1 = 1$, $d_2 = 3$, $d_3 = 3$. Thus the LCS formula holds, yielding $\phi_3 = 16$. Using (2.9) one may compute the value of ϕ_3 for $\mathbf{A}_2$ by row-reducing a certain 35×42 matrix. The result is $\phi_3 = 17$. (Note: the value $\phi_3 = 12$ for this arrangement reported in [6,7] is incorrect, and in fact contradicts (3.4).) So A_1 and A_2 have non-isomorphic one-minimal models, and thus cannot be isomorphic as rings.

A more interesting situation arises when both $\mathbf{A}_1$ and $\mathbf{A}_2$ are fiber-type, with the same exponents. Then $b_n^{(1)} = b_n^{(2)}$ for all n, and furthermore, since the LCS formula holds, we also have $\phi_n^{(1)} = \phi_n^{(2)}$ for all n. It is not known whether A_1 and A_2 are necessarily isomorphic.

EXAMPLE 4.2. Let $\mathbf{A}_1$ and $\mathbf{A}_2$ are defining equations $Q_1 = xz(x+z)(x-z)(y+z)(y-z)$ and $Q_2 = xyz(x+y)(x+z)(y-z)$. The pictures are

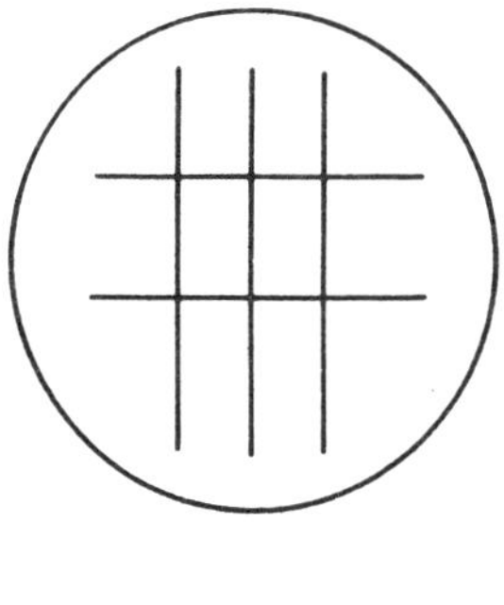

$\mathbf{A}_1$

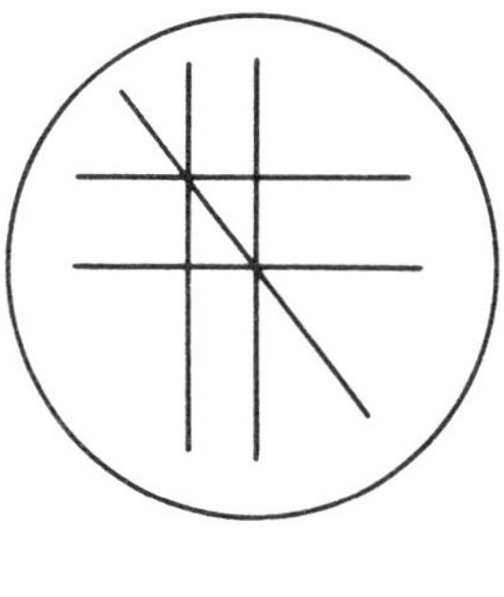

$\mathbf{A}_2$

Both $\mathbf{A}_1$ and $\mathbf{A}_2$ are fiber-type, with exponents $d_1 = 1$, $d_2 = 2$, $d_3 = 3$. The betti numbers are $b_0 = 1$, $b_1 = 6$, $b_2 = 11$, $b_3 = 6$. From the LCS formula ((2.6) or (3.4)) one computes $\phi_3 = 10$ for both arrangements. Let's look at the local elements in $V_3^{(1)}$ and $V_3^{(2)}$ (cf. (3.8)). Denote the quantity $\Sigma_{r\geqslant 3} c(r) N_n^{r-1}$ of (3.8) by ψ_n. The arrangement

$\mathbf{A}_1$ is a product, so (3.12) implies $\psi_3^{(1)} = \phi_3 = 10$. But for A_2 we have $c(3) = 4$ and $c(r) = 0$ for $r > 3$. Thus $\psi_3^{(2)} = 4N_3^2 = 4 \cdot 2 = 8$. So $V_3^{(2)}$ has two global elements and $V_3^{(1)}$ has none. We can see these elements of V_3 by examining the map $E^1 \otimes I^2 \to E^3$. First of all, local elements arising from a rank 2 lattice element $X = H_p \cap H_q \cap H_r$ look like $r_1 = (e_r - e_p)\partial e_{pqr}$ and $r_2 = (e_q - e_p)\partial e_{pqr}$ (notation of [13]). For $X = H_p \cap H_q \cap H_r \cap H_s$ of multiplicity 4 we get in addition $s_1 = (e_s - e_p)\partial e_{pqr} + (e_r - e_p)\partial e_{pqs}$ and $s_2 = (e_r - e_p)\partial e_{pqs} + (e_q - e_p)\partial e_{prs}$. These account for all of $V_3^{(1)}$—there is one X of multiplicity 3 and three more (independent) 3-fold intersections coming from the multiplicity 4 lattice element. In $V_3^{(2)}$, there are only eight local elements, two for each of the four 3-fold intersections in rank 2. The two global elements look like

$$t_1 = (e_5 - e_4)\partial e_{126} + (e_4 - e_2)\partial e_{135} + (e_5 - e_1)\partial e_{234} + (e_2 - e_1)\partial e_{456},$$

and

$$t_2 = (e_3 - e_4)\partial e_{126} + (e_4 - e_6)\partial e_{135} + (e_6 - e_1)\partial e_{135} + (e_3 - e_1)\partial e_{456}.$$

The reader may check that these expressions indeed vanish in E^3.

It remains an open question whether A_1 and A_2 are isomorphic in this case. (It is known that M_1 and M_2 are not homotopy equivalent.) The local-global considerations above seem to indicate an intrinsic difference in the one-minimal models of A_1 and A_2, but it is not clear whether the combinatorial invariants ψ_n are indeed algebraic invariants of the one-minimal models.

With these examples in mind we propose the following conjecture.

(i) The quantities ψ_n, $n \geqslant 3$, defined in (4.2) are invariants of the cohomology algebra.

(ii) In case $\ell = 3$, the cohomology algebras A_1 and A_2 are isomorphic if and only if $b_n^{(1)} = b_n^{(2)}$ for $0 \leqslant n \leqslant 3$, $\phi_3^{(1)} = \phi_3^{(2)}$, and $\psi_3^{(1)} = \psi_3^{(2)}$.

REMARK. Since this paper was submitted, the author has succeeded in proving conjecture (i). As a consequence, the algebras of Example 4.2 are not isomorphic. Also, the converse of (3.12) has been established for fiber-type arrangements.

REFERENCES

1. E. Brieskorn. Sur les groups de tresses (d'apres V.I. Arnol'd). Sem. Bourbaki 24[e] annee 1971/72, Springer Lecture Notes No. 317, Springer-Verlag, Berlin, 1973.

2. A. Bousfield and V. Guggenheim. On PL DeRham Theory and rational homotopy type. Mem. AMS Vol. 8, No. 179, Amer. Math. Soc., Providence (1976).

3. A. Bousfield and D. Kan. Homotopy limits, completions, and localizations. Springer Lecture Notes No. 304, Springer-Verlag, Berlin (1972).

4. M. Falk. The minimal model of the complement of an arrangement of hyperplanes. Preprint, Univ. of Iowa, 1985, to appear in Trans. AMS.

5. M. Falk and R. Randell, Braid groups and products of free groups, to appear in Proc. of Santa Cruz Conf. on Artin's Braid Group, 1987.

6. M. Falk and R. Randell. The lower central series of a fiber-type arrangement. Invent. Math. 82 (1985), 77-88.

7. M. Falk and R. Randell. The lower central series of generalized pure braid groups. Geometry and Topology, Lecture Notes on Pure and Applied Math. Series, Vol. 105, Marcel Dekker, Inc. (1986).

8. T. Kohno. Holonomy Lie algebras, logarithmic connections, and the lower central series of fundamental groups, in these Proceedings.

9. T. Kohno. On the holonomy Lie algebra and the nilpotent completion of the fundamental group of the complement of hypersurfaces. Nagoya Math. J. 92 (1983), 21-37.

10. T. Kohno. On the minimal algebra and $K(\pi,1)$ property of affine algebraic varieties. Xeroxed notes, 1983.

11. T. Kohno. Serie de Poincare-Koszul associee aux groupes de tresses pure. Invent. Math. 82 (1985), 57-75.

12. W. Magnus, A. Karras, D. Solitar. Combinatorial Group Theory. John Wiley, New York, 1966.

13. P. Orlik and L. Solomon. Combinatorics and topology of complements of hyperplanes. Invent. Math. 56 (1980), 167-189.

14. P. Orlik and L. Solomon. Unitary reflection groups and cohomology. Invent. Math. 59 (1980), 77-94.

15. D. Sullivan. Infinitessimal computations in topology. Publ. Math. IHES 47 (1977), 269-331.

16. H. Terao. Modular elements of lattices and topological fibration. Adv. in Math. 62 (1986), 135-154.

DEPARTMENT OF MATHEMATICS
NORTHERN ARIZONA UNIVERSITY
FLAGSTAFF, AZ 86011

Contemporary Mathematics
Volume **90**, 1989

THE MODIFIED ANALYTIC TRIVIALIZATION OF A FAMILY OF REAL ANALYTIC MAPPINGS

Toshizumi Fukui

§1. INTRODUCTION

It has been well-understood that finding a 'nice and natural' equivalence relation of real analytic object (functions, mappings, varieties, and so on) is one of the fundamental problems in the theory of real singularities. In [K_1] and [K_2], T.-C. Kuo introduced the notion of modified analytic trivialization for analytic families of analytic functions. This notion induces an equivalence relation in the set of all germs of real analytic functions, which is slightly weaker than bianaliticity and much stronger than topological equivalence. In his lecture on "blow-analytic equisingularity" given at the general meeting of Mathematical Society of Japan in October, 1984, using the same idea, he defined the notion of blow-analytic equivalence of real analytic mappings. According to his criterion, for instance, the family $W_t(x,y) = xy(x - y)(x - ty)$, $(t > 2)$, which is not C^1-trivial, admits a modified analytic trivialization. He proved some finite classification theorem of real analytic function germs. Then this equivalence relation seems to be nice.

In this note we treat real analytic mappings (or varieties) and give some sufficient condition for modified analytic trivialization. This is a continuation of [Y-F].

We now recall the notion of the modified analytic trivialization. Let $\pi : X \to \mathbb{R}^n$ be a proper analytic modification (in this note we only consider proper analytic modifications whose complexifications are also proper complex analytic modifications). Let S_π be defined by the smallest subset of X such that $\pi \mid (X - S_\pi)$ is an isomorphism. Let $f_t(x) := F(x;t)$ be an analytic family of real analytic mappings from $\mathbb{R}^n$ to $\mathbb{R}^p$ parametrized by $t = (t_1, ..., t_m) \in I$ where I is a compact cube $\times_{i=1}^m [a_i,b_i]$ in $\mathbb{R}^m$. Assume that $f_t(0) = F(0;t) = 0$.

DEFINITION (1.1). We say that f_t admits an *almost modified analytic trivialization* (*almost* π-*MAT*) *via* π *along* I if the following properties hold. There exist

a neighbourhood U of the origin of $\mathbb{R}^n$,
two neighbourhoods $\widehat{U}_1$, $\widehat{U}_2$ of $\pi^{-1}(0)$ in $\pi^{-1}(U)$,
two neighbourhoods U_1', U_2' of the origin of $\mathbb{R}^p$, and
two analytic families of analytic isomorphisms

$$H_t : \widehat{U}_1 \to \widehat{U}_2, \quad \text{and}$$
$$H'_{x,t} : U_1' \to U_2', \quad (x \in \widehat{U}_1),$$

such that $H_t(\pi^{-1}(0)) = \pi^{-1}(0)$, $H_t(S_\pi) = S_\pi$, $f_t(H_t(x)) = H'_{x,t}(f_0(x))$, for any $x \in \widehat{U}_1$; i.e., the following diagram is commutative.

$$\begin{array}{ccccc} \widehat{U}_1 & \xrightarrow{id_X \times f_0} & \widehat{U}_1 \times U_1' & \ni & (x,y) \\ \downarrow H_t & & \downarrow H_t' & & \downarrow \\ \widehat{U}_2 & \xrightarrow{id_X \times f_t} & \widehat{U}_2 \times U_2' & \ni & (H_t(x), H'_{x,t}(y)) \end{array}$$

where $H_t'(x,y) = (H_t(x), H'_{x,t}(y))$.

Moreover, if H_t induces a homeomorphism of $\pi(\widehat{U}_1)$ to $\pi(\widehat{U}_2)$, and "$\pi(x) = \pi(x') \Rightarrow H'_{x,t} = H'_{x',t}$", we say that f_t admits a *modified analytic trivialization* (*MAT*) *via* π *along* I; i.e., the following diagram is commutative.

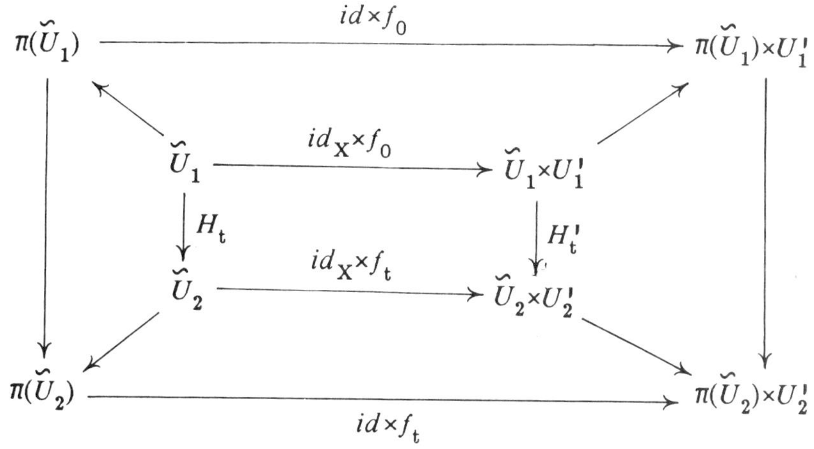

We say that f_t admits a *MAT* along I if f_t admits a MAT via π along I for some modification π.

This definition is an analogue of J. Mather's K equivalent (see [M]).

Set $V(f) := \{x \in \mathbb{R}^n \mid f(x) = 0\}$ and

$$\breve{V}(f) := \text{closure of } (f^{-1}(0) - f(S_\pi)) \quad \text{in} \quad X.$$

It is easy to show if f_t admits an almost MAT via π then $(X,\breve{V}(f_t), S_\pi)$ is analytic trivial near $f^{-1}(0)$, and if f_t admits a MAT then $(\mathbb{R}^n,V(f_t),0)$ is topological trivial near 0.

Next we recall Newton polygons.

Let $f = (f_1, ..., f_p) : \mathbb{R}^n \to \mathbb{R}^p$ be a real analytic mapping of n variables $x = (x_1, ..., x_n)$. Let $\Sigma\, a_{i,\nu}x^\nu$ $(\nu = (\nu_1, ..., \nu_n))$ be the Taylor's expansion of f_i $(i = 1, ..., p)$ at 0. Let us set $\Gamma_+(f_i)$ = convex hull of $\{\nu + \mathbb{R}^n_+ \mid a_{i,\nu} \neq 0\}$ (the *Newton polygon* of f_i),

$$\Gamma_f = (\Gamma_+(f_1), ..., \Gamma_+(f_p)),$$

and

$$\Gamma_+(f) = \Gamma_+(f_1) + \cdots + \Gamma_+(f_p) = \{\alpha + \mathbb{R}^n_+ \mid \alpha = \alpha_1 + \cdots + \alpha_p,\ \alpha_i \in \Gamma_+(f_i)\},$$

where $\mathbb{R}_+$ is the set of non-negative real numbers. Let $g = \Sigma\, b_\nu x^\nu$ be an analytic function and define $g_T = \Sigma_{\nu\in T} b_\nu x^\nu$ for any subset T of $\mathbb{R}^n_+$.

For $a = (a_1, ..., a_n) \in \mathbb{R}^n_+$, and $\alpha = (\alpha_1, ..., \alpha_n) \in (\mathbb{R}^n_+)^*$, define

$$\langle a,\alpha\rangle := a_1\alpha_1 + \cdots + a_n\alpha_n,$$
$$\ell_j(\alpha) = \min\{\langle a,\alpha\rangle \mid a \in \Gamma_+(f_j)\},$$
$$\tau_j(\alpha) = \{a \in \Gamma_+(f_j) \mid \langle a,\alpha\rangle = \ell_j(\alpha)\}, \quad \text{for } j = 1, ..., p,$$
$$\ell(\alpha) = \min\{\langle a,\alpha\rangle \mid a \in \Gamma_+(f)\},$$
$$\tau(\alpha) = \{a \in \Gamma_+(f) \mid \langle a,\alpha\rangle = \ell(\alpha)\},$$
$$\Gamma^*(f_j) := (\mathbb{R}^n_+)^*/\sim_j, \text{ where } \alpha \sim_j \alpha' \Leftrightarrow \tau_j(\alpha) = \tau_j(\alpha'), \text{ and}$$
$$\Gamma^*(f) := (\mathbb{R}^n_+)^*/\sim, \text{ where } \alpha \sim \alpha' \Leftrightarrow \tau(\alpha) = \tau(\alpha').$$

Then the following properties (2.1) - (2.3) hold, which will be proved in §2.

(2.1) $\quad \ell(\alpha) = \ell_1(\alpha) + \cdots + \ell_p(\alpha).$

(2.2) $\quad \tau(\alpha) = \tau_1(\alpha) + \cdots + \tau_p(\alpha),$

(2.3) $\quad \tau(\alpha) = \tau(\alpha') \Leftrightarrow \tau_1(\alpha) = \tau_1(\alpha'), ..., \tau_p(\alpha) = \tau_p(\alpha').$

DEFINITION (1.2). $\tau = (\tau_1, ..., \tau_p)$ is a *face* of Γ_f if there exists $\alpha \in (\mathbb{R}^n_+)^*$ such that $\tau_1 = \tau_1(\alpha), ..., \tau_p = \tau_p(\alpha)$. Let us define f_τ by $f_\tau := (f_{1,\tau_1}, ..., f_{p,\tau_p}$. A face $(\tau_1, ..., \tau_p)$ is *compact* if $\tau_1 + \cdots + \tau_p$ is a compact face of $\Gamma_+(f)$. A non-compact face $(\tau_1, ..., \tau_p)$ is *co-ordinate* if $\tau_1 + \cdots + \tau_p$ is a coordinate face of $\Gamma_+(f)$, up to parallel translation T with $T(\Gamma_+(f)) \subset (\mathbb{R}_+)^n$.

DEFINITION (1.3). We say that f is *non-degenerate* if the matrix Jf_τ has the maximal rank on the set $\{f_\tau = 0, x_1 ..... x_n \neq 0\}$, for any compact face τ of Γ_f.

Using the same trick in [Ku], we can prove that non-degeneracy is generic in the space of coefficients.

We can construct a proper analytic modification $\pi : X \to \mathbb{R}^n$ corresponding to $\Gamma_+(f)$, (See §2.)

Now, we state the main theorem.

THEOREM (1.4). *Let $f_t(x) := F(x;t)$ be an analytic family of non-degenerate real analytic mappings. Suppose that the Newton* *polygons* Γ_{f_t} *are independent of* $t \in I$. *Let* $\pi : X \to \mathbb{R}^n$ *be one of the proper analytic modifications corresponding to the* $\Gamma_+(f_t)$. *Then the family* $F(x;t)$ *admits an almost MAT via* π *along* I. *Moreover, if* $f_{t,\tau}$ *is independent of* t *for each non-compact and non-coordinate face* τ, *then* F *admits a MAT via* π *along* I.

The author is grateful to Professor Etsuo Yoshinaga and Takashi Nishimura for useful discussions.

§2. MODIFICATION

In this section, we prove (2.1) - (2.3), and give a summary on modifications. The detail of this modification is found in [Ka], [TE], [V], etc. .

LEMMA (2.1). $\ell(\alpha) = \ell_1(\alpha) + \cdots + \ell_p(\alpha)$.

PROOF. Assume that there exists $a_j \in \Gamma_+(f_j)$ such that $\ell_j(\alpha) = \langle a_j,\alpha\rangle$. Let us set $a = a_1 + \cdots + a_p$. Then

$$\ell_1(\alpha) + \cdots + \ell_p(\alpha) = \langle a,\alpha\rangle \geqslant \ell(\alpha).$$

Conversely, assume that there exists $a \in \Gamma_+(f)$ such that $\ell(\alpha) = \langle a,\alpha\rangle$. And write $a = a_1 + \cdots + a_p$, $a_j \in \Gamma_+(f_j)$, $k = 1, \ldots, p$. Then,

$$\ell(\alpha) = \langle a,\alpha\rangle = \langle a_1,\alpha\rangle + \cdots + \langle a_p,\alpha\rangle \geqslant \ell_1(\alpha) + \cdots + \ell_p(\alpha).$$

Q.E.D.

LEMMA (2.2). $\tau(\alpha) = \tau_1(\alpha) + \cdots + \tau_p(\alpha)$.

PROOF.

$$\begin{aligned} a \in \tau(\alpha) &\Leftrightarrow \langle a,\alpha\rangle = \ell(\alpha), \quad a = a_1 + \cdots + a_p, \quad a_i \in \Gamma_+(f_i) \\ &\Leftrightarrow \langle a_i,\alpha\rangle = \ell_i(\alpha), \quad a_i \in \Gamma_+(f_i), \quad i = 1, \ldots, p \\ &\Leftrightarrow a_i \in \tau(f_i), \quad i = 1, \ldots, p \end{aligned}$$

Q.E.D.

LEMMA (2.3). $\tau(\alpha) = \tau(\alpha') \Leftrightarrow \tau_i(\alpha) = \tau_i(\alpha')$, $i = 1, \ldots, p$.

PROOF. By (2.2), "$\Leftarrow$" part is trivial. We now prove "$\Rightarrow$" part. Assume that $a_i \in \tau_i(\alpha)$, $i = 1, \ldots, p$, and $a_1 \notin \tau_1(\alpha')$. Since $\langle a_1,\alpha'\rangle > \ell_1(\alpha')$, there exists $k \in \{2, \ldots, p\}$ such that $\langle a_k,\alpha'\rangle < \ell_k(\alpha')$. This contradicts the definition of ℓ_k. Q.E.D.

Lemma (2.3) asserts that $\Gamma^*(f)$ is a subdivision of $\Gamma^*(f_i)$.

Let Σ be a simplicial unimodular finite subdivision of $\Gamma^*(f)$. Let $\Sigma^{(n)}$ be the set of all n-simplecies of Σ. For any $\sigma \in \Sigma^{(n)}$, we define $a^1(\sigma), \ldots, a^n(\sigma) \in \mathbf{Z}^n$ by $\sigma \cap \mathbf{Z}^n = \mathbf{Z}_+a^1(\sigma) + \cdots + \mathbf{Z}_+a^n(\sigma)$. Define X by

$$X := \bigcup_{\sigma\in\Sigma^{(n)}} \mathbb{R}^n(\sigma)/\sim$$

where $\mathbb{R}^n(\sigma)$ is a copy of $\mathbb{R}^n$ with a coordinate system $y_\sigma = (y_{\sigma,1}, \ldots, y_{\sigma,n})$, and $y_\sigma \sim y_\tau$ ($y_\sigma \in \mathbb{R}^n(\sigma)$, $y_\tau \in \mathbb{R}^n(\tau)$, $\sigma,\tau \in \Sigma^{(n)}$) is defined by

$$y_\tau = {}^{a(\tau)^{-1}a(\sigma)}y_\sigma.$$

Here, $a(\sigma) = (a^1(\sigma), \ldots, a^n(\sigma))$, and

$$^{A}y := \left(y_1^{a_1^1} \cdots y_n^{a_1^n}, \ldots, y_1^{a_n^1} \cdots y_n^{a_n^n}\right)$$

for a matrix $A = (a_i^j) \in \mathrm{Mat}(n\times n, \mathbf{Z})$.
Then, X is a non-singular n-dimensional algebraic manifold and

$$\pi : X \to \mathbb{R}^n,\ \pi(y_\alpha) := {}^{a(\sigma)}y_\sigma\ (y_\sigma \in \mathbb{R}^n(\sigma))$$

is a proper analytic modification of $\mathbb{R}^n$. Let $\pi_\sigma := \pi \mid \mathbb{R}^n(\sigma)$.

Define $E_{\sigma,J}$ and $E^*_{\sigma,J}$ by

$$E_{\sigma,J} := \{y_\sigma \in \mathbb{R}^n(\sigma) \mid y_{\sigma,j} = 0,\ j \in J\},$$

and

$$E^*_{\sigma,J} := \{y_\sigma \in E_{\sigma,J} \mid y_{\sigma,j} \neq 0,\ j \notin J\},$$

where J is a subset of $\{1, \ldots, n\}$.

Note the following easy lemmata. (See [Y-F].)

LEMMA (2.4). $\pi_\sigma(E_{\sigma,J}) = \{0\}$ *if and only if* $\tau = \cap_{i\in J}\tau(a^i(\sigma))$ *is a compact face of* Γ_f

LEMMA (2.5). $\pi_\sigma|E^*_{\sigma,J}$ is injective if $\tau = \cap_{i\in J}\tau(a^i(\sigma))$ *is a coordinate face of* Γ_f

Let g denote an analytic function near 0 in $\mathbb{R}^n$, and let $\Sigma_\nu b_\nu x^\nu$ be the Taylor expansion of g. Define $\tilde{g}_\sigma$ by

$$g(\pi_\sigma(y_\sigma)) = \prod_{k=1}^{n} y_{\sigma,k}^{\ell_k} \cdot \tilde{g}_\sigma,$$

where ℓ_k $(k = 1, ..., n)$ is the minimal number of $\langle a^j(\sigma),\nu\rangle$ with $b_\nu \neq 0$. Then

$$\tilde{g}_\sigma \mid E_{\sigma,J} = g_{\tau_{\sigma,J}} \Big/ \prod_{k=1}^{n} y_{\sigma,k}^{\ell_k}. \tag{2.6}$$

The following trick plays an important role in our proof of Theorem (1.4).

TRICK (2.7). The function $\tilde{g}_\sigma$ has no zeros on $E^*_{\sigma,J}$ if $g_{\tau,J}$ has no zeros on $\{x_1 \cdots\cdots x_n \neq 0\}$.

PROOF. Since $\tilde{g}|E_{\sigma,J}$ is a function on $y_j (j \notin J)$, we obtain that

there exists $y_\sigma \in E^*_{\sigma,J}$ with $\tilde{g}_\sigma(y_\sigma) = 0$

$\Leftrightarrow$ there exists $y_\sigma \in \{y_{\sigma,1} \cdots\cdots y_{\sigma,n} \neq 0\}$ with $\tilde{g}_\sigma|E_{\sigma,J} = 0$

$\Rightarrow$ there exist $x \in \{x_1 \cdots\cdots x_n \neq 0\}$ with $g_{\tau,J}(x) = 0$. (By (2.6).)

§3. SINGULAR RIEMANNIAN METRIC

Let $I = [a,b]$. We shall define a singular Riemannian metric on $\mathbb{R}^n$ (i.e. a Riemannian metric on $(\mathbb{R}^n - \{x_1 \cdots\cdots x_n = 0\})$) by the following

$$\left[x_i \frac{\partial}{\partial x_i}, x_j \frac{\partial}{\partial x_j} \right] = \delta_{ij} .$$

Let $F = (F_1, ..., F_p) : \mathbb{R}^n \times I \to \mathbb{R}^p$ be a family of analytic mappings. Let grad F_i be the gradient vector of F_i with respect to the induced singular Riemannian metric on $\mathbb{R}^n \times I$ from the above metric. By the elementary computation, we obtain that

$$\text{(3.1)} \qquad \operatorname{grad} F_i = \Sigma_j\, x_j^2 \frac{\partial F_i}{\partial x_j} \frac{\partial}{\partial x_j} + \frac{\partial F_i}{\partial t} \frac{\partial}{\partial t}.$$

Then, we obtain that

$$\text{(3.2)} \qquad (\operatorname{grad} F_i, \operatorname{grad} F_j) = \Sigma_k x_k^2 \frac{\partial F_i}{\partial x_k} \frac{\partial F_j}{\partial x_k} + \frac{\partial F_i}{\partial t} \frac{\partial F_j}{\partial t} .$$

§4. KUO'S VECTOR FIELD

In this section, we discuss the Kuo's vector field. The analysis of the Kuo's vector field is a key step of the proof of our theorem.

Let U be a neighbourhood of 0 in $\mathbb{R}^n$ and $I = [a,b]$. Let (,) denote the induced singular Riemannian metric on $\mathbb{R}^n \times I$. Let S be an open dense subset of $U \times I$ such that there exist a direct sum decomposition $T \oplus N$ of the tangent space T_pS of $p \in S$. Here,

$$T = \{v \in T_pS \mid \langle \nabla_j, v \rangle = 0, \ \ j = 1, ..., p\},$$
$$N = T^{\perp},$$

and

$\nabla_i = \operatorname{grad} F_i$ (a column vector).

Let $p : T_pS \to T$ be the natural projection.

Define the meromorphic vector fields V' and V by

$$V' = p\left[\frac{\partial}{\partial t} \right],$$

and

$$V = \frac{1}{((\partial/\partial t),\ V')}\ V'.$$

By the definition of V, V is tangent to the level set of f and the t-component of V is 1. We call V a *Kuo's vector field.*

PROPOSITION (4.1)

$$V' = -\sum_{i=1}^{n} Q_i x_i \frac{\partial}{\partial x_i} + P \frac{\partial}{\partial t},$$

where

$$P = 1 - \sum_{k=1}^{p} \frac{D_{k,t}^{\ 2}}{D_{k-1} D_k} = \frac{J_p}{D_p},$$

$$Q_i = \sum_{k=1}^{p} \frac{D_{k,i} D_{k,t}}{D_{k-1} D_k},$$

$$D_k = \det\ ((\nabla_i, \nabla_j))_{i,j=1,\dots,k}\,,$$

$$J_k = \det \left[\sum_{k=1}^{n} x_k^{\ 2} \frac{\partial F_i}{\partial x_k} \frac{\partial F_j}{\partial x_k} \right]_{i,j=1,\dots,s},$$

$$D_{k,t} = \det \left[{}^t(\nabla_1, \dots, \nabla_k) \cdot \left[\nabla_1, \dots, \nabla_{k-1}, \frac{\partial}{\partial t} \right] \right],$$

$$D_{k,i} = \det \left[{}^t(\nabla_1, \dots, \nabla_k) \cdot \left[\nabla_1, \dots, \nabla_{k-1}, x_i \frac{\partial}{\partial x_i} \right] \right], \quad k = 1, \dots, p,$$
$$i = 1, \dots, n.$$

PROOF. Note that

$$V' = \frac{\partial}{\partial t} - \left(e_1, \frac{\partial}{\partial t} \right) e_1 - \dots - \left(e_p, \frac{\partial}{\partial t} \right) e_p,$$

where $e_1, \dots, e_p$ are orthonormal basis of N. By Gram-Schmid process, we can choose

$$e_1 = \nabla_1 / \|\nabla_1\|,$$
$$e_2 = \nabla_2' / \|\nabla_2'\|, \quad \nabla_2' = \nabla_2 - (\nabla_2, e_1) e_1,$$
$$\dots\dots$$

and

$$e_p = \nabla'_p / \|\nabla'_p\|, \quad \nabla'_p = \nabla_p - (\nabla_p, e_1)e_1 - \cdots - (\nabla_p, e_{p-1})e_{p-1}.$$

Let u_i be an analytic vector field with $u_i/\|u_i\| = e_i$. Then

$$V' = \frac{\partial}{\partial t} - \frac{\left[u_1, \dfrac{\partial}{\partial t}\right]}{(u_1, u_1)} u_1 - \cdots - \frac{\left[u_p, \dfrac{\partial}{\partial t}\right]}{(u_p, u_p)} u_p \, . \tag{4.2}$$

We can choose $u_1, \ldots, u_p$ as the following way:

$$u_1 = \nabla_1, \; u_k = \prod_{i=1}^{k-1} (u_i, u_i) \left[\nabla_k - \sum_{i=1}^{k-1} \frac{(\nabla_k, u_i)}{(u_i, u_i)} u_i\right], \quad k = 2, \ldots, p. \tag{4.3}$$

Since ${}^t(u_1, \ldots, u_k) = R_k {}^t(\nabla_1, \ldots, \nabla_k)$, where

$$R_k = \begin{bmatrix} 1_{(u_1,u_1)} & & 0 \\ & & \\ * & & \prod_{i=1}^{k-1} (u_i, u_i) \end{bmatrix},$$

we obtain that

$$\begin{aligned} \prod_{i=1}^{k} (u_i, u_i) &= \det((u_i, u_j))_{i,j=1,\ldots,k} \\ &= \det({}^t(u_1, \ldots, u_k) \cdot (u_1, \ldots, u_k)) \\ &= \det(R_k {}^t(\nabla_1, \ldots, \nabla_k) \cdot (\nabla_1, \ldots, \nabla_k)^t R_k) \\ &= (\det R_k)^2 \cdot \det((\nabla_i, \nabla_j))_{i,j=1,\ldots,k} \\ &= \prod_{i=1}^{k-1} (u_i, u_i)^{2k-2i} \det((\nabla_i, \nabla_j))_{i,j=1,\ldots,k} \, . \end{aligned}$$

Thus

$$(u_k, u_k) = \prod_{i=1}^{k-1} (u_i, u_i)^{2k-2i-1} \det((\nabla_i, \nabla_j))_{i,j=1,\ldots,k} \, . \tag{4.4}$$

By (4.3),

$$\left[u_k, \frac{\partial}{\partial t}\right] = \det \begin{bmatrix} (u_1,u_1) & & 0 & (u_1, \partial/\partial t) \\ & \ddots & & \\ 0 & & (u_{k-1},u_{k-1}) & (u_{k-1}, \partial/\partial t) \\ (\nabla_k,u_1) & \cdots & (\nabla_k,u_{k-1}) & (\nabla_k, \partial/\partial t) \end{bmatrix}$$

$$\begin{aligned} &= \det\left[{}^t(u_1, \ldots, u_{k-1},\nabla_k) \cdot \left[u_1, \ldots, u_{k-1}, \frac{\partial}{\partial t}\right]\right] \qquad (4.5)\\ &= \det\left[\begin{bmatrix} R_{k-1} & 0 \\ 0 & 1 \end{bmatrix} {}^t(\nabla_1, \ldots, \nabla_k) \cdot \left[\nabla_1, \ldots, \nabla_{k-1}, \frac{\partial}{\partial t}\right] {}^t\begin{bmatrix} R_{k-1} & 0 \\ 0 & 1 \end{bmatrix}\right] \\ &= (\det R_{k-1})^2 \cdot D_{k,t} \,. \end{aligned}$$

Similarly,

$$\left[u_k, x_i \frac{\partial}{\partial x_i}\right] = (\det R_{k-1})^2 \cdot D_{k,i} \,. \qquad (4.6)$$

Thus

$$\begin{aligned} \frac{\left[u_k, \frac{\partial}{\partial t}\right]^2}{(u_k,u_k)} &= \frac{(\det R_{k-1})^4 D_{k,t}^{\;2}}{(u_k,u_k)} \\ &= \frac{\prod_{j=1}^{k-2}(u_j,u_j)^{4(k-j-1)} D_{k,t}^{\;2}}{\prod_{j=1}^{k-1}(u_j,u_j)^{2k-2j-1} D_k} \quad \text{(by (4.4) and (4.5))} \\ &= \frac{\prod_{j=1}^{k-2}(u_j,u_j)^{2k-2j-3} D_{k,t}^{\;2}}{(u_{k-1},u_{k-1}) D_k} \qquad (4.7)\\ &= \frac{\prod_{j=1}^{k-2}(u_j,u_j)^{2k-2j-3} D_{k,t}^{\;2}}{\prod_{j=1}^{k-2}(u_j,u_j)^{2k-2j-3} D_{k-1} D_k} \quad \text{(by (4.4))} \\ &= \frac{D_{k,t}^{\;2}}{D_{k-1} D_k} \,. \end{aligned}$$

Similarly, we obtain that

$$\frac{\left[u_k, \frac{\partial}{\partial t}\right]\left[u_k, x_i \frac{\partial}{\partial x_i}\right]}{(u_k, u_k)} = \frac{D_{k,t} D_{k,i}}{D_{k-1} D_k}. \tag{4.8}$$

Therefore,

$$\begin{aligned} P &= t\text{-component of } V' \\ &= \left[V', \frac{\partial}{\partial t}\right] \\ &= \left[\frac{\partial}{\partial t}, \frac{\partial}{\partial t}\right] - \sum_{k=1}^{p} \frac{(u_k, \partial/\partial t)^2}{(u_k, u_k)} && \text{(by (4.2))} \\ &= 1 - \sum_{k=1}^{p} \frac{D_{k,t}{}^2}{D_{k-1} D_k}. && \text{(by (4.7))} \end{aligned}$$

And,

$$\begin{aligned} -Q_i &= x_i\text{-component of } V \\ &= x_i{}^2\left[V', \frac{\partial}{\partial x_i}\right] \\ &= x_i{}^2\left[\frac{\partial}{\partial t}, \frac{\partial}{\partial x_i}\right] - \sum_{k=1}^{p} x_i \frac{(u_k, \partial/\partial t)}{(u_k, u_k)}\left[u_k, x_i \frac{\partial}{\partial x_i}\right] && \text{(by (4.2))} \\ &= -\sum_{k=1}^{p} \frac{D_{k,i} D_{k,t}}{D_{k-1} D_k}. && \text{(by (4.8))} \end{aligned}$$

Next we show that $P = J_p/D_p$. Let $\mathrm{Vol}(\nabla_1, \ldots, \nabla_k)$ denote the k-dimensional volume of the paralleltope with edges $\nabla_1, \ldots, \nabla_k$. Then

$$\begin{aligned} P &= (V', V') \\ &= \frac{\mathrm{Vol}(\nabla_1, \ldots, \nabla_p, \frac{\partial}{\partial t})^2}{\mathrm{Vol}(\nabla_1, \ldots, \nabla_p)^2} \\ &= \frac{J_p}{D_p}. && \text{Q.E.D.} \end{aligned}$$

Therefore we obtain the following expression of Kuo's vector field.

$$V = -\sum_{i=1}^{n}\sum_{k=1}^{p} \frac{D_p D_{k,i} D_{k,t}}{J_p D_{k-1} D_k} x_i \frac{\partial}{\partial x_i} + \frac{\partial}{\partial t}. \tag{4.9}$$

§5. PROOF OF THEOREM

We assume that $I = [a,b]$. The following lemma is easy.

LEMMA (5.1). *Assume that there exist*

an analytic vector field $\tilde{\xi}$ *near* $\pi^{-1}(0)$ *in* X, *and*
an analytic vector field $\tilde{\eta}$ *near* $\pi^{-1}(0) \times 0$ *in* $X \times \mathbb{R}^p$,
such that

1) $\tilde{\xi}$ *is tangent to* $\pi^{-1}(0)$ *and* S_π,
2) $\tilde{\eta}$ *is tangent to* $X \times 0$,
3) X*-component of* $\tilde{\eta}$ *is* 0, *and*
4) $(df_t)(\tilde{\xi}(x)) + (df_t/dt)(\pi(x)) = \tilde{\eta}(x,f_0(x))$, *for any* x *near* $\pi^{-1}(0)$ *in* X.

Then f_t *admits an almost* MAT *via* π *along* I.

Let $f_t(x) = F(x;t)$ be an analytic family of non-degenerate mappings. Suppose that $\Gamma_+(f_t)$ is independent of t.

LEMMA (5.2). *There exists an analytic function* g_j *such that*

1) $\Gamma_+(\tilde{g}_j) = 2\Gamma_+(f_{t,j})$, *and*
2) g_j *is positive on* $\pi^{-1}(0)$,

where

$$g_j(\pi_\sigma(y_\sigma)) = \prod_{c=1}^{n} y_{\sigma,c}^{\,2\ell_j(a^c(\sigma))}\,\tilde{g}_j(y_\sigma).$$

PROOF. Let h_j be a non-degenerate function with $\Gamma_+(h_j) = \Gamma_+(f_{t,j})$. Then $g_j = \Sigma_{i=1}^{n}(x_i, (\partial h_j/\partial x_i))^2$ satisfies the desired properties. Q.E.D.

Let g_j be as in Lemma (5.2). Define A, B, and B' by

$$A = D_1, \dots, D_{p-1}J_p ,$$

$$B = \sum_{j=1}^{p} f_j^2 g_j^{p-j} \prod_{k\neq j} g_k^{p-k+1}, \quad \text{and}$$

$$B' = \sum_{j=1}^{p} f_j\, g_j^{p-j} \prod_{k\neq j} g_k^{p-k+1}\rho^* y_j .$$

Let us denote

$$\xi = \frac{A}{A+B}\left[V - \frac{\partial}{\partial t}\right], \quad \text{(where } V \text{ be as in §4), and}$$

$$\eta = \sum_{j=1}^{p} \frac{B'}{A+B}\left[\frac{df_t}{dt}\right]_j \frac{\partial}{\partial y_j}.$$

By (4.9), we obtain the following expression of ξ.

$$\xi = \frac{1}{A+B}\sum_{i=1}^{n}\sum_{k=1}^{p}\prod_{j\neq k,k-1} D_j D_p D_{k,i} D_{k,t}\, x_i \frac{\partial}{\partial x_i}. \tag{5.3}$$

Define π' by $(\pi\,|X - S_\pi)^{-1}$.

LEMMA (5.4). *The vector field* $d\pi'(\xi)$ (*resp.* $d(\pi' \times id_{\mathbb{R}^p})(\eta)$) *is analytically extendible to a small neighbourhood of* $\pi^{-1}(0)$ (*resp.* $\pi^{-1}(0) \times 0$).

Define $\breve{\xi}$ (resp. $\breve{\eta}$) the extension of $d\pi'(\xi)$ (resp. $d(\pi' \times id_{\mathbb{R}^p})(\eta)$). Then $\breve{\xi}$ and $\breve{\eta}$ satisfy the assumption of Lemma (5.1). Therefore the trajectory of $\breve{\xi} + \partial/\partial t$ (resp. $\breve{\eta} + \partial/\partial t$) gives the desired isomorphism $H_t : \breve{U}_1 \to \breve{U}_2$ (resp. $H_t' : \breve{U}_1 \times U_1' \to \breve{U}_2 \times U_2'$).

Moreover, if $f_{t,\tau}$ is independent of t for non-compact and non-coordinate face τ, then $\breve{\xi}$ (resp. $\breve{\eta}$) $= 0$ on S_π (resp. $S_\pi \times 0$). (Because $D_{k,t}$ is identically zero on S_π.) Because $D_{k,t}$ is identically zero on S_π.) Thus H_t (resp. H_t') induces a local homeomorphism of $(\mathbb{R}^n,0)$ (resp. $(\mathbb{R}^n \times \mathbb{R}^p,0)$).

PROOF OF LEMMA (5.4). Since

$$\Gamma_+(J_s) = \sum_{i=1}^{s} 2\Gamma_+(f_{t,j}),$$

$$\Gamma_+(D_s) = \sum_{i=1}^{s} 2\Gamma_+(f_{t,j}),$$

$$\Gamma_+(D_{s,i}) \subset \sum_{i=1}^{s-1} 2\Gamma_+(f_{t,j}) + \Gamma_+(f_{t,s_s}), \quad \text{and}$$

$$\Gamma_+(D_{s,t}) \subset \sum_{i=1}^{s-1} 2\Gamma_+(f_{t,j}) + \Gamma_+(f_{t,s}),$$

there are analytic functions $\widetilde{J}_s$, $\widetilde{D}_s$, $\widetilde{D}_{s,i}$, and $\widetilde{D}_{s,t}$ such that

$$J_s(\pi_\sigma(y_\sigma);t) = \prod_{c=1}^{n} y_{\sigma,c}^{\,2\Sigma_{d=1}^{s}\ell_d(a^c(\sigma))}\, \widetilde{J}_s(y_\sigma;t),$$

$$D_s(\pi_\sigma(y_\sigma);t) = \prod_{c=1}^{n} y_{\sigma,c}^{\,2\Sigma_{d=1}^{\sigma}\ell_d(a^c(\sigma))}\, \widetilde{D}_s(y_\sigma;t),$$

$$D_{s,i}(\pi_\sigma(y_\sigma);t) = \prod_{c=1}^{n} y_{\sigma,c}^{\,2\Sigma_{d=1}^{s-1}\ell_d(a^c(\sigma))+\ell_s(a^c(\sigma))}\, \widetilde{D}_{s,i}(y_\sigma;t),$$

and

$$D_{s,t}(\pi_\sigma(y_\sigma);t) = \prod_{c=1}^{n} y_{\sigma,c}^{\,2\Sigma_{d=1}\ell_d(a(\sigma))+\ell_s(a(\sigma))}\, \widetilde{D}_{s,t}(y_\sigma;t).$$

Define $\widetilde{f}_{t,j}$ by

$$f_{t,j}(\pi_\sigma(y_\sigma)) = \prod_{c=1}^{n} y_{\sigma,c}^{\,\ell_j(a^c(\sigma))}\, \widetilde{f}_{t,j}(y_\sigma).$$

By the assumption of non-degeneracy and trick (2.7).

$$\widetilde{J}_p = \widetilde{f}_{t,1} = \cdots = \widetilde{f}_{t,p} = 0 \quad \text{has no common solution on } \pi_\sigma^{-1}(0).$$

It is easy to show that $\widetilde{J}_p \neq 0$ implies $\widetilde{J}_s \neq 0$ and $\widetilde{D}_s \neq 0$ for $s = 1, \ldots, p$.

Since

$$d\pi'\left(x_i, \frac{\partial}{\partial x_i}\right) = \sum_{j=1}^{n} b_i^j(\sigma) y_{\sigma,j} \frac{\partial}{\partial y_{\sigma,j}},$$

where $(b_i^j(\sigma))$ is the inverse matrix of $(a_i^j(\sigma))$, $d\pi'(\xi)$ is analytically extendible on some small neighbourhood of $\pi^{-1}(0)$. By (5.2), $d(\pi' \times id_{\mathbb{R}P})(\eta)$ is analytically extendible on some small neighbourhood of $\pi^{-1}(0) \times 0$. Q.E.D.

§6. EXAMPLE

EXAMPLE (6.1). Let $f_t = (f_{t,1}, f_{t,2}) : (\mathbb{R}^3,0) \to (\mathbb{R}^2,0)$ be a map-germ defined by

$$f_{t,1}(x_1,x_2,x_3) = x_1^2 + x_2^2 + x_3^2, \text{ and}$$

$$f_{t,2}(x_1,x_2,x_3) = a_1x_1^3 + a_2x_2^3 + a_3x_3^3,$$

where $a_i = a_i(x,t)$ is an analytic function such that $a_i(0,t) \neq 0$ on I. Assume that $a_i(0,t) \neq a_j(0,t)$ for $i \neq j$, $t \in I$. Let π: $X \to \mathbb{R}^n$ be the blowing-up at $\{0\}$. Then f_t admits a MAT via π along I.

EXAMPLE (6.2). Let $f_t = (f_{t,1}, f_{t,2}) : (\mathbb{R}^3,0) \to (\mathbb{R}^2,0)$ be a map-germ defined by

$$f_{t,1}(x_1,x_2,x_3) = x_1^2 + x_2^2 + x_3^2, \text{ and}$$

$$f_{t,2}(x_1,x_2,x_3) = a_1x_1^2 + a_2x_2^3 + a_3x_3^4,$$

where $a_i = a_i(x,t)$ is an analytic function with $a_i(0,t) \neq 0$. Then f_t admits a MAT.

REFERENCES

[Ka] A. Kaneko: Newton polygon, singularity, oscillating integral. Math. Lect. Note Sophia Univ. **11**, (1981) (in Japanese).

[TE] G. Kempf, F. Knudsen, D. Mumford, B. Saint-Donat: Toroidal Embeddings, I. Lect. Notes Math., Vol. 339, Berlin, Heidelberg, New York, Springer, 1973.

[K_1] T.-C. Kuo: The modified analytic trivialization of singularities. J. Math. Soc. Japan **32**, pp. 605-614 (1980).

[K_2] T.-C. Kuo: Une classification des singulautes reelles. C.R. Acad. Sci. Paris Ser. A-B **288**, pp. A805-A812 (1979).

[Ku] A. G. Kushinirenko: Polyedres de Newton et Nombres de Milnor. Invent. Math. **32**, pp. 1-32 (1976).

[M] J. Mather: Stability of C^∞-mappings III. Inst. Hautes Etudes Sci. Publ. Math. **37**, pp. 1-32 (1976).

[V] A. N. Varcenko: Newton polyhedra and estimation of oscillating integrals. Funct. Anal. Appl. **10**, pp. 175-196 (1977).

[Y-F] T. Fukui, E. Yoshinaga: The modified analytic trivialization of family of real analytic functions. Invent. Math. **82**, pp. 467-477 (1985).

DEPARTMENT OF MATHEMATICS
FACULTY OF SCIENCE
TOKYO METROPOLITAN UNIVERSITY
2-1-1 FUKAZAWA, SETAGAYA-KU,
TOKYO 158 JAPAN

Contemporary Mathematics
Volume **90**, 1989

ON THE TOPOLOGICAL TYPES OF QUASI-ORDINARY SURFACE GERMS

Yih-Nan Gau

INTRODUCTION

A quasi-ordinary surface germ is a covering of the complex plane which is branched over the coordinate axes, (see (1.1) for definition). Essentially because the fundamental group of the complement of the axes in $\mathbf{C}^2$ is $\mathbf{Z} \times \mathbf{Z}$, quasi-ordinary germs can be 'parametrized' by fractional power series, (1.2). For example, the surface germ (at the origin) $z^n = x^a y^b$ (in particular, the classical $A_{n,q}$ singularity $z^n = xy^{n-q}$) is quasi-ordinary and can be written as $\Pi_i(z - \omega^i x^{a/n} y^{b/n}) = 0$.

$A_{n,q}$, whose normalization is the cyclic quotient $\mathbf{C}^2/\mathbf{Z}_n$, [H-S, p. 233] is typical as far as normalization is concerned: the normalization of any quasi-ordinary hypersurface germ is a cyclic quotient singularity, see [L_1, (3.1.1)]. In the *surface* case the links of the normalizations are then 3-dim lens spaces. In [C], Costa describe the lens spaces thus obtained (in terms of the parametrization) and the identification needed to recover the links of the quasi-ordinary surface singularities.

In this paper we give another description of the links and their normalizations (section 2) and derive from it a necessary and sufficient condition for the *links* of two quasi-ordinary surface singularities to be homeomorphic, see (3.6). In section 4, we consider the same question for the quasi-ordinary surface germs

0271-4132/89 $1.00 + $.25 per page

(instead of their links). We are successful only in the most "topologically singular" case, see prop. (4.3). The answer to the remaining case is related to "whether there is a "topological h-cobordism" between two homotopy equivalent lens spaces", see (4.5). Also a question of Lê-Oka is verified in the quasi-ordinary case, see (2.6.6).

We note in passing that the *embedded* topological types of quasi-ordinary hypersurface singularities has been classified in $[G_2]$, using results from $[L_1]$.

I appreciate the comments of J. Lipman and K. Oh.

SECTION 1. DEFINITIONS AND THE MAIN QUESTION

Let F be an analytic surface in some open subset U of $\mathbf{C}^3$:

$$F = \{(a,b,c) \in U: f(a,b,c) = 0, f \text{ is holomorphic in } U\}.$$

Assume that the origin 0 is a singular point of F and F is irreducible* at $0 = (0,0,0)$.

(1.1) DEFINITION. The germ $(F,0)$ is *quasi-ordinary* if there is a finite map of analytic germs:

$$\pi: (F,0) \to (\mathbf{C}^2,0)$$

which induces a surjection on Zariski tangent spaces and whose discriminant locus D (the curve over which either F is singular or D is not a submersion) either is smooth or has an ordinary double point at 0.

REMARK (1.1.1). In the above definition, if D is smooth at 0 then F is equisingular along $\pi^{-1}(D)$, $[G_1$, (2.8.4)]. In this case, F is topologically trivial along $\pi^{-1}(D)$ near 0, [T, p. 598].

*We are interested in the topology of F near 0 (1.4). Because homeomorphisms preserves irreducible components [G-L, (A.8), p. 172], it suffices to consider the irreducible case.

Quasi-ordinary germs can be parametrized by fractional power series, just as plane curve germs can be parametrized by Puiseux series. To state this result, we introduce some notation. Let ζ be a fractional power series with bounded denominator, i.e. $\zeta = H(X^{1/n}, Y^{1/n})$ for some $H \in \mathbf{C}[[X,Y]]$, the formal power series ring. We assume that H is convergent. Let

$$Q_n = \{q > 0: nq \text{ is an integer}\},$$

then ζ can be written as $\Sigma\, c_{ab}X^aY^b$ with $c_{ab} \in \mathbf{C}$ and $(a,b) \in Q_n \times Q_n$.

(1.2) DEFINITION (Cf. [L_2, (1.5)], [G_1, (2.1.4)]. A fractional power series $\zeta = \Sigma\, c_{ab}X^aY^b$ is *quasi-ordinary* if there are pairs of rational numbers (λ_i,μ_i), $1 \leqslant i \leqslant s$, which satisfy:

1. $c_{\lambda_i,\mu_i} \neq 0$.
2. $\lambda_i \leqslant \lambda_{i+1}$, $\mu_i \leqslant \mu_{i+1}$ for $0 \leqslant i < s$. (Here we take $\lambda_0 = \mu_0 = 0$.)
3. $(\lambda_t,\mu_t) \notin \mathbf{Z} \times \mathbf{Z} + \Sigma_{1 \leqslant i < t}\mathbf{Z}(\lambda_i,\mu_i)$ for $1 \leqslant t \leqslant s$.
4. if $c_{ab} \neq 0$, then $(a,b) \in \mathbf{Z} \times \mathbf{Z} + \Sigma_{(\lambda_i,\mu_i) \leqslant (a,b)} \mathbf{Z}(\lambda_i,\mu_i)$.

The monomials $M_i := X^{\lambda_i}Y^{\mu_i}$ are called the *characteristic monomials* of ζ.

(1.2.1) Let G be the galois group $\mathrm{Gal}[E(\zeta)/E]$, where E is the quotient field of $\mathbf{C}[[x,y]]$. (Although G is abelian, we will use multiplicative notation.) It is easy to check that $\Pi_{g\in G}(z - g\zeta) \in \mathbf{C}[[x,y]][z]$.

(1.3) PROPOSITION. An analytic surface germ $(F,0)$ is *quasi-ordinary* iff, up to an analytic change of coordinates, F is defined (near 0) by:

$$f(x,y,z) := \Pi_{g\in G}(z - g\zeta) = 0,$$

for some quasi-ordinary fractional power series ζ. (In this case, the projection in (1.1) is given by π: $(x,y,z) \to (x,y)$.)

PROOF. See [G_1, (2.1.1), (2.1.4)].

Each $g\zeta$ in (1.3) is said to be a *parametrization* of $(F,0)$.

(1.4) QUESTION: When are two quasi-ordinary germs $(F,0)$ and $(F',0)$ topologically equivalent (i.e. a neighborhood of 0 in F is homeomorphic to a neighborhood of 0 in F')?

Let ζ and ζ' be parametrizations of $(F,0)$ and $(F',0)$ respectively. In trying to answer the above question, we can and shall assume that ζ is the sum of its characteristic monomials:

$$\zeta = M_1 + M_2 + \dots + M_s. \tag{1.4.1}$$

(Same thing can be assumed of ζ'.) This is due to the fact that the characteristic monomials determine the local topology of $(\mathbf{C}^3,F)$ at 0, see [L_2, p. 167], [G_1, (2.2.3)].

Moreover, as far as (1.4) is concerned, we can assume that ζ is "reduced", i.e.

$$\lambda_1 > 0 \quad \text{and} \quad \mu_1 > 0. \tag{1.4.2}$$

This follows from [L_1, prop. (6.3)].

SECTION 2. LINKS AND THEIR NORMALIZATIONS

In this section we relate the "algebra" of the fractional power series ζ to the "geometry"of F near 0, especially the "polydisk" link of $(F,0)$. (From now on we shall omit the word "polydisk.") We assume that ζ is *reduced* and is the sum of its characteristic monomials, (1.4.1), (1.4.2).

(2.1) First we introduce some notations and basic definitions.

1. For the λ_i, μ_i (i = 1, ..., s) in (1.2), let (m_i,n_i) and (m_i',n_i') be integer pairs defined inductively by:

$$\lambda_i = m_i/n_1n_2 \dots n_i, \quad (m_i,n_i) = 1$$

$$\mu_i = m_i'/n_1'n_2' \dots n_i',\ (m_i',n_i') = 1.$$

Let $n_x := n_1 n_2 \dots n_s$ and $n_y := n_1'n_2' \dots n_s'$.

2. If $\zeta = H(x^{1/n},y^{1/n})$ and $\omega = \exp(2\pi i/n)$ (for some n), let $\alpha,\beta \in G$ (1.2.1) be defined by:

$$\alpha(\zeta) = H(\omega x^{1/n},y^{1/n})$$
$$\beta(\zeta) = H(x^{1/n},\omega y^{1/n}).$$

(2.1.1) REMARK. Since $(re^{i\theta})^{1/n} \to \omega r^{1/n} = \alpha(x^{1/n})|_{x=r}$, as $\theta \to 2\pi^-$, we have:

$$a^j\zeta(r_1e^{i\theta},y) \to \alpha^{j+1}\zeta(r_1,y) \quad \text{as } \theta \to 2\pi^-$$

Similarly, $\beta^i\zeta(x,r_2e^{i\theta}) \to \beta^{i+1}\zeta(x,r_2)$ as $\theta \to 2\pi^-$. (The fractional power of a complex number $z = r\exp(i\theta)$, $r \geq 0$, $0 \leq \theta < 2\pi$, is defined to be: $z^{m/n} = [r^{1/n}\exp(i\theta/n)]^m$. See convention (0.4) in [G_1] for details.)

3. Let $H_x := \langle\alpha\rangle$, $H_y := \langle\beta\rangle$ be the subgroups of G generated by α and β. Let $\mathbf{c}$, $\mathbf{d}$ be the indices of H_x, H_y in G. It can be checked that α, β have orders n_x, n_y respectively [G_1, (2.4.1), (2.4.2)]. Thus

$$|G| = \mathbf{c}n_x = \mathbf{d}n_y.$$

We define $\mathbf{p}$ to be:

$$\mathbf{p} = n_x/\mathbf{d} = n_y/\mathbf{c} = |G|/\mathbf{cd}.$$

Since G is generated by α and β [G_1, (2.1.2)], the natural homomorphism

$$(*) \qquad H_x \to G/H_y$$

is onto and has the subgroup $\langle\alpha^{\mathbf{d}}\rangle$ as its kernel. Thus $G = \cup_{0\leq i<\mathbf{d}} \alpha^i H_y$ and $\alpha^{\mathbf{d}} \in H_y$. Similarly, $\beta^{\mathbf{c}} \in H_x$.

4. Let δ ($<n_y$), γ ($<n_x$) be the smallest positive integer defined by:

$$\alpha^{\mathbf{d}} = \beta^{\delta}, \quad \beta^{\mathbf{c}} = \alpha^{\gamma}.$$

Let $\mathbf{r} := \gamma/\mathbf{d}$, $\mathbf{s} := \delta/\mathbf{c}$.

(2.2) LEMMA. $\mathbf{p}$ *and* $\mathbf{r}$, $\mathbf{s}$ (*both* $<\mathbf{p}$) *are integers and* $\mathbf{rs} = 1$ (mod $\mathbf{p}$).

PROOF. Since α^{γ} and α^{n_x} are both in H_y, it follows from (*) that $\mathbf{d}|n_x$ and $\mathbf{d}|\gamma$. Similarly $\mathbf{c}|\delta$. To prove the second assertion, raise $\beta^{\mathbf{c}} = \alpha^{\gamma}$ to the $(\delta/\mathbf{c})$-th power:

$$\alpha^{(\gamma\delta/\mathbf{c})} = \beta^{\delta} = \alpha^{\mathbf{d}}.$$

Thus $\gamma\delta/\mathbf{c} \equiv \mathbf{d} \pmod{n_x}$. Divide the congruence by $\mathbf{d}$ to get $\mathbf{rs} \equiv \gamma\rightarrow \pmod{\mathbf{p}}$.

EXAMPLE. The surface $z^7 = x^3y^5$ is quasi-ordinary with parametrization $\zeta = x^{3/7}y^{5/7}$. In this case:

$$n = n_x = n_y = 7, \quad \mathbf{c} = \mathbf{d} = 1, \quad \mathbf{p} = 7, \quad \mathbf{s} = \delta = 2, \quad \mathbf{r} = \gamma = 4.$$

It can be checked that these same values are obtained for the surface $z^7 = x^2y^1$. We will see latter that the links (2.4.3) of these two surfaces are homeomorphic: they are the $(\mathbf{p},\mathbf{r}) = (7,4)$ lens spaces (2.6.3), (2.7).

Geometric interpretation of $\mathbf{c}$ *and* $\mathbf{d}$. Let $L(y_0)$ denote the complex plane (in $\mathbf{C}^3$) defined by $Y = y_0$. ($|y_0|$ will always be small but nonzero.) The plane section $F \cap L(y_0)$ of F is a plane curve containing the point $(0,y_0,0)$. *Denote* the germ of $F \cap L(y_0)$ at $(0,y_0,0)$ by $\mathbf{F}(y_0)$.

(2.3) PROPOSITION. *The number of irreducible components of* $\mathbf{F}(y_0)$, *denoted by* irc $\mathbf{F}(y_0)$, *is* $\mathbf{c}$. *Similarly*, $\mathbf{d} =$ *irc* $\mathbf{F}(x_0)$.

PROOF. Let $C_j(y_0)$, $0 \leqslant j < \mathbf{c}$, be the plane curve in $L(y_0)$which is parametrized by $z = \beta^j\zeta(x,y_0)$;

(2.3.1) $$C_j(y_0) = \{(x,y_0,z)\colon z = g\zeta(x,y_0),\ x \in \mathbf{C},\ g \in \beta^j H_x\}.$$

It can be checked that the germs of $C_j(y_0)$ (at 0, y_0, 0)) are exactly the (irreducible) components of $\mathbf{F}(y_0)$. Similarly the components of $\mathbf{F}(x_0)$ are represented by:

(2.3.2) $D_j(x_0) = \{(x_0,y,z)\colon z = g\zeta(x_0,y),\ y \in \mathbf{C},\ g \in \alpha^j H_y\},\ 0 \leqslant j < d.$

See [G_1, (2.4.6), (2.6.1)] for details.

(2.4) Next we describe the (polydisk) link of $(F,0)$. For $r = (r_1,r_2,r_3)$, let

$$D_{xyz}(r) := \{(x,y,z)\colon |x| \leqslant r_1,\ |y| \leqslant r_2,\ |z| \leqslant r_3\},$$
$$S_{xyz}(r) := \partial D_{xyz}(r).$$

$D_{xz}(r_1,r_3)$, $S_y(r_2)$ etc. are similarly defined. We shall identify, e.g. S_y with $0xS_yx0$. From now on, r_i will always be small enough and satisfy the following conditions (see [G_1, (3.0), (2.5.1)]):

(*) $|\zeta(x,y)| < r_3$ provided $|x| < r_1$, $|y| < r_2$. In this case $F \cap (S_z \times D_{xy})$ is empty.

(**) For $(x,y) \in D_{xy}(r_1,r_2)$, $g\zeta(x,y) = 0$ iff $xy = 0$; and $g\zeta(x,y) = g'\zeta(x,y)$ iff either $g = g'$ or $xy = 0$.

(2.4.1) NOTE. It follows from (*), (**) that, $F \cap D_{xyz}(r)$ is a $|G|$-sheeted branched covering of $D_{xy}(r_1,r_2)$, the discriminant locus (in $\mathbf{C}^2$) of which is contained in $xy = 0$. In a small neighborhood of 0, the singular locus of F is thus contained in the X-, Y-axes. See [G_1, (2.5), (2.8)].

(2.4.2) NOTE. The conic structure of F at 0 [B-V] can be exhibited explicitly. In fact, the map:

$$t \times (x,y,g\zeta(x,y)) \to (tx,ty,g\zeta(tx,ty)),\quad 0 \leqslant t \leqslant 1,\quad (x,y) \in S_{xy},$$

shows that $F \cap D_{xyz}(r)$ is a cone over $F \cap S_{xyz}(r) = K(r)$ with 0 as the cone point. It is easy to see that the conic structure respects the X-, Y-axes.

CONVENTION. We shall sometimes omit the radius r. Also,

since we will be interested mainly in what happens inside D_{xyz}, we shall write F, $C_j(y)$ etc. for $F \cap D_{xyz}$, $C_j(y) \cap y \times D_{xz}$ etc.

(2.4.3) By a (polydisk) *link* $K = K(r)$ of $(F,0)$ we mean $F \cap S_{xyz}(r)$ $(= \partial F)$, with r small enough. Let

$$K_1 = F \cap (S_y \times D_{xz})$$
$$K_2 = F \cap (S_x \times D_{yz})$$

In view of condition (*) in (2.4), $K = K_1 \cup K_2$ and $K_1 \cap K_2 = \partial K_1 = \partial K_2$.

(2.5) PROPOSITION. *K_1 is a fibre bundle over S_y whose fibre over y_0 is:*

$$\cup_{0 \leqslant j < c} C_j(y_0),$$

which is topologically the one-point-union (= disjoint union with one interior point from each identified) of **c** *2-disks. Similarly, K_2 fibres over S_x whose fibre is topologically the one-point-union of* **d** *2-disks.*

PROOF. Let $P = (x, y_0, g\zeta(x, y_0)) \in K_1 \cap L(y_0)$, the fibre over y_0, and let I be a small neighborhood of y_0 in S_y. It can be checked that the mapping:

$$I \times (K_1 \cap L(y_0)) \ni (y,P) \to (x,y,g\zeta(x,y)) \in K_1$$

is well defined (i.e. independent of choice of g) and gives a homeomorphism onto $K_1 \cap (I \times D_{xz})$; hence the local triviality of the natural projection $K_1 \to S_y$.

From the proof of proposition (2.3) we see that the fibre:

$$K_1 \cap L(y_0) = F \cap L(y_0) = \cup_{0 \leqslant j < c} C_j(y_0).$$

Claim. $C_j(y)$ is topologically a 2-disk.

PROOF OF CLAIM. Since the germ $C_j(y)$ at $(0,y,0)$ is irreducible ($\Rightarrow$ topologically unibranch [M_1, p. 117]), $\partial C_j(y)$, the link of $C_j(y)$, is a closed 1-manifold, hence a circle.

Another proof: from (2.3.1) we have:

$$(2.5.1) \qquad \partial C_j(y) = \{(x,y,z) \colon |x| = r_1,\ z = \alpha^i\beta^j\zeta(x,y),\ 0 \leq i < n_x\}.$$

Since $\alpha^{n_x} = \alpha^0 = id$ (2.1)-3, $\partial C_j(y)$ is a circle.

The assertion about D_j can be similarly proved. Q.E.D.

(2.5.2) REMARK. S_y is a deformation retract of K_1. In fact, by the choice of radius r (2.4), we can deform $C_j(y)$ to the point $(0,y,0)$ through:

$$(t,x,y,z) \to (tx,y,g\zeta(tx,y)),$$

where $z = g\zeta(x,y)$. See [G_1, (3.1)].

(2.5.3) REMARK. Let $P = (x,y,z) \in F$. If $xy \neq 0$, then F is smooth (in particular analytically irreducible) at P, note (2.4.1). If $x = 0$, $y \neq 0$, then $z = g\zeta(0,y) = 0$, and $P \in Y$-axis. In this case, F has **c** analytically irreducible components at P, whose intersections with the plane $L(y)$ are the irreducible components $C_j(y)$ of the plane section $F \cap L(y)$. (This can be argued from the topological triviality of F along Y-axis at P and proposition (2.3), see [G_1, proof of (4.3.1)] for details.) Similarly, F has **d** analytically irreducible components at $(x,0,0)$.

(2.6) We now consider the normalization $\mathcal{F}$ of $(F,0)$:

$$\nu : \mathcal{F} \to F.$$

For each $P \in F$, $\nu^{-1}(P)$ consists of k points, which correspond to the k irreducible components of the analytic germ (F,P) [W, Th. 3A, p. 256]. Now, F is analytically irreducible at $P \in K_1 - S_y$, but has **c** irreducible components at $P \in S_y$, (2.5.3). Thus, in view of (2.5), $\mathcal{K}_1 := \nu^{-1}(K_1)$ fibres over S_y with fibre the disjoint union of **c** 2-disks. Moreover the monodromy action is cyclic: as y traverse the circle S_y once, $C_j(r_2)$ is identified with $C_{j+1}(r_2)$ [here it is understood that $C_{\mathbf{c}} = C_0$], remark (2.1.1). Thus:

(2.6.1) K_1 is a solid torus with $\partial C_0(r_2)$ as a meridian. The map ν: $\mathrm{K}_1 \to K_1$ is an identification map: it is one-to-one except at the core $\nu^{-1}(S_y)$ of K_1, whereas $\nu^{-1}(S_y)$ is a $\mathbf{c}$-fold cyclic covering over S_y. In other words, ν^{-1} "pulls" the $\mathbf{c}$ branches of K_1 at $P \in S_y$ apart.

(2.6.2) NOTE. Since S_y (resp. $\nu^{-1}(S_y)$) is a deformation retract of K_1 (resp. K_1) (2.5.2), the map ν_*:

$$\langle[\nu^{-1}(S_y)]\rangle = \mathbf{Z} = H_1(\mathrm{K}_1) \to H_1(K_1) = \mathbf{Z} = \langle[S_y]\rangle,$$

is just "multiplication by $\mathbf{c}$". ([ ] means "the homology class of").

Similarly, with the roles of x, y interchanged, the "normalization" $\mathrm{K}_2 := \nu^{-1}(K_2)$ of K_2 is a solid torus with its core identified in a d-to-one fashion. Thus K can be described as the union of two solid tori with their boundary suitably identified; i.e. K is a Lens space [R, p. 233]. To sum up:

(2.6.3) PROPOSITION. $\mathrm{K} = \mathrm{K}_1 \cup \mathrm{K}_2$ *is a Heeggard splitting of a lens space* (*cf.* (2.7)) *and the link* K *of* F *is the identification space*:

$$K = \mathrm{K}/\sim,$$

where $\sim$ *identifies points of the core* S_1 *of* K_1 (*resp.* S_2 *of* K_2) *which are in the same orbit under the* $\mathbf{Z}_\mathbf{c}$ (*resp.* $\mathbf{Z}_\mathbf{d}$) *cyclic* (*rotation*) *action.*

For the rest of this subsection we consider the special case $\mathbf{c} = \mathbf{d} = 1$.

If $\mathbf{c} = \mathbf{d} = 1$, then K is homeomorphic to K, hence by proposition (2.6.3), K is a topological manifold.

To see that the *converse* is also true, consider again the map:

$$\nu : (\mathrm{K}_1, \nu^{-1}(S_y)) \to (K_1, S_y).$$

Let $P \in S_y$ and $P_0, \ldots, P_{\mathbf{c}-1}$ be the points on the core $\nu^{-1}(S_y)$ of K_1 with $\nu(P_j) = P$, $0 \leqslant j < \mathbf{c}$. We parametrize $\nu^{-1}(S_y)$ by $e^{i\theta}$, $0 \leqslant \theta < 2\pi$, so that $P_j = e^{i(2j\pi/\mathbf{c})}$. Then a small neighborhood V of P (in K_1) is obtained from $\nu^{-1}(V)$ (= disjoint union of (3-disk) neighborhoods V_j

(in K_1) of P_j) by identifying the arc $\nu^{-1}(S_y) \cap V_j$ with the arc $\nu^{-1}(S_y) \cap V_{j+1}$:

$$e^{i\theta} \in \nu^{-1}(S_y) \cap V_j \text{ is identified with } e^{i(\theta+2\pi/c)} \in \nu^{-1}(S_y) \cap V_{j+1}.$$

(As usual, $V_c = V_0$.) Thus near $P \in S_y$, K_1 is topologically the disjoint union of **c** 3-disks with **c** arcs (one from each 3-disk) identified into one arc (in S_y). Since the complement of an arc in a 3-disk is connected, $V_j - (\nu^{-1}(S_y) \cap V_j)$ is connected. Thus $V - (S_y \cap V)$ (which is homeomorphic to the union of $V_j - (\nu^{-1}(S_y) \cap V_j)$, $0 \leqslant j < \mathbf{c}$, (2.6.1)) has **c** connected components. It follows that V is not a topological manifold at P if $\mathbf{c} > 1$. In view of (2.4.1) we have:

K_1 is a topological manifold (with boundary) at P iff $\mathbf{c} = 1$.
Similarly, K_2 is a topological manifold iff $\mathbf{d} = 1$.

In view of proposition (2.6.3), we can now state the following characterization of quasi-ordinary surface germs which are homeomorphic to their own normalizations:

(2.6.4) LEMMA (Cf. lemma 3.2). *The following are equivalent*:

1. $\mathbf{c} = 1$, $\mathbf{d} = 1$.
2. *K is homeomorphic to* $\mathfrak{K}$.
3. *K is a topological manifold.*

We can also characterize those quasi-ordinary surface germs which are topologically trivial (meaning that it has a neighborhood homeomorphic to a 4-cell). First we recall Mumford's result [M_2, p. 244]): if a normal surface germ is topologically trivial, then it is in fact smooth.

(2.6.5) PROPOSITION. *Let $(F,0)$ be as in section 2 (in particular, ζ is reduced (1.4.2)) with normalization $(\mathfrak{F},0)$. The following are equivalent:*

1. *$(F,0)$ and $(\mathfrak{F},0)$ are homeomorphic germs and $(\mathfrak{F},0)$ is smooth.*
2. *$(F,0)$ is topologically trivial.*

3. *The singular homology group* $H_1(F,F\backslash 0) = H_1(K)$ *is trivial.*

4. $(F,0)$ *is a smooth germ.*

If we do not assume that ζ *is reduced* (1.4.2), *then* 4 *should be replaced by*:

4'. $(F,0)$ *is either smooth or equisingular along the singular locus of* F *(which is a smooth curve).*

PROOF. $1 \to 2$ is clear. $2 \to 3$ because $H_1(F,F\backslash 0)$ is topologically invariant; $H_1(F,F\backslash 0) = H_1(K)$ follows from the conic structure. $3 \to 4/4'$ holds true for quasi-ordinary *hypersurface* singularities and is shown in [L_1, (6.4)] (see also [L_1, (1.1)]. $4/4' \to 2$ is clear, see (2.1.2).

$2 \leftrightarrow 1$ (which is true for all surface germs) follows from: (a) if an analytic set S in U is analytically irreducible at each point in U, then S is homeomorphic to its normalization, [W, p. 254, proof of 2F]; (b) a reducible analytic germ can not be topologically trivial. (This follows from [B-M, (3.2), (4.3)] and the topological invariance of Borel-Moore homology groups.)

(2.6.6) REMARK. It was asked if the implication $1 \to 4'$ is true for all surface germs [LDT].

(2.7) In this subsection we show that K is the $(\mathbf{p},\mathbf{r})$ lens space ($\mathbf{p}$, $\mathbf{r}$ are defined in (2.1)) by giving explicitly the meridians m_i and longitudes l_i of the torus K_i $(i = 1,2)$, which satisfy:

(2.7.1) $$[m_2] = \mathbf{p}[l_1] + \mathbf{r}[m_1]$$

(considered as homology classes in $H_1(K_1 \cap K_2)$), see [R, p. 234].

Let $m_1 = \partial C_0(r_2)$ (2.6.1), (2.5.1). Similarly a meridian curve m_2 of K_2 is:

(2.7.2) $$m_2 = \partial D_0(r_1) = \cup_{0 \leq j < n_y}\{(r_1,r_2e^{i\theta},z) \mid z = \beta^j\zeta(r_1,r_2e^{i\theta}),\ 0 \leq \theta < 2\pi\}$$
$$:= \cup_{0 \leq j < n_y}\gamma_j \,.$$

A longitude of K_1. First note that, from note (2.6.2), any longitude l_1 satisfies:

(2.7.3) $$\nu_*[l_1] = c[S_y] \in H_1(K_1).$$

So we consider the following simple arc (contained in $\partial D_0(r_1)$ (2.3.2)) which is the lifting of c copies of S_y:

© $$\cup_{0 \leqslant j < c}\{(r_1,y,z) \mid y = r_2 e^{i\theta},\ 0 \leqslant \theta < 2\pi,\ z = \beta^j \zeta(r_1,y)\}.$$

(2.7.4) NOTE. This arc "starts" at $P = (r_1,r_2,\zeta(r_1,r_2))$ and "ends" at:

$$P' = (r_1,r_2,\beta^c\zeta(r_1,r_2)) = (r_1,r_2,\alpha^\gamma\zeta(r_1,r_2)).$$

Thus P' is obtained from P by a rotation of the circle $\partial C_0(r_2)$ (considered as a circle with length 2π) by $2\pi\gamma/n_x$ in the positive direction. (The orientation on K is the one induced from the complex structure on F.)

Now in order to obtain l_1, we close up the arc © by joining P' to P through the *negatively* oriented arc on the circle $\partial C_0(r_2)$.

We now prove (2.7.1).

Consider the map ν_*: $H_1(K_1) \to H_1(K_1)$ in (2.6.2). Under the deformation retraction $K_1 \to S_y$ (2.5.2), m_2 (which lies in $\partial(K_1)$) is deformed (as $x \to 0$) onto S_y in an "n_y-to-one" fashion. Thus

$$\nu_*([m_2]) = n_y[S_y]$$

In view of (2.7.3), this implies that:

(2.7.5) $$[m_2] = (n_y/c)[l_1] = p[l_1] \in H_1(K_1).$$

Next we disect the meridian $m_2 = \partial D_0(r_1)$ (2.7.2) into n_y/c simple arcs (the first one is exactly the arc ©):

$$\cup_{0 \leqslant j < c}\gamma_j,\ \cup_{c \leqslant j < 2c}\gamma_j,\ \ldots,\ \cup_{n_y - c \leqslant j < n_y}\gamma_j$$

Note that, in view of note (2.7.4), each arc is a twist of its predecesor by $(\gamma/n_x)2\pi$ in the m_1-direction. It follows that m_2 winds $(n_y/c)(\gamma/n_x)$ times around the meridian direction of K_1.

The assertion (2.7.1) now follows from (2.7.5), $\mathbf{c}n_x = \mathbf{d}n_y$ (2.1.2) and:

$$\gamma n_y/\mathbf{c}n_x = \gamma n_y/\mathbf{d}n_y = \gamma/\mathbf{d} = \mathbf{r}.$$

We can now state:

(2.7.6) PROPOSITION. *The link* $\bar{K}$ *of the normalization* $\bar{F}$ *of* F *is a* $(\mathbf{p},\mathbf{r})$ *lens space, where* $\mathbf{p} = n_x/\mathbf{d} = n_y/\mathbf{c}$ *and* $\mathbf{r} = \gamma/\mathbf{d}$.

(2.7.7) REMARK. If we interchange the roles of x, y, then (2.7.6) gives: $\bar{K}$ is a $(\mathbf{p},\mathbf{s})$ lens space, where $\mathbf{s} = \delta/\mathbf{c}$. Note that $\mathbf{rs} \equiv 1 \pmod{\mathbf{p}}$, (2.2).

SECTION 3. TOPOLOGICAL CLASSIFICATION OF LINKS

Let $(F,0)$, $(F',0)$ be quasi-ordinary surface germs. In this section we give necessary and sufficient conditions for the link $K := K(r)$ of $(F,0)$ to be homeomorphic to the link $K' := K'(r)$.

We assume that hypotheses (1.4.1), (1.4.2) *are satisfied for both* F *and* F'. Let $\mathbf{c}$, $\mathbf{d}$, n_x, n_y, γ, δ be defined for (F,ζ) as in (2.1). Let $\mathbf{c}'$, $\mathbf{d}'$, n_x', n_y', γ', δ' be the corresponding data for (F',ζ').

(3.1) Let $\Sigma(K)$ be the *topological singular locus* of K, i.e. $\Sigma(K)$ is the set of points in K near which K is not a topological manifold. It is clear that $\Sigma(K)$ is topologically invariant, i.e.

(3.1.1) if $h\colon K \to K'$ is a homeomorphism then $h(\Sigma(K)) = \Sigma(K')$.

The discussion in (2.6.3) can now be formulated in terms of $\Sigma(K)$:

(3.2) LEMMA. $\Sigma(K) = S_x \cup S_y$, iff $\mathbf{c} > 1$, $\mathbf{d} > 1$;

$= S_y$, iff $\mathbf{c} > 1$, $\mathbf{d} = 1$;

$= \Phi$, iff $\mathbf{c} = 1$, $\mathbf{d} = 1$.

(3.3) We introduce another topological term. We say that K has *topological multiplicity* m at P (in symbols: $\mu_P(K) = m$), if P has a small enough ϵ-neighborhood V in K so that $V - V \cap \Sigma(K)$ has m connected components and the closure of each component contains P. (Cf. [M_1, p. 43, p. 117], [H_2, Prop. 1, p. 97].)

In view of (3.1.1), $\mu_P(K)$ is a topological invariant:

(3.3.1) $$\mu_P(K) = \mu_{h(P)}(K').$$

From (2.6.3), we have:

(3.3.2) $$\text{if } P \in S_y, \text{ then } \mu_P(K) = \mathbf{c}.$$

Similarly, if $P \in S_x$, then $\mu_P(K) = \mathbf{d}$.

(3.4) PROPOSITION. *If the links K, K' of $(F,0)$ and $(F',0)$ are homeomorphic, then $|G| = |G'|$ and $\{\mathbf{c},\mathbf{d}\} = \{\mathbf{c}',\mathbf{d}'\}$. Moreover, up to an interchange of the roles of x and y in F when $\mathbf{c} \neq \mathbf{d}$, we have:*

$$(\mathbf{c},\mathbf{d}) = (\mathbf{c}',\mathbf{d}'), \quad (n_x,n_y) = (n_x',n_y') \quad \textit{and} \quad \mathbf{p} = \mathbf{p}'.$$

PROOF. It is shown in [G_1, (3,4,1)] or [L_1, Th. (6.1)] that the first integral homology group $H_1(K)$, has order $|G| = \mathbf{c}n_x = \mathbf{d}n_y$. Since $\mathbf{p} = n_y/\mathbf{c}$ (2.1)-4, we need only show that, up to an interchange of x, y in F, $\mathbf{c} = \mathbf{c}'$, $\mathbf{d} = \mathbf{d}'$.

If $\mathbf{c} = 1$, $\mathbf{d} = 1$, then it follows from lemma (2.6.4) that $\mathbf{c} = 1$, $\mathbf{d} = 1$.

If $\mathbf{c} > 1$ and $\mathbf{d} > 1$, then $\Sigma(K) = S_x \cup S_y$. By invoking (3.1.1) and (3.3.1), we have $\{\mathbf{c},\mathbf{d}\} = \{\mathbf{c}',\mathbf{d}'\}$. If $\mathbf{c} \neq \mathbf{d}$ and $\mathbf{c} = \mathbf{d}'$, then interchange x, y in F.

If $\mathbf{c} > 1$, $\mathbf{d} = 1$, then $\Sigma(K) = S_y$ (3.2). By (3.1.1), $\Sigma(K')$ has one component. Interchange x and y if necessary, we have: $\Sigma(K') = S_y$ and $\mathbf{d} = 1$, (3.2). Moreover, if $P \in S_y$ then:

$$\begin{aligned} \mathbf{c} &= \mu_P(K) && (3.3.2), \\ &= \mu_{h(P)}(K') && (3.3.1), \\ &= \mathbf{c}' && (3.3.2). \end{aligned}$$

Q.E.D.

We shall need the facts (A.1)-(A.4) (about lens spaces) in the appendix.

(3.5) PROPOSITION. *If* h: $K \to K'$ *is a homeomorphism then, up to an interchange of the roles of* x, y *in* F, *we have*:

(3.5.1) $\qquad$ either $(\mathbf{r},\mathbf{s}) = (\mathbf{r}',\mathbf{s}')$ or $(\mathbf{r},\mathbf{s}) = (\mathbf{p} - \mathbf{r}', \mathbf{p} - \mathbf{s}')$,

or equivalently

(3.5.2) $\qquad$ either $(\gamma,\delta) = (\gamma',\delta')$ or $(\gamma,\delta) = (n_{\mathrm{x}}' - \gamma', n_{\mathrm{y}}' - \delta')$.

PROOF. In view of (3.4), we can divide the proof into cases according to the values of $\mathbf{c}$ and $\mathbf{d}$.

(1) $\underline{\mathbf{c} = \mathbf{d}}$. First consider the case $\underline{\mathbf{c} = \mathbf{d} = 1}$. In this case, K (resp. K') is homeomorphic to $\tilde{K}$ (resp. $\tilde{K}'$) (2.6.4). Hence, by (2.7), the Lens spaces $L(\mathbf{p},\mathbf{r})$ and $L(\mathbf{p}',\mathbf{r}')$ are homeomorphic. Thus, in view of (A.1) and lemma (2.2):

$$\mathbf{p} = \mathbf{p}' \text{ and } \mathbf{r}' \equiv \text{either } \pm\mathbf{r} \text{ or } \pm\mathbf{s} \pmod{\mathbf{p}}.$$

If $\mathbf{r}' = \pm\mathbf{s} \pmod{\mathbf{p}}$, interchange the roles of x and y in F to conclude that $\mathbf{r} = \pm\mathbf{r}' \pmod{\mathbf{p}}$, (2.7.7). Since $0 \leqslant \mathbf{r} < \mathbf{p}$, $0 \leqslant \mathbf{r}' < \mathbf{p}'$, $\mathbf{rs} = \mathbf{r}'\mathbf{s}' = 1 \pmod{\mathbf{p}}$, the assertion (3.5.1) follows.

Now assume $\underline{\mathbf{c} = \mathbf{d} > 1}$. By (3.2) and (3.1.1):

$$h(S_{\mathrm{x}} \cup S_{\mathrm{y}}) = S_{\mathrm{x}}' \cup S_{\mathrm{y}}'.$$

Since h maps the connected components of $V - \Sigma(K)$ in (3.3) to the connected components of $h(V - \Sigma(K)) = h(V) - \Sigma(K')$ (2.6.1), h can be lifted to a homeomorphism $\tilde{h}$: $\tilde{K} \to \tilde{K}'$. Now the argument for the $\mathbf{c} = \mathbf{d} = 1$ case also applies here.

(2) $\underline{\mathbf{c} \neq \mathbf{d}}$. In this case we first interchange x and y if necessary, so that $\mathbf{c} = \mathbf{c}'$, $\mathbf{d} = \mathbf{d}'$ (3.4). Assume first that $\underline{\mathbf{c},\mathbf{d} > 1}$. As in the case "$\mathbf{c} = \mathbf{d} > 1$", we have a homeomorphism $\tilde{h}$: $\tilde{K} \to \tilde{K}'$. Moreover since $h(S_{\mathrm{x}}) = S_{\mathrm{x}}'$ and $h(S_{\mathrm{y}}) = S_{\mathrm{y}}'$, ((3.3.1) and (3.3.2)), $\tilde{h}$ maps the cores to the cores:

$$\hbar(\nu^{-1}(S_x)) = \nu^{-1}(S_x'), \quad \hbar(\nu^{-1}(S_y)) = \nu^{-1}(S_y').$$

Thus it follows that $\mathbf{r} = \pm\mathbf{r}'(\text{mod } \mathbf{p})$, (A.3). Now argue as in (1).

We are left with the case $\mathbf{c} > 1 = \mathbf{d}$. (The case $\mathbf{d} > 1 = \mathbf{c}$ is similar.) In this case, as above, we have a homeomorphism $\hbar$ from K to K', with $\hbar(\nu^{-1}(S_y)) = \nu^{-1}(S_y')$. (3.5.1) then follows from (A.4).

Finally the equivalence of (3.5.1) and (3.5.2) follows directly from the definition of $\mathbf{r}$, $\mathbf{s}$ in (2.1)-4 and (3.4). Q.E.D.

We now show that the necessary conditions in propositions (3.4) and (3.5) are also sufficient. Recall that G is the galois group of ζ (1.2.1) and $|G| = \mathbf{c}n_x = \mathbf{d}n_y$ (2.1)-3.

(3.6) THEOREM. *The links K, K' of $(F,0)$, $(F',0)$ are homeomorphic iff, up to a possible interchange of x and y in F':*

$$(|G|,\mathbf{c},\mathbf{d}) = (|G'|,\mathbf{c}',\mathbf{d}') \text{ and either } (\mathbf{r},\mathbf{s}) = (\mathbf{r}',\mathbf{s}') \text{ or}$$
$$(\mathbf{r},\mathbf{s}) = (\mathbf{p} - \mathbf{r}', \mathbf{p} - \mathbf{s}').$$

PROOF. The implication "$\rightarrow$" has been shown in (3.4), (3.5).

To prove the converse we note that, in view of (A.2) and $\mathbf{p} = |G|/\mathbf{cd}$ (2.1)-3, we have a homeomorphism:

$$\hbar : K = L(\mathbf{p},\mathbf{r}) \rightarrow L(\mathbf{p}',\mathbf{r}') = K',$$

which maps the core $\nu^{-1}(S_x)$, resp. $\nu^{-1}(S_y)$, to the core $\nu'^{-1}(S_x')$, resp. $\nu'^{-1}(S_y')$. To complete the proof, we show that there is a homeomorphism

$$k: K \rightarrow K',$$

which respects the identification maps ν: $K \rightarrow K$, ν': $K' \rightarrow K'$ (2.6.3). (k then induces, through ν, ν', a homeomorphism between K and K'). To be precise:

Claim. There is a homeomorphism $\hbar$ from K to K' which maps the cores $\nu^{-1}(S_x)$, $\nu^{-1}(S_y)$ to the cores $\nu'^{-1}(S_x')$, $\nu'^{-1}(S_y')$ and for $P \in S_x$ (resp. S_y),

(3.6.1) $\quad k\{\nu^{-1}(P)\} = \{\nu'^{-1}(P')\}$, for some $P' \in S'_x$ (resp. S'_y).

Note that maps from the cores of K to the cores of K' which satisfies (3.6.1) clearly exists. We choose one such map k_0 which has the same orientation as the restriction of $\hbar$ to the cores of K (denoted by $\hbar_0$), i.e.

$$\text{degree } \hbar_0 = \text{degree } k_0 \ (= \pm 1) \text{ on each core.}$$

To prove the claim, we need a lemma.

LEMMA. *Let* **T** *be a solid torus* $C \times D^2$ *where* $D^2 = \{(x,y) \in R^2, x^2 + y^2 \leqslant 1\}$ *and* $C = [0,2\pi]/0 \sim 2\pi$. *Suppose* $\iota\colon C \to C$ *is isotopic to the identity map, then there exists a homeomorphism*:

$$\tau\colon \mathbf{T} \to \mathbf{T}, \textit{ with } \tau|_{\partial \mathbf{T}} = \text{id} \quad \textit{and } \tau|_{C\times\{o\}} = \iota.$$

PROOF OF LEMMA. Let $J\colon C \times [0,1] \to C$ be the isotopy between $id = J_0$ and $\iota = J_1$.

Define $\mathbf{J}\colon \mathbf{T} \times [0,1] \to \mathbf{T}$ to be:

$$\mathbf{J}\colon (\theta,x,y,t) \to (J_{t(1-r)}(\theta),x,y), \text{ where } \theta \in C \text{ and } r = (x^2 + y^2)^{1/2}.$$

Let $\tau = \mathbf{J}|_{t=1}$. One checks easily that τ is a homeomorphism. (In fact, $\mathbf{J}$ is an isotopy between id and τ.)

PROOF OF CLAIM. We apply the lemma with $\mathbf{T} = K_i$, $i = 1,2$, and $\iota = \hbar_0^{-1} \circ k_0$. (Since $\hbar_0^{-1} \circ k_0$, a homeomorphism, has degree one, it can be isotoped into the identity.) The two homeomorphisms thus obtained from the lemma are both the identity map on $K_1 \cap K_2$ and hence combine to form a homeomorphism:

$$\tau\colon K \to K \quad \text{with } \tau|_{\text{cores}} = \hbar_0^{-1} \circ k_0 \quad \text{and} \quad \tau|_{\partial(K_1)} = id.$$

It can be checked that the composition $k := h\circ\tau$ is the desired homeomorphism. (From the proof of the lemma, $\hbar$ is in fact isotopic to k.) Q.E.D.

SECTION 4. TOPOLOGICAL TYPES OF QUASI-ORDINARY SURFACE GERMS

In section 3 we give necessary and sufficient condition for two links of quasi-ordinary surface germs to be homeomorphic. In this section we shift the emphasis from the links to the surface germs themselves. An application of the conic structure (2.4.2) immediately gives sufficient condition for two quasi-ordinary surfaces to have the same topological type (4.1.1). We are able to show that the condition is also necessary if $\mathbf{c}, \mathbf{d} > 1$, hence obtain a classification of the topological types in the case $\mathbf{c}, \mathbf{d} > 1$ (4.3). The unanswered cases depend on a question about lens spaces: (4.5).

(4.1) Let $(F,0)$, $(F',0)$ be as in section 3. We say that F and F' *have the same topological type at* 0, denoted by $(F,0) \approx (F',0)$ if there is a homeomorphism h between a small neighborhood U of 0 in F and a neighborhood U' of 0 in F' which maps the origin 0 to 0.

(4.1.1) PROPOSITION. (sufficient condition) $(F,0) \approx (F',0)$ provided, up to an interchange of X and Y in F, we have:

$$(|G|,\mathbf{c},\mathbf{d}) = (|G'|,\mathbf{c}',\mathbf{d}') \text{ and either } (\mathbf{r},\mathbf{s}) = (\mathbf{r}',\mathbf{s}') \text{ or } (\mathbf{r},\mathbf{s}) = (\mathbf{p}-\mathbf{r}', \mathbf{p}-\mathbf{s}').$$

PROOF. This follows immediately from theorem (3.6) and the conic structure of analytic sets (2.4.2).

The necessary condition in proposition (3.4) continues to hold here:

(4.2) PROPOSITION. *If* $(F,0) \approx (F',0)$ *then up to an interchange of the roles of* X *and* Y *in* F *(when* $\mathbf{c} \neq \mathbf{d}$*), we have:*

$$|G| = |G'|,\ (\mathbf{c},\mathbf{d}) = (\mathbf{c}',\mathbf{d}'),\ (n_x,n_y) = (n_x',n_y') \text{ and } \mathbf{p} = \mathbf{p}'.$$

PROOF. The basic idea is that of the proof of (3.4), but with the following modifications.

(1) If $(F,0) \overset{\sim}{\approx} (F',0)$ then their links are homotopy equivalent, hence have isomorphic homology groups.
(This is true for general analytic germs, see [Lo, p. 114]).

(2) Along X- and Y-axes, but not at the origin, F and F' are equisingular and hence topologically trivial, see (1.1.1).

With (2), some of the results about K continue to hold for F:

(3) The *topological* singular locus (3.1) of $F \cap U$ is:

$$\begin{aligned} \Sigma(F \cap U) &= X^{\mathrm{U}} \cup Y^{\mathrm{U}} && \text{iff } \mathbf{c} > 1,\ \mathbf{d} > 1: \\ &= Y^{\mathrm{U}} && \text{iff } \mathbf{c} > 1,\ \mathbf{d} = 1; \\ &= \{0\} && \text{iff } \mathbf{c} = 1,\ \mathbf{d} = 1; \end{aligned}$$

where U is a small open set (containing 0) and X^{U} denotes "X-axis $\cap\ U$".

REMARK. For the singular locus of F, see [G, (2.8.3)], [L_1, Th. (7.3)].

PROOF. We know that $\Sigma(F)$ contains $\{0\}$, (2.6.5), and is contained in $X^{\mathrm{U}} \cup Y^{\mathrm{U}}$, note (2.4.1). The assertion follows from lemma (3.2) because, in view of (2), F is locally homeomorphic to the product of K and a 2-disk neighborhood of P ($\neq 0$) in X^{U} or Y^{U}.

(4) Using (2) and (3.3.2) we can show that:

$$\text{if } 0 \neq P \in X^{\mathrm{U}}, \text{ then } \mu_{\mathrm{P}}(F) = \mathbf{d}; \text{ if } 0 \neq P \in Y^{\mathrm{U}}, \text{ then } \mu_{\mathrm{P}}(F) = \mathbf{c}.$$

Next we show that proposition (3.5) also carry over to F in the case when $\mathbf{c}$, $\mathbf{d}$ are both > 1.

(4.3) PROPOSITION. *Assume* $\mathbf{c} > 1$, $\mathbf{d} > 1$. If $(F,0) \overset{\sim}{\approx} (F',0)$ *then up to an interchange of the roles of* X *and* Y *in* F, *either* $(\mathbf{r},\mathbf{s}) = (\mathbf{r}',\mathbf{s}')$ *or* $(\mathbf{r},\mathbf{s}) = (\mathbf{p} - \mathbf{r}', \mathbf{p} - \mathbf{s}')$.

PROOF. Since $\mathbf{p} = \mathbf{p}'$ (4.2), we need only show that, up to an interchange of the roles of X and Y in F:

$$r \equiv \pm r' \pmod{p}, \ s \equiv \pm s' \pmod{p}.$$

(Since $0 < r, s < p$, $rs \equiv 1 \pmod{p}$ and the same for r', s', we have either $(r,s) = (r',s')$ or $(r,s) = (p - r', p - s')$.)

Let h, U, U' be as in (4.1). Then $\Sigma(F \cap U) = X^U \cup Y^U$, $\Sigma(F' \cap U') = X^{U'} \cup Y^{U'}$ (4.2), (4.2.3). Since h maps $\Sigma(F)$ to $\Sigma(F')$ (3.1.1). we have, up to an interchange of x and y in F, $h(X^U) = X^{U'}$ and $h(Y^U) = Y^{U'}$.

Let U, $U^\wedge$ be small enough polydisks D_{xyz} around 0 so that $h(F \cap U)$ is contained in $F' \cap U^\wedge$ and $K = \partial(F \cap U)$, $K' = \partial(F' \cap U^\wedge)$ are links of $(F,0)$, $(F',0)$ respectively. N.B. $S_x = \partial(X^U)$ (2.4).

It follows from the conic structure (2.4.2) that:

(*) the inclusion $(K,S_x,S_y) \subset (F^U,X^U,Y^U)$ is a homotopy equivalence.

(**) h: $(F^U,X^U,Y^U) \to (F'^{U^\wedge},X^{U^\wedge},Y^{U^\wedge})$ is a homotopy equivalence, [Lo, p. 114].

Recall the map ν: $\tilde{K} \to K$ from (2.6.1). Denote $\nu^{-1}(S_x)$, $\nu^{-1}(S_y)$ by C_1, C_2. We have the following commutative diagram:

$$\begin{array}{lccccc}
(0) & H_2(\tilde{K}-C_1, \tilde{K}-C_1 \cup C_2) & \to & H_1(\tilde{K}-C_1 \cup C_2) & \to & H_1(\tilde{K}-C_2) \\
 & \downarrow & & \downarrow & & \downarrow \\
(1) & H_2(K-S_x, K-S_x \cup S_y) & \to & H_1(K-S_x \cup S_y) & \to & H_1(K-S_y) \\
 & \downarrow & & \downarrow & & \downarrow \\
(2) & H_2(F^U-X^U,F^U-X^U \cup Y^U) & \to & H_1(F^U-X^U \cup Y^U) & \to & H_1(F^U-Y^U) \\
 & \downarrow & & \downarrow & & \downarrow \\
(3) & H_2(h(F^U-X^U),h(F^U-X^U \cup Y^U)) & \to & H_1(h(F^U-X^U \cup Y^U)) & \to & H_1(h(F^U-Y^U)) \\
 & \downarrow & & \downarrow & & \downarrow \\
(4) & H_2(F'^{U^\wedge}-X^{U^\wedge},F'^{U^\wedge}-X^{U^\wedge} \cup Y^{U^\wedge}) & \to & H_1(F'^{U^\wedge}-X^{U^\wedge} \cup Y^{U^\wedge}) & \to & H_1(F'^{U^\wedge}- Y^{U^\wedge}) \\
 & \uparrow & & \uparrow & & \uparrow \\
(5) & H_2(K'-S'_x,K'-S'_x \cup S'_y) & \to & H_1(K'-S'_x \cup S'_y) & \to & H_1(K'-S'_y)
\end{array}$$

NOTE. The first map in each (nonexact) sequence is the boundary homomorphism, the second is induced by inclusion.

Those "vertical" maps between the sequences (0) and (1) are induced by ν; those between (1) and (2), (3) and (4), (5) and (4) are induced by inclusions.

The idea of proof is to show:

CLAIM. * the congruence class $\pm\mathbf{r}$ (mod $\mathbf{p}$) is determined by sequence (1) and, similarly $\pm\mathbf{r}'$ (mod $\mathbf{p}$) by sequence (5).

** (1), ..., (5) are all functorially isomorphic (in symbols: (1) = (2) = $\cdots$.)

PROOF of **. (1) = (2) by (*), (2) = (3) because h is a homeomorphism, (3) = (4) by (**). Similarly, for F', (5) = (4).

PROOF OF *. Consider the commutative diagram from above (induced by ν):

$$\begin{array}{llllll}
(0) & H_2(K-C_1, K-C_1 \cup C_2) & \to & H_1(K-C_1 \cup C_2) & \to & H_1(K-C_2) \simeq \mathbf{Z} \\
& \downarrow & & \downarrow & & \downarrow \\
(1) & H_2(K-S_x, K-S_x \cup S_y) & \to & H_1(K-S_x \cup S_y) & \to & H_1(K-S_y) \simeq \mathbf{Z}
\end{array}$$

(Note that since K is a $(\mathbf{p},\mathbf{r})$ lens space, the proof of (A.4) shows that $\pm\mathbf{r}$ (mod $\mathbf{p}$) is determinable from the sequence (0)).

The first two "downward" maps are isomorphisms: (a) $K-C_1 \cup C_2$ is homeomorphic to $K-S_x \cup S_y$ (2.6.3) and (b) the transverse class $\tau(C_2)$ in $H_2(K-C_1, K-C_1 \cup C_2)$ is mapped by ν_* into a generator of $H_2(K-S_x, K-S_x \cup S_y) \simeq \mathbf{Z}$, cf. proof of (A.3). On the other hand, the third one has degree $\mathbf{d}$ (Proof: the class of the core $C_1 = \nu^{-1}(S_x)$ which generate $H_1(K-C_2)$ is mapped into $\mathbf{d}\,[S_x]$ (2.6.3).)

The way $\pm\mathbf{r}$ (mod $\mathbf{p}$) is determined from the sequence (0) in the proof of (A.3) can now be easily adopted for the sequence (1). Q.E.D.

The above lemma together with propositions (4.1), (4.2) gives:

(4.4) THEOREM. *Assume* $\mathbf{c}, \mathbf{d} > 1$. *Then* $(F,0) \cong (F',0)$ *iff, up to a possible interchange of X and Y*:

$$(|G|,\mathbf{c},\mathbf{d}) = (|G'|,\mathbf{c}',\mathbf{d}') \text{ and either } (\mathbf{r},\mathbf{s}) = (\mathbf{r}',\mathbf{s}') \text{ or } (\mathbf{r},\mathbf{s}) = (\mathbf{p}-\mathbf{r}', \mathbf{p}-\mathbf{s}').$$

In view of theorem (3.6) we have:

COROLLARY. *When* $\mathbf{c}, \mathbf{d} > 1$, $(F,0) \cong (F',0)$ *iff* K *is homeomorphic to* K'.

We do not know if the assertion of (4.4) still holds in the remaining cases. When $\mathbf{c}, \mathbf{d} = 1$ ($\Rightarrow \mathbf{c}', \mathbf{d}' = 1$) the conic structure implies that the inclusion of $h(F^U)$ in $F'^{U^\wedge}$ is a homotopy equivalence [Lo, p. 124]. It follows that K and K' are homotopy equivalent. Hence, by (2.6.4), (2.7) and a classical result [H-W]:

(4.4.1) $\quad \mathbf{p} = \mathbf{p}'$ and $\pm\mathbf{r}\mathbf{r}'$ is a quadratic residue (mod $\mathbf{p}$).

If the answer to the following question is positive, then:

when $\mathbf{c} = \mathbf{d} = 1$, $(F,0) \cong (F',0)$ iff (4.4.1) holds.

(4.5) Let L, L' be two lens spaces which are homotopy equivalent but not homeomorphic, say (7,1) and (7,2) lens spaces. Is it true that the cones over L, L' have the same topological type (4.1) at the cone points?

APPENDIX

Let $\mathbf{T}_1$ and $\mathbf{T}_2$ be solid tori with longitudes l_i and meridians m_i, $i = 1,2$. Let T_1, C_1 resp. T_2, C_2 be the boundary and core (center) of $\mathbf{T}_1$ resp. $\mathbf{T}_2$. Let t: $T_2 \to T_1$ be a homeomorphism with:

$$t_*[m_2] = p[l_1] + r[m_1].$$

Then $\mathbf{T}_1 \cup_t \mathbf{T}_2$ is the (p,r) lens space $L = L(p,r)$.

Let $L' = L'(p',r')$ be similarly defined, with everything "primed". It is known that (see e.g. [H-W, p. 224]):

(A1). if L is homeomorphic to L', then $p = p'$ and $r' \equiv \pm r^{\pm 1}$ (mod p).

(A2). if $p = p'$ and $r' \equiv \pm r$ (mod p), then there is a homeomorphism between L and L' which takes the core C_1' to C_1,

C_2' to C_2.

The converse to (A.2) is also true, (also cf. [H_1, p. 21]):

(A3). If there is a homeomorphism **h** between L and L' which takes the core C_1 to C_1', C_2 to C_2', then $r' \equiv \pm r \pmod p$.

PROOF. First note that the inclusions:

$$T_1 \to L - C_1 \cup C_2, \quad \mathbf{T}_1 \to L - C_2, \quad \mathbf{T}_2 \to L - C_1$$

are all deformation retractions, hence are homotopy equivalences. (Proof: points in $\mathbf{T}_1 - C_1$ are deformed into T_1 which is then identified with T_2. Similarly for $\mathbf{T}_2 - C_2$.) It follows that:

$$-H_2(L-C_1, L-C_1 \cup C_2) \cong H_2(\mathbf{T}_2, \mathbf{T}_2 - C_2) \cong H_0(C_2) \cong \mathbf{Z}$$

is generated by the transverse class $\nu(C_2)$ [D, p. 315].

$$-H_1(L-C_1 \cup C_2) \cong H_1(T_1) \cong H_1(T_2) \cong \mathbf{Z} \oplus \mathbf{Z} \cong \langle [l_1] \rangle \oplus \langle [m_1] \rangle,$$

$$-H_1(L-C_2) \cong H_1(\mathbf{T}_1) \cong \mathbf{Z} \cong \langle [C_1] \rangle.$$

Consider the map $\mathbf{j}\colon A \to B$ defined as the composition $i_* \circ \partial$:

$$(*) \quad A := H_2(L-C_1, L-C_1 \cup C_2) \to H_1(L-C_1 \cup C_2) \to B := H_1(L-C_2),$$

where the ∂ is the boundary homomorphism and i_* is induced by inclusion.

The congruence class of $\pm r$ modulo p is determined by $\mathbf{j} = i_* \circ \partial$ in the following way (since **h** induces a functorial isomorphism between (*) and the corresponding sequences for L', the assertion follows):

The map **j** has degree $\pm p$: there are generators $[a]$, $[b]$ of A, B so that:

$$\mathbf{j}[a] = p[b].$$

Choose any $x \in i_*^{-1}[b]$, then $\pm r \pmod p$ is determined from:

$$\partial[a] - (px) = \pm(r + kp)[m],$$

where $[m] \in H_1(L - C_1 \cup C_2)$ is a generator of the infinite cyclic group $\mathrm{Ker}(i_*)$, and the integer k depends on the choice of x).

We now check the details with $[a] = [\nu(C_2)]$, $[b] = [C_1]$ and $[m] = [m_1]$. With generators suitably chosen we have:

$$\partial[\nu(C_2)] = h_*[m_2] = p[l_1] + r[m_1],$$
$$i_*[l_1] = [C_1], \qquad i_*[m_1] = 0.$$

Thus $\mathrm{Ker}(i_*) = \langle[m_1]\rangle$. Next, p is the degree of the composition j : :

$$\mathrm{j}[\nu(C_2)] = i_*(p[l_1] + r[m_1]) = p[C_1].$$

For any $x \in i_*^{-1}[C_1] = \{[l_1] + k[m_1]\colon k \in \mathbf{Z}\}$ we have:

$$\partial[\nu(C_2)] - px = p[l_1] + r[m_1] - p([l_1] + k[m_1])$$
$$= (r + kp)[m_1]\colon\ k \in \mathbf{Z}. \qquad \text{Q.E.D.}$$

(A4). If there is a homeomorphism h between L and L' which takes the core C_1 to C_1', then $r' = \pm r \pmod p$.

PROOF. In order to apply (A3), we need only show that there is a self homeomorphism of L' which maps C_1' to C_1' and $h(C_2)$ to C_2'.

We can assume that $\mathbf{h}(\mathbf{T}_2)$ is contained in $\mathbf{T}_2'$ (shrink $\mathbf{T}_1'$ if necessary). Since $\mathbf{T}_1' \subset \mathbf{h}(\mathbf{T}_1)$ are two solid tori with common core C_1', $\mathbf{T}_1'$ can be isotoped onto $\mathbf{h}(\mathbf{T}_1)$ (i.e. there is a homotopy $H_t\colon \mathbf{T}_1' \to \mathbf{h}(\mathbf{T}_1)$, such that $H_t|_{C_1'} = id$, $H_0 = id$, H_t are homeomorphisms). $\mathbf{h}(\mathbf{T}_2)$ can then be isotoped onto $\mathbf{T}_2'$. Thus $\mathbf{h}(C_2)$ is also a core of $\mathbf{T}_2'$. It follows that there is a self homeomorphism of $\mathbf{T}_2'$ which is the identity on the boundary and maps $h(C_2)$ to C_2'. Patch this homeomorphism with the identity map on $\mathbf{T}_1'$ gives the desired map. Q.E.D.

REMARK. Another proof of (A3), (A4) can also be extracted from [G-H, p. 298].

REFERENCES

[B-V] D. Burghelea and A. Verona: Local homological properties of analytic sets. Manuscript Math. 7 (1972), pp. 55-66.

[C] A. F. Costa: On the link of quasi-ordinary singularities. Géométrie algébrique et applications, Vol. II, Travaux en cours 23. Hermann, Paris (1987).

[D] A. Dold: Lectures on algebraic topology. Springer, Berlin (1972).

[G_1] Y. N. Gau: Topology of quasi-ordinary surface singularities. Topology 25 (1986), pp. 495-519.

[G_2] Y. N. Gau: Topological classification of quasi-ordinary hypersurface singularities. Memoirs of the AMS, No. 388 (1988).

[G-L] Y. N. Gau and J. Lipman: Differential invariance of multiplicity on analytic varieties. Invent. Math. 73 (1983), pp. 165-188.

[G-H] M. J. Greenberg and J. R. Harper: Algebraic Topology, A first course. Benjamin/Cummings Publishing Company. London, (1981).

[H-S] P. De La Harpe and P. Siegfried: Singularites de Klein. Monograph. Enseig. Math. Univ. Geneve, Geneva.

[H_1] J. Hemple: 3-Manifolds. Annals of Mathematics Studies. Princeton University Press, Princeton, New Jersey (1976).

[H_2] H. Hervé: Several Complex Variables. Tata Studies in Math. 82, Springer (1963).

[H-W] P. J. Hilton and S. Wylie: Homology Theory. Cambridge University Press (1960).

[LDT] Lê Dũng Tráng: Questions raised in "Problem session in knots, braids and singularities" (plans-sur-bex, 1982), Monograph. Enseig. Math. 31, Univ. Geneve, Geneva (1983).

[L_1] J. Lipman: Topological invariants of quasi-ordinary singularities. Memoirs of the AMS, No. 388 (1988).

[L_2] J. Lipman: Quasi-ordinary singularities of surfaces in C^3. Proc. Symp. Pure Math., Vol. **40** (1983), Part 2, Amer. Math. Soc., Providence, R.I., pp. 187-230.

[L_3] J. Lipman: Quasi-ordinary singularities of embedded surfaces. Thesis, Harvard University (1965).

[Lo] E. J. N. Looijenga: Isolated Singular Points on Complete Intersections. London Math. Soc., Lecture Notes Series No. 77. Cambridge Univ. Press (1984).

[M_1] D. Mumford: Algebraic Geometry I: Complex Projective Varieties. Springer, Berlin (1978).

[M_2] D. Mumford: The topology of normal singularities of an algebraic surface and a criterion for simplicity, Publ. Math. Inst. Hautes Etudes Scientifiques, No. 9 (1961).

[R] D. Rolfson: Knots and Links. Publish or Perish Inc., Berkeley (1976).

[T] B. Teissier: Introduction to equisingularity problems. Proc. Symp. Pure Math., Vol. 29, Amer. Math. Soc., Providence, R. I. (1975), pp. 593-632.

[W] H. Whitney: Complex Analytic Varieties. Addison-Wesley, Menlo Park, CA (1972).

CALIFORNIA STATE UNIVERSITY
LONG BEACH, CALIFORNIA

Contemporary Mathematics
Volume **90**, 1989

REPORT ON THE PROBLEM SESSION

H. Hauser and R. Randell

The problems and questions listed below were mostly raised during the problem session at the end of the conference. About one third was sent to the organizers afterwards. They are grouped thematically. Some of the problems were already announced at other occasions and do not claim originality. The various comments to the problems were added by the corresponding authors.

1. TOPOLOGY OF A VARIETY (*Y. N. Gau*)

A. Let (X,x) be an analytic hypersurface germ in $\mathbb{C}^n$. Does the (embedded) local topological type of $(\mathbb{C}^n \times \mathbb{C}^d, X \times \mathbb{C}^d, x \times 0)$ determine the local topological type of $(\mathbb{C}^n,X,x)$?

REMARK. It is well known that there are manifolds M_1, M_2 with $M_1 \times \mathbb{R} \cong M_2 \times \mathbb{R}$, (homeomorphic to) but $M_1 \not\cong M_2$.

B. Let X be a local analytic surface (in $\mathbb{C}^3$). Suppose that X has a nonisolated singular point at x and X is topologically smooth at x. Does it follow that $\Sigma(X)$ is smooth at x and X is equisingular along $\Sigma(X)$?

REMARK. Another form of this question has been asked by Lê [1]. An affirmative answer will follow, if the following conjecture of Lê is true: If $(\mathbb{C}^2,0) \xrightarrow{\pi} (\mathbb{C}^3,0)$ is injective and $\pi(\mathbb{C}^2)$ is a complex surface (singular at 0) then π has rank one.

REFERENCES

[1] Problem session in knots, braids and singularities (Plans-sur-Bex, 1982), Monograph. Ens. Math. **31**, Univ. Genève, Geneva (1983).

[2] Lê, D. T., Une application d'un théorème d'A'Campo à l'équisingularité. Indag. Math. **35** (1973), 403-409.

2. TOPOLOGY OF A VARIETY (*J. Lipman*)

A. *Classify quotient singularities topologically.* Quotient singularities are germs of the form $(\mathbb{C}^n,0)/G$, with G a finite group; so the problem is related to that of classifying linear representations of finite groups topologically, cf. [1], [2], [3].

B. Let S be a semigroup of pairs of positive integers. Let X_S be the analytic germ whose local ring (of germs of holomorphic functions) is

$$\mathbb{C}\langle\{u^{\alpha}v^{\beta}\}_{(\alpha,\beta)\in S}\rangle \subset \mathbb{C}\langle u,v\rangle.$$

Given another such semigroup T, say "S is topologically equivalent to T" is the germs X_S and X_T are homeomorphic.

Classify semigroups under this equivalent relation. We do not assume here that X_S is normal. The problem arises in the context of toroidal embeddings [4, p. 4] and is also related, via Zariski's "saturation", to the next problem on quasi-ordinary singularities (cf. [5,§2]). The question can of course be posed in any dimension.

C. There is a classical relation between the topology of a plane curve singularity and the characteristic pairs of a Puiseux parametrization (via compound toroidal knots [6]). As explained in Gau's talk at this conference, the embedded topology of a quasi-ordinary singularity [5] and the characteristic monomials of any suitably normalized fractional power series parametrization determine each other (cf. [7] for the two-dimensional case, and [8] for the general case).

Find a nice knot-theoretic interpretation of the characteristic monomials. For example, consider the surface in $\mathbb{C}^3$ given by $z^n = x^a y^b$ where $(n,a) = (n,b) = 1$. The link L of the origin is a lens space (= 3-sphere modulo $\mathbf{Z}/n$-action) sitting inside a 5-sphere S^5. *How can one recover the integers a,b from the topological pair $L \subset S^5$?* (According to Gau [7, Thm. (3.4.1)] n is the order of the homology group $H_1(L,\mathbf{Z})$. But a,b cannot depend just on L, because for any $p,q \geqslant 0$, L is homeomorphic to the link of the origin on $z^n = a^{a+pn}y^{b+qn}$.)

REFERENCES

[1] D. Prill, Local classification of quotients of complex manifolds by discontinuous groups, Duke Math. J. **34** (1967), 375-386.

[2] W.-C. Hsiang, W. Pardon, When are topologically equivalent orthogonal transformations linearly equivalent? Invent. Math. **68** (1982), 275-316.

[3] S. Cappell, J. Shaneson, The topological rationality of linear representations, Publ. Math. I.H.E.S. **56** (1982), 101-128.

[4] G. Kempf, F. Knudsen, D. Memford, B. Saint-Donat, Toroidal Embeddings I, Lecture Notes in Math. **339**, Springer-Verlag, 1973.

[5] J. Lipman, Quasi-ordinary singularities of surfaces in $\mathbb{C}^3$, in *Singularities*, Proc. Symp. Pure Math. **40** Amer. Math. Soc., Providence, 1983, Part 2, pp. 161-171.

[6] J. Reeve, A summary of results in the topological classification of plane algebroid singularities, Rend. Sem. Mat. Univ. e Politec, Torino **14** (1954-55), 159-187.

[7] Y.-N. Gau, Topology of quasi-ordinary surface singularities, Topology (to appear).

[8] Y.-N. Gau, Embedded topological classification of quasi-ordinary singularities, preprint.

3. TWO-DIMENSIONAL SINGULARITIES (*H. Laufer*)

Let Γ be the weighted dual graph of the exceptional set of a normal two-dimensional singularity (V,p).

A. Give necessary conditions on Γ for (V,p) to be a hypersurface singularity.

B. Give sufficient conditions on Γ for a hypersurface singularity (V,p) to exist.

4. μ-CONSTANT FAMILIES AND MILNOR FIBRATIONS (*M. Oka*)

A. Let $f_t(z_1,\dots,z_n)$ be an analytic family of μ-constant analytic functions defined in a neighborhood of the origin.

(i) Does the topological invariance follow from the μ-constancy?

(ii) Do there exist a base change and a family of analytic changes of the coordinates $z(t)$ such that $z(0) = z$ and the family $f_t(z(t))$ has a uniform radius for its Milnor fibrations?

REMARK. (i) is true for $n \neq 3$ by Lê and Ramanujam [5]. The case $n = 3$ is claimed in [2] but the proof is not clear. (ii) implies (i). Without changes of coordinates there is a counter-example (cf. [3] and [1]).

B. Let $f_t(z_1,\ldots,z_n)$ be as above and assume that f_0 has a non-degenerate Newton boundary which is convenient. Do there exist a base change and a family of changes of coordinates $z(t)$ such that $z(0) = z$ and the Newton boundary of $f_t(z(t))$ is equal to that of f_0?

REMARK. If $n = 2$ this is true by Oka [7]. If f_0 is homogeneous, this is proved by [6]. In case f_0 is a weighted homogeneous polynomial the assertion may be possibly reduced to Theorem 2 of [4] but it is not an obvious corollary.

REFERENCES

[1] J. G. Timourain, The invariance of Milnor's number implies topological triviality, Am. J. Math. **99**, No. 2 (1977), 437-446.

[2] B. Perron, "μ-Constant" implique "type topologique constant" en dimension trois, C.R. Acad. Sci. Paris, t. 295, (1982), 735-738.

[3] M. Oka, On the weak simultaneous resolution of a negligible truncation of the Newton boundary, These Proceedings.

[4] A. N. Varchenko, A lower bound for the codimension of μ-constant in terms of mixed Hodge structures, Vestnik Moskovskogo Universiteta. Mathematika, Vol. 37, No. 6, (1982), 28-31.

[5] Lê and Ramanujam, The invariance of the Milnor numbers implies the invariance of the topological type, Am. J. Math. **98** (1976), 67-78.

[6] Gabrielov and Kouchnirenko, Description of deformations with constant Milnor number for homogeneous functions, Funct. Anal. Appl., Vol. 9, No. 4, (1975), 67-68.

[7] M. Oka, On the stability of the Newton boundary, Proc. of Symp. in Pure Math., Vol. 40, (1983), 259-268.

5. μ-CONSTANT FAMILIES AND ZARISKI'S CONJECTURE (*D. O'Shea*)

A. *Setting.* One strategy for approaching Zariski's question as to whether multiplicity is a topological invariant of isolated hypersurface singularities is to ask whether multiplicity must be constant in a μ-constant family, and then to ask whether any two topologically equivalent hypersurface singularities can be embedded in a μ-constant family. Neither of these questions has been answered. What follows are related questions about which almost nothing seems to be known. Before stating them, we need a definition.

Let $\{f_t\}$, f_t: C^n, $0 \to C$, 0 be a family of isolated singularities which depend smoothly on $t \in [0,1]$. We say that $\{f_t\}$ is *embedded* in a family $\{F_t\}$, F_t: C^{n+k}, $0 \to C$, 0 if: 1) there exists a smooth family of injective linear maps r_t: $C^n \hookrightarrow C^{n+k}$ such that $F_t \cdot r_t = f_t$ and 2) $\mu^{(j)}(f_t,0) = \mu^{(j)}(F_t,0)$ for all $1 \leqslant j \leqslant n+1$ and all $t \in [0,1]$.

The Actual Questions.

a) Choose a family $\{f_t\}$ which is not equimultiple (that is, which is such that $\mu^{(1)}(f_0) \neq \mu^{(1)}(f_t)$ for small $t \neq 0$. Is it possible to embed $\{f_t\}$ in a μ-constant family $\{F_t\}$? More generally, what, if any, changes in $\mu^*(f_t)$ mandate changes in $\mu^{(i)}(F_t)$, $i > n+1$, for any family F_t in which f_t embeds?

b) Suppose that f,g: C^n, $0 \to C$, 0 have isolated singularities at the origin 0. If f and g have the same topological type, does there exist a μ-constant family f_t: C^n, $0 \to C$, 0 such that $f_0 = f$ and $f_1 = g$? What (if any) are the obstructions to the existence of such a family?

c) Let f, g be as above. Does there exist a μ-constant family F_t: C^{n+k}, $0 \to C$, 0 such that $F_0|_{C^n\times 0} = f$, $F_1|_{C^n\times 0} = g$ for some (and, hence, any) sufficiently large k?

REMARKS. Concerning a): apart from a constraint that follows easily from Teissier's "Minkowski-type inequalities" for multiplicity sequences (*Ann. Math.* **106** (1977), 38-44), I know of no other general conditions on the behaviour of $\mu^*(F_t)$. In particular, I know of no *a priori* reason why one cannot embed an arbitrary family of isolated singularities into a μ-constant family of one dimension higher provided only that μ is sufficiently large. This raises the disturbing (and, one feels, unlikely) possibility that the multiplicity fails to be topologically invariant in the worse possible way.

Concerning b) and c): questions b) and c) can be caricatured as "same topological type $\Rightarrow$ μ-homotopic?" and "same topological type $\Rightarrow$ stably μ-homotopic?", respectively.

B. (*Giblin/O'Shea*).

Setting. A special case of what follows arose in the course of investigating whether, at every point of a smooth, compact real hypersurface (in Euclidean space), there is a sphere (or hyperplane) which is tangent to the hypersurface at that point and at some other point. This is known to be the case if the dimension of the hypersurface is one or two and unknown if the dimension is three or greater.

The actual question. Let f,g: $R^n, 0 \to R, 0$ be homogeneous polynomials of the same degree and suppose that g has an isolated minimum at 0 and f an isolated singularity at 0. Let $S = \{x \in R^n: |x| = 1\}$ and let h be the restriction of the function f/g to S. Is it possible to find a manifold M with $\partial M = S$ and a differentiable function H on M such that

a) H extends h; that is, $H|_{\partial M} = h$; and

b) H has no critical points on $M - \partial M$?

REMARKS. We do not even know the answer to this question when f and g are of second degree and $n \geqslant 3$.

6. THE SPECTRUM OF A SINGULARITY AND ZARISKI'S CONJECTURE (*J. Steenbrink*)

A. Is the spectrum of an isolated hypersurface singularity an invariant of its embedded topological type?

COMMENTS: See [1] for the definition of spectrum. The spectrum is constant under μ-constant deformations [2].

B. Does the spectrum of an isolated hypersurface singularity determine its multiplicity?

COMMENTS: For quasihomogeneus singularities the spectrum determines the weights (if uniquely determined) and hence the multiplicity. One might be able to use the semi-continuity of the spectrum [3] to give sharp estimates for the multiplicity by deformation from/into quasi-homogeneous singularities. An affirmative answer to A. and B. would imply Zariski's conjecture....

REFERENCES

[1] V. I. Arnol'd: On some problems in singularity theory. Geometry and Analysis, Papers dedicated to the memory of V. K. Patodi, Bombay 1981, 1-10.

[2] A. N. Varchenko: The complex exponent of a singularity docs not change along strata μ = const. Funct. An. Appl. **16**, (1982), 1-10.

[3] J. H. M. Steenbrink: Semicontinuity of the singularity spectrum. Invent. Math. **79**, (1985), 557-565.

7. ALGEBRAICITY OF VARIETIES (*R. Pellikaan*)

A germ of an analytic space $(V,0)$ in $(\mathbb{C}^n,0)$ is called algebraic if there exist local coordinates $x_1,\ldots,x_n$ such that the ideal defining $(V,0)$ is generated by polynomials in $x_1,\ldots,x_n$. Whitney [9] gave an example of a germ which is not algebraic, e.g., $f(x,y,z) = xy(x + y)(x + (z + 2)y)(x + 3e^z y)$.

Artin [1] and Tougeron [8] proved that in case $(V,0)$ has an isolated singularity at 0 then $(V,0)$ is algebraic.

Bochnak [2] proved that in case the components of $(V,0)$ are complete intersections with isolated singularities, which intersect transversally outside of the origin, $(V,0)$ is also algebraic. He conjectures that the same is true without assuming the components to be complete intersections. For functions Mather [4] proved a finite determinacy theorem for isolated singularities, which implies that they are right-equivalent with a polynomial. This can be generalized to non-isolated singularities by methods of Siersma [6], Izumiya and Matsuoka [3] and Pellikaan [5]. Siersma has the following question: Suppose f has a one dimensional singular locus Σ with transversal to Σ only simple singularities. Is f right-equivalent to a polynomial?

Van Straten [7] conjectures that weakly normal surface singularities are algebraic. He asks the following question. Suppose $(V,0)$ is a germ such that for all $p \in V - \{0\}$ we have that (V,p) is algebraic. Is then $(V,0)$ algebraic?

REFERENCES

[1] M. Artin: Algebraic approximation of structures over complete local rings, Publ. Math. I.H.E.S. **36** (1969), 23-58.

[2] J. Bochnak: Algebraicity versus analyticity, Rocky Mountain J. Math. **14** (1984), 863-880.

[3] S. Izumiya and S. Matsuoka: Notes on smooth function germs on varieties, Proc. Am. Math. Soc. **97** (1986), 146-150.

[4] J. Mather: Stability of C^∞-mappings III. Finitely determined map germs, Publ. Math. I.H.E.S. **35** (1968), 127-156.

[5] R. Pellikaan: Hypersurface singularities and resolutions of Jacobi modules, Thesis, R. U. Utrecht, 1985.

[6] D. Siersma: Isolated line singularities, Proc. Symp. Pure Math. **40**, Singularities at Arcata., ed. P. Orlik, Vol. 2, 485-496.

[7] D. van Straten: Weakly normal surface singularities and their improvements, Thesis, R. U. Leiden, 1986.

[8] J. C. Tougeron: Solution d'un système d'équations analytiques réelles et applications, Ann. Inst. Fourier **26** (1976), 109-135.

[9] H. Whitney: Local properties of analytical varieties, Differential and Combinatorial Topology, Princeton Univ. Press, 1965, 205-244.

8. APPROXIMATION PROBLEM (*H. Hauser, G. Müller*)

For coordinates u, t, y, z on $\mathbb{C}^m$, $\mathbb{C}^n$, $\mathbb{C}^p$ and $\mathbb{C}^q$ respectively assume given a polynomial $h(u,t,y,z) \in \mathbb{C}[u,t,y,z]$, a parameter system $t(x) = (t_1(x), ..., t_n(x))$, $t_i(x) \in \mathbb{C}\{x\}$, formal series $\bar{y}(u) = (\bar{y}_1(u), ..., \bar{y}_p(u))$, $\bar{y}_i(u) \in \mathbb{C}[[u]]$, and $\bar{z}(u,x) = (\bar{z}_1(u,x), ..., \bar{z}_q(u,x))$, $\bar{z}_i(u,x) \in \mathbb{C}[[u,x]]$ such that

$$h(u,t(x),\bar{y}(u),\bar{z}(u,x)) = 0.$$

Do there exist convergent $y(u) \in \mathbb{C}\{x\}^p$ and $z(u,x) \in \mathbb{C}\{u,x\}^q$ approximating $\bar{y}(u)$ and $\bar{z}(u,x)$ up to given degree and such that

$$h(u,t(x),y(u),z(u,x)) = 0?$$

TYPICAL EXAMPLE: $h(u,t,y,z) = g(u,y,z) - t_1$ with $g(u,y,z) \in \mathbb{C}[u,y,z]$ and $t_1(x)$ is a polynomial in x_1 with coefficients in $\mathbb{C}\{x_2, ..., x_n\}$ and $t_i(x) = x_i$ for $i \geqslant 2$.

REMARKS. (1) If $t(x)$ is not a parameter system of $\mathbb{C}\{x\}$ the answer is in general negative as show examples of Gabrielov [G] and Becker [B].

(2) If $h(u,t,y,z)$ is linear in z and $t(x)$ a parameter system one can show that the answer is by the affirmative.

(3) If the $t_i(x)$ are algebraic series and not necessarily a parameter system of $\mathbb{C}\{x\}$ the answer is yes by Theorem 3.7 of [P].

(4) The relevance of the problem to singularity theory is the following: Many constructions related to reductive group actions on analytic varieties can be done formally, whereas it is very hard to make them directly convergent. An affirmative answer to the above question would fill this gap in several situations. For instance, Saito's characterization of quasi-homogeneous hypersurface singularities would immediately extend to the non-isolated case (cf. [S]). More generally, one could resolve the Levi-subgroup problem for analytic algebras (cf. [M] and [H-M]).

REFERENCES

[B] J. Becker: Exposé on a conjecture of Tougeron. Ann. Inst. Fourier, Grenoble, **27** (1977), 9-27.

[G] A. M. Gabrielov: Formal relations between analytic functions. Izv. Akad. Nauk. SSR, **37** (1973), 1056-1088.

[H-M] H. Hauser, G. Müller: Algebraic singularities have maximal reductive automorphism groups. Nagoya J. Math. **113** (1989)..

[M] G. Müller, Reduktive Autmorphismengruppen analytischer ℂ-Algebra, Journ. Reine u. Angew. Math., **364** (1986), 26-34.

[P] D. Popescu: General Neron desingularization and Approximation. Nagoya Jour. Math., **104** (1986), 85-116.

[S] K. Saito: Quasihomogene Singularitäten von Hyperflächen. Invent. Math., **14** (1971), 121-142.

9. ARRANGEMENTS (*M. Falk*)

Let $A = \{H_1, ..., H_n\}$ be a central arrangement in $\mathbb{C}^\ell$. Let $M = \mathbb{C}^\ell - \cup_i H_i$ and let Q be the defining equation for the singular variety $\cup_i H_i$.

A. Find a combinatorial algorithm for computing the first homology of the Milnor fiber of Q, or prove that there is none.

B. Find a reasonable (e.g. combinatorial) algorithm for distinguishing the cohomology rings $H^*(M)$ of free or fiber-type arrangements with identical sets of exponents.

REMARKS. A. The Milnor fiber F is a cyclic n-fold cover of the orbit space $M/\mathbb{C}^*$, which is homeomorphic to the complement of an affine arrangement in $\mathbb{C}^{\ell-1}$. If $\ell = 3$ and the hyperplanes have real defining equations, there is a presentation of $\pi_1(M/\mathbb{C}^*)$ given by Randell [R1]. Then $\pi_1(F)$ can be obtained by a Reidermeister-Schreier rewriting process. The presentation of $\pi_1(M/\mathbb{C}^*)$ is not combinatorial, and it is not known whether the group itself is combinatorial. Randell [R2] has some local computations which sometimes yield bounds on the first betti number of F, based on combinatorial data.

B. Let A_1 and A_2 be free arrangements with identical sets of exponents. Then the cohomology rings $H^*(M_1)$ and $H^*(M_2)$ will be isomorphic as groups. In some cases, the 1-minimal models will serve to distinguish the ring structures. But if A_1 and A_2 are both fiber-type, then all the numerical invariants arising from the 1-minimal models agree. In fact, all the corresponding subquotients in the lower central series of $\pi_1(M_1)$ and $\pi_1(M_2)$ are isomorphic. It is unlikely that $H^*(M_1)$ and $H^*(M_2)$ are necessarily isomorphic as rings. On the other hand, it is possible that, for fiber-type arrangements, $H^*(M)$ determines the homotopy type of M.

General references may be found in [F].

REFERENCES

[F] M. Falk, The cohomology and fundamental group of a complement of hyperplanes, in this Proceedings.

[R1] R. Randell, The fundamental group of the complement of a union of hyperplanes: correction, Invent. Math. **80** (1985), 467-468.

[R2] R. Randell, Some topology of Zariski surfaces, Proc. 1979 Siegen Topology Symp., Springer Lecture Notes 788, 145-164, Springer, New York.

10. ARRANGEMENTS (*P. Orlik*)

In order to make this problem set self-contained, all definitions are included. Let V be a complex vector space of dimension ℓ. A *hyperplane* in V is a vector subspace of codimension 1. An *arrangement* $A = (A,V)$ in V is a finite set of hyperplanes. Let $L(A)$ be the set of intersections of elements of A. Partially order $L(A)$ by reverse inclusion so V is its minimal element, A is its set of atoms, and $\cap \{H \mid H \in A\}$ is its maximal element. We say that a property of A is *combinatorially determined* if it depends on $L(A)$. The complement of A is the smooth 2ℓ-manifold $M = M(A) = V - \cup \{H \mid H \in A\}$. It is known [4] that the cohomology ring $H^*(M,Z)$ is combinatorially determined.

A. PROBLEMS ON THE TOPOLOGY OF M.

A.1. Is $\pi_1(M)$ combinatorially determined?

A.2 Are the higher homotopy groups of M combinatorially determined?

A.3. Is the homotopy type of M combinatorially determined?

A.4. Is the homeomorphism type of M combinatorially determined?

A.5. Is the diffeomorphism type of M combinatorially determined?

Choose a basis $\{x_i\}$ in V^*. Let $S = \mathbb{C}[x_1, ..., x_\ell]$. For each $H \in A$ choose a linear form $\alpha_H \in V^*$ with $\ker \alpha_H = H$. Call A a *real* arrangement if each α_H has real coefficients. The polynomial $Q = \Pi\, \alpha_H$ is a *defining polynomial* for A. Let Der_S be the S-module of derivations of S. A nonzero element $\theta \in \mathrm{Der}_S$ is homogeneous of degree b if $\theta(S_p) \subset S_{p+b}$. This makes Der_S a graded S-module. Define the *derivations tangent to A* by $D_S(A) = \{\theta \in \mathrm{Der}_S \mid \theta Q \in QS\}$. Then $D_S(A)$ is a graded submodule of Der_S. The arrangement A is called *free* [8] if $D_S(A)$ is a free S-module. If A is free then there exists a basis for $D_S(A)$ as S-module consisting of ℓ homogeneous elements: $\{\theta_1, ..., \theta_\ell\}$. It is known that arrangements are generically not free.

B. PROBLEMS ON FREE ARRANGEMENTS.

B.1. Are there computable conditions on the coefficients of the linear forms α_H of a real arrangement which make A free?

B.2. Are there such conditions for all arrangements?

B.3. Is the property that A is free combinatorially determined?

B.4. Let $X \in L(A)$. The *restriction* of A to X is an arrangement in X defined by $A^X = \{H \cap X \mid H$ does not contain $X\}$. If A is free, is A^X free?

B.5. If A is free, is there a computable algorithm to produce a basis $\theta_1, \ldots, \theta_\ell$ from the defining equation Q?

Let $G \subset GL(V)$ be a finite group generated by reflections [7]. Let A be the set of reflecting hyperplanes. Then A is a *reflection* arrangement. It is known that every reflection arrangement is free [9]. However, not every restriction of a reflection arrangement is a reflection arrangement.

C. PROBLEMS ON REFLECTION ARRANGEMENTS.

C.1. Is every restriction of a reflection arrangement free?

C.2. If A is a reflection arrangement is $M(A)$ a $K(\pi,1)$-space? This is known to be true for Coxeter groups [1], for Shephard groups [5] for the monomial groups [3], and for rank 2 groups. Using the classification [7] only six groups remain.

COMMENTS. These are perhaps the hardest problems in arrangements. I decided to include them in part to put them on record and in part to show you the motivation for some of the work on arrangements. There are many more problems in [2]. An interesting new result [6] seems to give a positive answer to (A.3) for real arrangements.

REFERENCES: See also the references in [2].

[1] P. Deligne, Les immeubles des groupes de tresses généralisés, Invent. Math. 17 (1972), 273-302.

[2] M. Falk and R. Randell, On the homotopy type of arrangements, Advanced Studies in Pure Mathematics **8** (1986), 101-124.

[3] T. Nakamura, A note on the $K(\pi,1)$ property of the orbit space of the unitary reflection group $G(m,\ell,n)$, Sci. Papers College of Arts and Sciences, Univ. Tokyo **33** (1983), 1-6.

[4] P. Orlik and L. Solomon, Combinatorics and topology of complements of hyperplanes, Invent. Math. **56** (1980), 167-189.

[5] P. Orlik and L. Solomon, Discriminants in the invariant theory of reflection groups, Nagoya Math. J. **109** (1988), 23-45.

[6] M. Salvetti, Topology of the complement of real hyperplanes in $\mathbb{C}^N$, Invent. Math. **88** (1987), 603-618.

[7] G. C. Shephard and J. A. Todd, Finite unitary reflection groups, Can. J. Math. **6** (1954), 274-304.

[8] H. Terao, Arrangements of hyperplanes and their freeness I, J. Fac. Sci. Univ. Tokyo, Sect. IA Math. **27** (1980), 293-312.

[9] H. Terao, Free arrangements of hyperplanes and unitary reflection groups, Proc. Japan Acad. Ser. A, **56** (1980), 389-392.

11. ARRANGEMENTS (*R. Randell*)

Let us say that an arrangement A is a $K(\pi,1)$ arrangement if $M(A)$ (notation of 10.) is a $K(\pi,1)$ space. In [1] there is an example of a $K(\pi,1)$ arrangement which is not free.

A. Is every free arrangement a $K(\pi,1)$ arrangement?

This question appears in [2], and is apparently originally due to K. Saito.

Next, Falk has shown that the lattice of the arrangement gives information sufficient to construct the Sullivan one-minimal model of $M(A)$ (see [1]). That is, the one-minimal model is combinatorially determined.

B. If the one-minimal model of $M(A)$ is isomorphic to the minimal model, then is A a $K(\pi,1)$ arrangement?

Again, examples [1] show that the converse fails.

REFERENCES

[1] M. Falk, R. Randell: On the homotopy theory of arrangements, Adv. Studies in Pure Math. **8** (1986), 101-124.

[2] H. Terao: Arrangements of hyperplanes and their freeness I, Jour. Fac. Sci. Tokyo, **27** (1980), 293-320.

Contemporary Mathematics
Volume 90, 1989

TWO DIMENSIONAL SINGULARITIES WITH BOUNDED PLURI-GENERA δ_m ARE Q-GORENSTEIN SINGULARITIES

Shihoko Ishii

INTRODUCTION

A normal isolated singularity (X,x) of dimension n ($\geqslant 2$) is called a Q-Gorenstein singularity, if there exists a non-vanishing r-ple n-form which is holomorphic on $X - \{x\}$ for some positive integer r. We call the minimal value of such r's the index of a Q-Gorenstein singularity (X,x). It is well known that a Gorenstein singularity is equivalent to a Cohen-Macaulay Q-Gorenstein singularity of index one. In particular if $n = 2$, a Gorenstein singularity is equivalent to a Q-Gorenstein singularity of index one, since a 2-dimensional normal singularity is necessarily a Cohen-Macaulay singularity.

It seems important to find a good sufficient condition for a normal singularity to be a Q-Gorenstein singularity. Because the existence of a non-vanishing form sometimes makes a calculation easier.

Now let us mention our conjecture: A normal isolated singularity (X,x) of $\kappa_\delta(X,x) = 0$ is a Q-Gorenstein singularity.

Partially supported by Japan Association for Mathematical Science and Foundation for women mathematicians.

Here, κ_δ means the Kodaira dimension defined by the growth order of pluri-genera $\delta_m(X,x)$ ([5]).

In this paper, we will show that the conjecture is true in 2-dimensional cases. As a corollary of our theorem, we obtain that a 2-dimensional singularity with $\kappa_\delta = 0$ satisfies $\delta_m \leq 1$ for every $m \in \mathbf{N}$ and $\delta_m = 1$ for some m. More precisely, we get that they are either simple elliptic singularities, cusp singularities or their finite cyclic quotients.

Normal isolated singularities with bounded δ_m are divided into two cases, $\kappa_\delta = -\infty$ and $\kappa_\delta = 0$. The former is a rational case, which was already proved to be Q-Gorenstein singularities by M. Artin ([1]). Therefore, together with our main theorem, it shows the statement of the title.

The author has announced that Y. Koyama also proved the special case of our theorem; 2-dimensional purely elliptic singularities are Gorenstein singularities. He used the classification theory of the graph of the minimal resolution of surface singularities, while our proof in this article depends on the Riemann-Roch theorem of Morales [4] and Grauert-Riemenschneider vanishing theorem.

§1. DEFINITIONS AND BASIC FACTS

Let (X,x) be a normal isolated singularity of dimension $n \geq 2$. We will say that a resolution of a singularity is a good resolution if the reduced fiber of the singular point is a divisor of normal crossings.

DEFINITION 1.1 (Watanabe [5]). Define a pluri-genera $\{\delta_m\}_{m\in\mathbf{N}}$ by

$$\delta_m(X,x) = \dim_{\mathbf{C}}\Gamma(X - \{x\}, \mathcal{O}(mK_X))/L^{2/m}(X - \{x\}),$$

where $L^{2/m}(X - \{x\})$ denotes the set of all $L^{2/m}$-integrable m-ple holomorphic n-forms on $X - \{x\}$.

PROPOSITION 1.2 ([5]). *The pluri-genus* $\delta_m(X,x)$ *is represented as*

$$\delta_m(X,x) = \dim_{\mathbf{C}}\Gamma(\tilde{X} - E, \mathcal{O}(mK_{\tilde{X}}))/\Gamma(\tilde{X}, \mathcal{O}(mK_{\tilde{X}} + (m-1)E)),$$

where $f\colon \tilde{X} \to X$ *is a good resolution of the singularity and* $E = f^{-1}(x)_{\mathrm{red}}$.

DEFINITION 1.3. We define the Kodaira dimension κ_δ for a normal isolated singularity (X,x) as follows. Let $\kappa_\delta(X,x) = -\infty$ if $\delta_m(X,x) = 0$ for every $m \in \mathbf{N}$, while $\kappa_\delta(X,x)$ = the growth order of the function $\sup_{k\leqslant m}\delta_k(X,x)$ in m if $\delta_m(X,x) \neq 0$ for some $m \in \mathbf{N}$.

PROPOSITION 1.4. *Let* (X,x) *be an n-dimensional normal isolated singularity of* $\kappa_\delta(X,x) \leqslant 0$. *Then every m-ple holomorphic n-form on* $\tilde{X} - E$ *has poles of order at most m, where* $f\colon \tilde{X} \to X$ *is a good resolution of the singularity* (X,x) *and* $E = f^{-1}(x)_{\mathrm{red}}$.

PROOF. Suppose $\theta_1 \in \Gamma(\tilde{X} - E, \mathcal{O}(mK_{\tilde{X}}))$ has the pole of order $m + s$ $(s > 0)$ at a component E_1 of E. Take an element $\theta_2 \in \Gamma(\tilde{X}, \mathcal{O}(mK_{\tilde{X}}))$ and let r be the order of zero of θ_2 at E_1. Let ν be the valuation at E_1. Then $\nu(\theta_1^{M-j}\theta_2^j) = -(m+s)(M-j) + rj = -mM - sM + j(m+s+r)$ for $j = 0,1,2, \ldots, M$. Here, note that they are distinct values for each j. We can easily check that the inequality $\nu(\theta_1^{M-j}\theta_2^j) \leqslant -Mm$ holds if and only if $0 \leqslant j \leqslant (s/m+s+r)M$.

Now we have linearly independent elements,

$$\theta_1^M,\ \theta_1^{M-1}\theta_2,\ \ldots,\ \theta_1^{M-[\frac{s}{m+s+r}M]}\theta_2^{[\frac{s}{m+s+r}M]}$$

in $\Gamma(\tilde{X} - E, \mathcal{O}(MmK_{\tilde{X}})/\Gamma(\tilde{X}, \mathcal{O}(MmK_{\tilde{X}} + (Mm - 1)E))$. The number of such elements grows more and more, if M grows, which contradicts to $\kappa_\delta \leqslant 0$.

PROPOSITION 1.5. *If a normal* **Q**-*Gorenstein singularity* (X,x) *satisfies* $\kappa_\delta(X,x) \leqslant 0$, *then* $\delta_m(X,x) \leqslant 1$ *for every* $m \in \mathbf{N}$.

PROOF. Let f and E be as in Proposition 1.4, and r be the index of (X,x). Then there exists a non-vanishing r-ple holomorphic n-form ω on $\tilde{X} - E$. By Proposition 1.4, ω^m has poles of order at most rm on E. Take any element θ in $\Gamma(\tilde{X} - E, \mathcal{O}(mrK_{\tilde{X}}))$. Since ω^m does not vanish at $\tilde{X} - E \simeq X - \{x\}$, the fraction θ/ω^m is a holomorphic function on $X - \{x\}$, which is extendable onto X by the normality of X. Therefore, we can consider θ/ω^m as a function on $\tilde{X}$ with a constant value $k \in \mathbf{C}$ on E. Then $\theta - k\omega^m$ has the pole of order less than the one of ω^m at each component of E, which means $\theta - k\omega^m \in \Gamma(\tilde{X}, \mathcal{O}(mrK_{\tilde{X}} + (mr-1)E))$. Hence, $\Gamma(\tilde{X} - E, \mathcal{O}(mrK_{\tilde{X}})/\Gamma(\tilde{X}, \mathcal{O}(mrK_{\tilde{X}} + (mr-1)E)$ is generated by the class of ω^m. Thus, $\delta_{mr}(X,x) \leqslant 1$ for every m, which asserts also $\delta_m(X,x) \leqslant 1$ for every m.

DEFINITION 1.6. For a normal isolated singularity (X,x), call the minimal value m such that $\delta_m(X,x) \neq 0$ the δ-index of (X,x).

§2. RIEMANN-ROCH THEOREM FOR 2-DIMENSIONAL SINGULARITIES

In this section, we restrict our discussion to the cases $n = 2$. We fix notation. Let (X,x) be a normal 2-dimensional singularity and $f: \tilde{X} \to X$ a good resolution with $E = f^{-1}(x)_{red}$. Decompose E into irreducible components $E_1, E_2, \ldots, E_s$, and denote by ν_i the valuation of E_i $(i = 1, \ldots, s)$.

LEMMA 2.1. *Let $\theta \in \Gamma(\tilde{X} - E, \mathcal{O}(mK_{\tilde{X}}))$ has a non-singular zero-locus on $\tilde{X} - E$.*

Then,

$$R^1 f_* \mathcal{O}_{\tilde{X}}\left[mK_{\tilde{X}} - \sum_{i=1}^{s} \nu_i'(\theta)E_i\right] = 0,$$

where $\nu_i'(\theta)$ means $\ulcorner \nu_i(\theta) - \nu_i(\theta)/m \urcorner$.

PROOF. Denote by D_θ the closure on $\tilde{X}$ of zero locus of $\theta|_{\tilde{X}-E}$.

If $D_\theta \cup E$ is of normal crossings, then the statement is true by Esnault-Viehweg-Kawamata's covering method (cf. for example, Lemma 1.11, [2]).

If $D_\theta \cup E$ is not of normal crossings, take a suitable blowing up $g\colon \widetilde{X}' \to \widetilde{X}$ with centers on E such that $D_\theta' \cup E'$ is of normal crossings, where D_θ' is the proper transform of D_θ on $\widetilde{X}'$ and $E' = g^{-1}(E)_{\mathrm{red}}$. Decompose E' into irreducible components $E_1', E_2', \ldots, E_s', E_{s+1}', \ldots, E_t'$, where E_i' $(i = 1, \ldots, s)$ are the proper transforms of components E_i and E_i' $(i = s+1, \ldots, t)$ are the exceptional divisors for g. Denote the valuation at E_i' by μ_i. We note that

$$R^1(f \cdot g)_* \ \mathcal{O}\left[mK_{\widetilde{X}} - \sum_{i=1}^{t} \mu_i'(\theta)E_i'\right] = 0,$$

since $D_\theta' \cup E'$ is of normal crossings, where $\mu_i'(\theta) = \mu_i(\theta) - \ulcorner\mu_i(\theta)/m\urcorner$. Here, of course, $\nu_i'(\theta) = \mu_i'(\theta)$ for $1 \leqslant i \leqslant s$. Therefore we have the inclusion on $\widetilde{X}$,

$$g_* \ \mathcal{O}_{\widetilde{X}'}\left[mK_{\widetilde{X}'} - \sum_{i=1}^{t} \mu_i'(\theta)E_i'\right] \subset \mathcal{O}_{\widetilde{X}}\left[mK_{\widetilde{X}} - \sum_{i=1}^{s} \nu_i'(\theta)E_i\right],$$

whose cokernel has the support of dimension zero. So, for our lemma, it turns out to be sufficient to show

$$R^1 f_* g_* \ \mathcal{O}_{\widetilde{X}'}\left[mK_{\widetilde{X}'} - \sum_{i=1}^{t} \mu_i'(\theta)E_i'\right] = 0.$$

To this end, consider the spectral sequence,

$$E_2^{p,q} = R^p f_* R^q g_* L \Rightarrow E^{p+q} = R^{p+q}(f \cdot g)_* L,$$

where L is the sheaf $\mathcal{O}_{\widetilde{X}'}(mK_{\widetilde{X}'} - \Sigma_{i=1}^{t}\mu_i'(\theta)E_i')$.

Then, by the trivial edge sequence, $R^1 f_* g_* L = E_2^{1,0}$ is contained in $R^1(f \cdot g)_* L$ which was already checked to vanish.

LEMMA 2.2. *For a general element* $\theta \in \Gamma(\widetilde{X}, \mathcal{O}_{\widetilde{X}}(mK_{\widetilde{X}} + mE))$,

$$f_* \ \mathcal{O}\left[mK_{\widetilde{X}} - \sum_{i=1}^{s} \nu_i'(\theta)E_i\right] = f_* \ \mathcal{O}(mK_{\widetilde{X}} + (m-1)E), \text{ and}$$

$$R^1 f_* \mathcal{O}\left[mK_{\tilde{X}} - \sum_{i=1}^{s} \nu_i'(\theta)E_i\right] = 0.$$

PROOF. By the generality of θ, we may assume that its zero locus on $\tilde{X} - E$ is non-singular. Then, the second equality follows from Lemma 2.1. For the first assertion, remember the definition of $\nu_i'(\theta)$ we have:

(i) if $0 < \nu_i(\theta)$, then $0 \leqslant \nu_i'(\theta) < \nu_i(\theta)$,
(ii) if $-m+1 \leqslant \nu_i(\theta) \leqslant 0$, then $\nu_i'(\theta) = \nu_i(\theta)$, and
(iii) if $\nu_i(\theta) \leqslant -m$, then $\nu_i(\theta) < \nu_i'(\theta) \leqslant 0$.

Thus, the inequalities $\nu_i(\theta) \geqslant -m$ yield $\nu_i'(\theta) \geqslant -(m-1)$ for any i (cf. Lemma 1.12 of [2]). Therefore we have

$$f_* \mathcal{O}(mK_{\tilde{X}} - \Sigma\, \nu_i'(\theta)E_i) \subset f_* \mathcal{O}(mK_{\tilde{X}} + (m-1)E).$$

For the converse, we have to note that, in a given space of forms, ν_i attains the minimal value at general elements for every i. Now, take a general element ω of $\Gamma(\tilde{X}, \mathcal{O}(mK_{\tilde{X}} + (m-1)E))$. Here, if $\nu_i(\theta) = -m$, then $\nu_i'(\theta) = -m + 1$ which is less than or equal to $\nu_i(\omega)$. On the other hand, if $\nu_i(\theta) > -m$, we have $\nu_i'(\theta) \leqslant \nu_i(\theta)$, where the right hand side is less than or equal to $\nu_i(\omega)$ by the generality of θ in $\Gamma(\tilde{X}, \mathcal{O}(mK_{\tilde{X}} + m\, E))$. In any case, we have $\nu_i'(\theta) \leqslant \nu_i(\omega)$ for every i. Hence,

$$\begin{aligned} &f_* \mathcal{O}(mK_{\tilde{X}} + (m-1)E) \\ &\quad = f_* \mathcal{O}(mK_{\tilde{X}} - \Sigma\, \nu_i(\omega)E_i) \subset f_* \mathcal{O}(mK_{\tilde{X}} - \Sigma\, \nu_i'(\theta)E_i). \end{aligned}$$

LEMMA 2.3. *Let θ_m be a general element of $\Gamma(\tilde{X}, \mathcal{O}(mK_{\tilde{X}} + mE))$. Then,*

$$\delta_m(X,x) = -\frac{1}{2}\left[mK_f - \sum_{i=1}^{s} \nu_i'(\theta_m)E_i\right]\left[(m-1)K_f - \sum_{i=1}^{s} \nu_i'(\theta_m)E_i\right] + \epsilon(m),$$

where $\epsilon(m)$ is bounded and K_f is the **Q**-*divisor with the support on E which is numerically equivalent to $K_{\tilde{X}}$ on E.*

PROOF. First, we quote Riemann-Roch formula of Morales (1.4 of [4]); for a Cartier divisor L on $\tilde{X}$,

$$\chi_f(L) = \chi_f(\mathcal{O}_{\tilde{X}}) + (1/2)L_f(L_f - K_f) - d(L - \ulcorner L \urcorner) - (1/2)e(L - \ulcorner L \urcorner),$$

where L_f is the Q-divisor with the support on E which is numerically equivalent to L. Here, we note that the terms $d(\ )$ and $e(\ )$ are bounded (1.3.1, 1.3.2 or [4]).

Applying the formula to the invertible sheaf $L = \mathcal{O}(mK_{\tilde{X}} - \Sigma \nu_i'(\theta_m)E_i)$, we obtain

$$\dim j_*j^*f_*L/f_*L + \dim R^1f_*L = -\frac{1}{2}L_f\cdot(L_f - K_f) + \epsilon(L),$$

where, $\epsilon(L)$ is bounded, $j: X - \{x\} \to X$ is the inclusion, and the left hand side is $-\chi_f(L)$ by the definition. By lemma 2.2, $\dim j_*j^*f_*L/f_*L = \delta_m(X,x)$ and $R^1f_*L = 0$. This complete the proof.

§3. THE MAIN THEOREM

In this section, we use the same notation as in §2. Let ν_i denote the valuation at a component E_i of $E = f^{-1}(x)_{\text{red}}$.

THEOREM 3.1. *Let (X,x) be a 2-dimensional normal singularity of δ-index r with $\kappa_\delta(X,x) = 0$.*

Then, the singularity (X,x) is a **Q**-*Gorenstein singularity of index r.*

PROOF. Let us take a general element $\theta_m \in \Gamma(\tilde{X} - E, \mathcal{O}(mrK_{\tilde{X}}))$ for each $m \in \mathbf{N}$. Then, by Proposition 1.4, $\nu_i(\theta_m) \geqslant -rm$ for any i and $\nu_i(\theta_m) = -rm$ for some i, since $\kappa_\delta(X,x) = 0$. Put $E^m = \Sigma (\nu_i'(\theta_m)/mr)E_i$ for $m > 1$. Then we can easily check that the coefficients $\{\nu_i'(\theta_m)/mr\}_{m\in\mathbf{N}}$ are bounded below and above for each i. By Lemma 2.3,

$$\delta_{mr}(X,x) = -\frac{1}{2}\left[mrK_f - \sum_i \nu_i'(\theta_m)E_i\right]\left[(mr-1)K_f - \sum_i \nu_i'(\theta_m)E_i\right] + \epsilon(m)$$

$$= -\frac{1}{2}(mr)^2(K_f - E^m)^2 + 0(m),$$

where $\limsup_{m\to\infty} \frac{0(m)}{m^2} = 0$ because of the boundedness of E^m mentioned above.

Now, we denote the real number $\inf_{m\in\mathbf{N}} \left\{\frac{\nu_i(\theta_m)}{mr}\right\}$ by n_i for every i. It actually exists, since $\left\{\frac{\nu_i(\theta_m)}{mr}\right\}_{m\in\mathbf{N}}$ are also bounded below and above. Denote the **R**-divisor $\sum_i n_iE_i$ by E^∞.

Here we claim that there exists a subsequence $\{m_j\}_{j=1}^\infty$ of **N** such that

$$(1) \qquad \lim_{j\to\infty} (K_f - E^{m_j})^2 = (K_f - E^\infty)^2.$$

To show the claim, it is sufficient to show $\lim_{j\to\infty} \nu_i'(\theta_{m_j})/m_jr = n_i$ for a suitable subsequence $\{m_j\}_{j=1}^\infty$ of **N**. By the definition,

$$\nu_i'(\theta_m) = \nu_i(\theta_m) - \frac{\ulcorner \nu_i(\theta_m) \urcorner}{mr},$$

where the second term of the right hand side is bounded by all m. Therefore we have only to show $\lim_{j\to\infty} \nu_i(\theta_{m_j})/m_jr = n_i$ for a suitable subsequence $\{m_j\}_{j=1}^\infty$ of **N**. For any $\epsilon > 0$, there is at least one number m such that

$$(2) \qquad \frac{\nu_i(\theta_m)}{mr} - n_i < \epsilon,$$

by the definition of n_i. Let it be m_ϵ. Then for any $m = m_\epsilon d$ ($d \in$ **N**), $(\theta_{m_\epsilon})^d$ is an element of $\Gamma(\tilde{X} - E, \mathcal{O}(mrK_{\tilde{X}}))$. By the generality of θ_m, we have $d\nu_i(\theta_{m_\epsilon}) \geqslant \nu_i(\theta_m)$. Thus, all $m = m_\epsilon d$ satisfy the inequality (2). We have just gotten that for any $\epsilon > 0$ there exist infinite number of positive integers m which satisfy the inequality (2), which means the existence of subsequence $\{m_j\}$ of $\mathbf{N}$ such that $\lim_{j\to\infty} \nu_i(\theta_{m_j})/m_j r = n_i$.

Let $\{m_j\}_{j=1}^{\infty}$ be the subsequence satisfying (1). The boundedness of δ_m yields

$$-\frac{1}{2}(K_f - E^\infty)^2 = \lim_{j\to\infty} \frac{\delta_{m_j r}(X,x)}{(m_j r)^2} = 0.$$

By the negative definiteness of the intersection matrix of E, we have

$$K_f = E^\infty. \tag{3}$$

Let $J = \{i \mid n_i = \nu_i(\theta_1)/r\}$. In general, $-1 \leqslant n_i \leqslant \nu_i(\theta_1)/r$ for every i. By the assumption of κ_δ, there exists a suffix i such that $\nu_i(\theta_1)/r = -1$, which means $J \neq \phi$. Put $E_J = \Sigma_{i\in J} E_i$. Let D be the closure on $\tilde{X}$ of the zero locus of $\theta_1|_{\tilde{X}-E}$. Then $rK_{\tilde{X}} = D + \Sigma\, \nu_i(\theta_1)E_i$. By (3) we have an equality $K_{\tilde{X}} \cdot E_J = E^\infty \cdot E_J$. The left hand side is

$$\frac{1}{r} D\cdot E_J + \sum_{i\in J} \frac{\nu_i(\theta_1)}{r} E_i\cdot E_J + \sum_{i\notin J} n_i E_i\cdot E_J\,.$$

while the right hand side is

$$\sum_{i\in J} n_i E_i\cdot E_J + \sum_{i\notin J} n_i E_i\cdot E_J\,.$$

Noting that $\nu_i(\theta_1)/r \geqslant n_i$ for any i and equality holds if and only if $i \in J$, we have $D\cdot E_J = 0$ and

$$\sum_{i\notin J} \frac{\nu_i(\theta_1)}{r} E_i\cdot E_J = \sum_{i\notin J} n_i E_i\cdot E_J\,.$$

If there exists a component E_i, $i \notin J$, then, among such suffices i's, there is at least one suffix i satisfying $E_i \cap E_J \neq \phi$ by the connectedness of E. The second equality above yields $\nu_i(\theta_1)/r = n_i$ for such i, which is a contradiction of the definition of J. Therefore, we have $E_J = E$ and $D \cdot E = 0$. If we delete X sufficiently, θ_1 is a non- vanishing r-ple holomorphic 2-form on $X - \{x\}$, which means (X,x) is a **Q**-Gornstein singularity of index $\leqslant r$. Here, let s be the index of (X,x) and ω the non-vanishing s-ple holomorphic 2-form on $X - \{x\} \cong \tilde{X} - E$. Then, for any i, ν_i attains the minimal value at ω^m in $\Gamma(\tilde{X} - E, \mathcal{O}(msK_{\tilde{X}}))$. Therefore $\nu_i(\omega^m) = -sm$ for some m and i, because $\kappa_\delta(X,x) = 0$. So $\nu_i(\omega) = -s$ for some i, which means $\delta_s(X,x) \neq 0$. By the definition of δ-index, s must coincide with r.

COROLLARY 3.2. *If 2-dimensional normal singularity* (X,x) *satisfies* $\kappa_\delta(X,x) = 0$, *then* $\delta_m(X,x) \leqslant 1$, *for every* $m \in \mathbf{N}$.

PROOF. By the theorem, the singularity (X,x) is a **Q**-Gorenstein singularity. Then apply the Proposition 1.5.

COROLLARY 3.3. *If 2-dimensional normal singularity* (X,x) *satisfies* $\kappa_\delta(X,x) = 0$, *then the singularity* (X,x) *is either a simple elliptic singularity, a cusp singularity, or one of their finite cyclic quotient.*

PROOF. If the index is one, (X,x) is a purely-elliptic Gorenstein singularity, which turns out to be either simple elliptic or cusp ([5]). If the index is greater than one, take the canonical cover $\pi : (Y,y) \to (X,x)$. Then, we have (Y,y) is a purely-elliptic Gorenstein singularity, by applying the log-ramification formula to the equivariant morphism between good resolutions (cf. Prop. 1.7 of [3]).

REFERENCES

[1] Artin, M., On isolated rational singularities of surfaces. *Amer. J. Math.*, **88**, 129-136 (1966).

[2] Ishii, S., Small deformations of normal singularities. *Math. Ann.*, **275**, 139-148 (1986).

[3] Ishii, S., Isolated Q-Gorenstein singularities of dimension three. *Advanced Studies in Pure Math.*, **8**, Complex Analytic Singularities 1986, 165-198.

[4] Morales, M., Calcul de quelques invariants des singularities de surface normale. Monographie No. 31, *L'enseignement Mathematique*, 191-203.

[5] Watanabe, K., On plurigenera of normal isolated singularities I. *Math. Ann.*, **250**, 65-94 (1980).

DIVISION OF MATHEMATICS
SANGYO-IKA UNIVERSITY
YAHATA-NISHI
KITAKYUSHU 807, JAPAN

Current address:

DEPARTMENT OF MATHEMATICS
KYUSHU UNIVERSITY
FUKUOKA 812, JAPAN

Contemporary Mathematics
Volume **90**, 1989

ARRANGEMENTS OF HYPERPLANES AND BROKEN-CIRCUITS

Michel Jambu and Hiroaki Terao

§0. INTRODUCTION

Let $\mathcal{A}$ be an arrangement in $\mathbf{C}^{\ell}$, i.e., a finite collection of hyperplanes through the origin in $\mathbf{C}^{\ell}$. Then the set of all intersections of hyperplanes of $\mathcal{A}$ becomes a geometric lattice $L = L(\mathcal{A})$ ordered by the reverse inclusion. Let $M(\mathcal{A})$ be the complement of $\mathcal{A}$: $M(\mathcal{A}) = \mathbf{C}^{\ell} \setminus \cup_{H \in \mathcal{A}} H$. The algebra $A(L)$ was first introduced by P. Orlik and L. Solomon in [6] in order to give a combinatorial presentation of the cohomology algebra $H^*(M(\mathcal{A}),\mathbf{C})$. They proved in [6; 5.2] that the $\mathbf{C}$-algebra $A(L)$ is isomorphic to $H^*(M(\mathcal{A}),\mathbf{C})$. The definition of $A(L)$ is purely combinatorial. Actually the algebra $A(L)$ is defined in [7] over any field for any geometric lattice L. In [7; 3.14] it is said that "it might be useful to define $A(L)$ and study its properties over any commutative ring $\mathbf{K}$, in particular over $\mathbf{K} = \mathbf{Z}$". This is exactly what we will do in this note.

We sketch our main tool for the study. Let L be a geometric lattice and $\mathbf{K}$ be a commutative ring with 1. Let $\mathcal{A}(L)$ be the set of atoms of L. Define $E(L)$ be the exterior algebra of the $\mathbf{K}$-free module with a basis consisting of the symbols e_H in one-to-one correspondence with the atoms $H \in \mathcal{A}(L)$. Then the $\mathbf{K}$-algebra $A(L)$ is defined to be a factor algebra of $E(L)$ by a certain ideal $I(L)$:

0271-4132/89 $1.00 + $.25 per page

$A(L) = E(L)/I(L)$. The main tool in this note is the existence of a well-behaved **K**-submodule $C(L)$ of $E(L)$ satisfying $E(L) = C(L) \oplus I(L)$. In order to find such a submodule $C(L)$ we utilize the concept of broken-circuit [8]. Let $\mathcal{A}(L)$ be linearly ordered. An increasing tuple $(H_1, ..., H_p)$ of $\mathcal{A}(L)$ is said to be χ-*independent* [4] if it contains no broken-circuit. Then $C(L)$ is defined as a **K**-free module with a basis $\{e_{H_1} \cdots e_{H_p} \in E(L) \mid (H_1, ..., H_p)$ is χ-independent$\}$ and is called the *broken-circuit module*. One of our main results is

$$E(L) = C(L) \oplus I(L). \tag{2.5}$$

In other words, there exists a basis for $A(L)$ in one-to-one correspondence to χ-independent tuples of $\mathcal{A}(L)$. In case that **K** is a field, this was proved by D. Leborgne and the first author in [5] using dimension count argument.

Here is an outline of this note. In §1, we review the definition and properties of the algebra $A(L)$ after [6]. We give the definition (2.4) of $C(L)$ in §2. The theorem (2.5) asserting $E(L) = C(L) \oplus I(L)$ is proved by using the acyclicity (2.11) of a complex $(C(L), \partial)$. In §3, we prove the exactness of a sequence

$$0 \to A(L') \to A(L) \to A(L'') \to 0 \tag{3.8}$$

of **K**-linear maps. Here L' and L'' are the deletion lattice and the contraction lattice of L by a fixed atom H_0: $L' = L \backslash H_0$, $L'' = L/H_0$. The exactness has been proved, in case **K** is a field, by using the dimension count in [7; 3.13], where it is said that "(when **K** is a commutative ring) one can no longer argue (the injectivity of $A(L') \to A(L)$) with a dimension count as we have done; one is forced to prove it directly. If this is done then Brylawski's theorem follows from existence of the exact sequence (3.8)." We will implement this: As an application of existence of the broken-circuit module $C(L)$ the exactness (3.8) is proved without dimension count. In §4, we study the intersection lattice $L = L(\mathcal{A})$ of an arrangement $\mathcal{A}$ in $\mathbf{C}^\ell$. Let $R(\mathcal{A})_{\mathbf{Z}}$ be the $\mathbf{Z}$-algebra generated by the rational 1-forms $d\alpha_H/\alpha_H$, $H \in \mathcal{A}$, where α_H is a linear form on $\mathbf{C}^\ell$ with $\ker(\alpha_H) = H$. Let $A(\mathcal{A})_{\mathbf{Z}} = A(L(\mathcal{A}))_{\mathbf{Z}}$ be the algebra over $\mathbf{Z}$. It is known by E. Brieskorn [2; Lemma 5] that the de Rham correspondence $R(\mathcal{A}) \to H^*(M(\mathcal{A}),\mathbf{Z})$ is an isomorphism. In (4.2) we verify that there exists a canonical isomorphism $A(\mathcal{A})_{\mathbf{Z}} \rightrightarrows R(\mathcal{A})_{\mathbf{Z}}$ by using the broken-circuit module $C(\mathcal{A})_{\mathbf{Z}} = C(L(\mathcal{A}))_{\mathbf{Z}}$ over $\mathbf{Z}$.

This paper is a revised version of a preliminary one of the first author who presented it at the Conference on Singularities in Kawaguchiko, Japan on February 1986. The first author is grateful to D. Leborgne for many discusions during his stay at the University of Wuhan. While we were preparing this paper, we learned that (2.5) was obtained, by a different method, independently by Pierre Cartier who gave a talk on it at this Conference.

§1. THE ALGEBRA $A(L)$

Throughout this note, let $\mathbf{K}$ denote a commutative ring with the identity 1. In this section we define an algebra $A(L)$ over $\mathbf{K}$ for any geometric lattice L. Although $A(L)$ was originally defined only over the complex number field $\mathbf{C}$ in [6], our construction of $A(L)$ is completely parallel to the original one. No essentially new results can be found in this section.

(1.1) DEFINITION. A finite (partially) ordered set L is said to be a (finite) *geometric lattice* if it satisfies the following three conditions: a) L is a lattice (hence has the top element, 1_L and the bottom element 0_L), b) every $x \in L$ is a join of atoms (an atom is an element covering 0_L), and c) if $x,y \in L$ cover $x \wedge y$, then $x \vee y$ covers both x and y. Let $x = H_1 \vee \cdots \vee H_r$ where each H_i is an atom of L. Then we define the rank function $r : L \to \mathbf{Z}$ of L by $r(x) = \inf\{r \mid x = \vee_{i=1}^r H_i\}$.

Our most important example of geometric lattice is the intersection lattice of an arrangement:

(1.2) EXAMPLE. Let V be a vector space over a field F. Let $\mathcal{A}$ be an arrangement in V, that is, $\mathcal{A}$ is a finite collection of subspaces of codimension one. Let $L(\mathcal{A})$ be the set of intersections of elements of $\mathcal{A}$:

$$L(\mathcal{A}) = \left\{ \bigcap_{H \in B} H \mid B \subseteq \mathcal{A} \right\}.$$

Agree that $V \in L(\mathcal{A})$. Define $X < Y$ if and only if $X \supset Y$ for any $X,Y \in L(\mathcal{A})$. Then $L(\mathcal{A})$ is a geometric lattice whose set of atoms is exactly $\mathcal{A}$. Moreover $r(X) = \operatorname{codim}_V X$ for any $X \in L(\mathcal{A})$. Also we have

$$X \wedge Y = \bigcap_{\substack{H \in \mathcal{A} \\ X \cup Y \subseteq H}} H, \quad X \vee Y = X \cap Y$$

for any $X,Y \in L(\mathcal{A})$. This lattice $L(\mathcal{A})$ is called the *intersection lattice* of an arrangement $\mathcal{A}$.

Let L be a geometric lattice and $\mathcal{A} = \mathcal{A}(L)$ the set of its atoms. Let $\underline{S}_p = \underline{S}_p(L)$ denote the set of all p-tuples $S = (H_1, \ldots, H_p)$ of atoms and let $\underline{S} = \underline{S}(L) = \cup_{p \geqslant 0} \underline{S}_p$. In case $p = 0$ we agree that $S = (\ )$ is the empty tuple. Let $\oplus_{H \in \mathcal{A}} \mathbf{K}e_H$ be a free module over a commutative ring $\mathbf{K}$ which has a $\mathbf{K}$-basis consisting of elements e_H in one-to-one correspondence with atoms, and let

$$E = E(L) = \Lambda\left(\bigoplus_{H \in \mathcal{A}} \mathbf{K}e_H \right)$$

be its exterior algebra. We write uv rather than $u \wedge v$ for the multiplication in E. If $S = (H_1, \ldots, H_p) \in \underline{S}$, let

$$e_S = e_{H_1} \cdots e_{H_p} \in E.$$

If $S = (\ )$, agree that $e_S = 1 \in \mathbf{K}$. Introduce an arbitrary linear order $\prec$ on $\mathcal{A}$ and say $S = (H_1, \ldots, H_p) \in \underline{S}$ is *standard* if $H_1 \prec H_2 \prec \ldots \prec H_p$. Then E has a $\mathbf{K}$-basis consisting of all e_S with S standard. Define a $\mathbf{K}$-linear map

$$\partial : E \to E \tag{1.3}$$

by $\partial 1 = 0$, $\partial e_H = 1$ $(H \in A)$ and for a standard p-tuple $S = (H_1, ..., H_p) \in \underline{S}_p$,

$$\partial e_S = \sum_{k=1}^{p} (-1)^{k-1} e_{H_1} \cdots \hat{e}_{H_k} \cdots e_{H_p}.$$

Then $\partial^2 = 0$ and if $T \in \underline{S}$,

$$\partial(e_S e_T) = (\partial e_S)e_T + (-1)^p e_S(\partial e_T).$$

We write $\vee S = H_1 \vee \cdots \vee H_p$ for $S = (H_1, ..., H_p) \in \underline{S}_p$. We say that $S \in \underline{S}_p$ is *dependent* (or *independent*) if $r(\vee S) < p$ (resp. $r(\vee S) = p$).

(1.4) DEFINITION (Orlik-Solomon [6; p. 171]). Let $I = I(L)$ be the ideal of the $\mathbf{K}$-algebra $E = E(L)$ generated by all elements ∂e_S where S is dependent. The algebra $A = A(L)$ over $\mathbf{K}$ is defined by

$$A = E/I.$$

For any $X \in L$, define $E_X = E_X(L) = \Sigma_{\vee S = X} \mathbf{K} e_S$ and $A_X = E_X/(I \cap E_X)$. Then A_X is a subalgebra of A. Although the following two propositions are over the commutative ring $\mathbf{K}$, the proofs are completely parallel to those over $\mathbf{C}$ given as [6; 2.10] and [6; 2.18]. So they are omitted.

(1.5) PROPOSITION. $A = \bigoplus_{X \in L} A_X$. □

(1.6) PROPOSITION. *Let* $1 = 1_L$ *and* $0 = 0_L$ *be the top element and the bottom element of* L. *Put* $r = r(L) = r(1)$. *Then the* $\mathbf{K}$*-linear map* $\partial : E \to E$ *induces an exact sequence*

$$0 \to A_1(L) \xrightarrow{\partial} \bigoplus_{r(Y)=r-1} A_Y(L) \xrightarrow{\partial} \bigoplus_{r(Y)=r-2} A_Y(L) \xrightarrow{\partial} \cdots \xrightarrow{\partial} A_0(L) \to 0. \quad \square$$

Let $X \in L$. Define a sublattice L_X by

$$L_X = \{Y \in L \mid Y \leqslant X\}.$$

Note $A(L_X) \subseteq A(L)$ and we can naturally assume $E(L_X)$ is a subalgebra of $E(L)$.

(1.7) PROPOSITION. *The induced map* $A(L_X) \to A(L)$ *from the inclusion map* $E(L_X) \hookrightarrow E(L)$ *is a monomorphism. This induces an isomorphism*

$$A_Y(L_X) \xrightarrow{\sim} A_Y(L)$$

for $Y \leqslant X$.

PROOF. The first part is equivalent to [7; 2.34] when $\mathbf{K}$ is a field. The proof can be generalized to the case over the commutative ring $\mathbf{K}$ without changes. Since $E_Y(L_X) = E_Y(L)$, the second part is obtained from the first part. □

Let $Y \in L$ and $\mu : L \times L \to \mathbf{Z}$ be the Möbius function.

(1.8) PROPOSITION. $\mu(0,Y) = (-1)^{r(Y)} \operatorname{rank}_{\mathbf{K}} A_Y(L)$.

PROOF. Replacing L by L_X, $X \in L$, in (1.6), we get

$$\sum_{Y \in L_X} (-1)^{r(Y)} \operatorname{rank}_{\mathbf{K}} A_Y(L_X) = 0.$$

By (1.7) we have $\operatorname{rank}_{\mathbf{K}} A_Y(L_X) = \operatorname{rank}_{\mathbf{K}} A_Y(L)$ if $Y \leqslant X$. Put $f(Y) = (-1)^{r(Y)} \operatorname{rank}_{\mathbf{K}} A_Y(L)$ for $Y \in L$. Then $\Sigma_{Y \in L_X} f(Y) = 0$. We also have $f(0) = 1$ because $A_0(L) = \mathbf{K}$. Therefore $f(Y) = \mu(0,Y)$. □

Let $P(A(L),t) = \Sigma_{p \geqslant 0} (\operatorname{rank}_{\mathbf{K}} A_p(L)) t^p$ be the Poincaré series of the graded $\mathbf{K}$-algebra $A(L)$. Since $A(L) = \oplus_{p \geqslant 0} A_p(L) = \oplus_{p \geqslant 0} \oplus_{r(X)=p} A_X(L)$, the following is an immediate result from (1.8).

(1.9) COROLLARY. $P(A(L),t) = \Sigma_{X \in L} \mu(0,X)(-t)^{r(X)}$. □

The characteristic polynomial of a geometric lattice is by definition [1; p. 155]

$$\chi(L,t) = \sum_{X \in L} \mu(0,X)t^{r(L)-r(X)}.$$

Then (1.9) can be written as

$$P(A(L),t) = (-t)^{r(L)}\chi(L, -t^{-1}). \tag{1.11}$$

§2. THE BROKEN-CIRCUITS

Let L be a geometric lattice with the linearly ordered set $A = A(L)$ of its atoms. In this section we define the broken-circuit module $C(L)$ over a commutative ring $\mathbf{K}$ and prove $E(L) = C(L) \oplus I(L)$ in (2.5).

(2.1) DEFINITION. We say that a p-tuple $S = (H_1,..., H_p) \in \underline{S}(L)$ is a *circuit* (of L) if S is a minimal dependent tuple, that is, S is dependent and $(H_1, ..., \hat{H}_k, ..., H_p)$ is independent for $k = 1, ..., p$.

Let $S = (H_1, ..., H_p) \in \underline{S} = \underline{S}(L)$ and H be an atom of L. We abuse notation slightly and write (S,H) instead of $(H_1, ..., H_p,H) \in \underline{S}$. Denote the maximum element of $S = (H_1, ..., H_p)$ by max S. If S is standard, then max $S = H_p$.

(2.2) DEFINITION (Rota [8]). We say that a standard p-tuple $S \in \underline{S}$ is a *broken-circuit* (of L) if there exists $H \in A$, max $S \prec H$, such that $(S,H) \in \underline{S}$ is a circuit of L.

(2.3) DEFINITION (Brylawski [4]). A standard p-tuple $S \in \underline{S}_p$ is said to be χ-*independent* (in L) if it does not contain any broken-circuit of L.

(2.4) DEFINITION. Let $C = C(L)$ be the free graded $\mathbf{K}$-module $\oplus_{p \geq 0} C_p(L)$, where $C_0(L) = \mathbf{K}$, $C_p(L)$ is the free $\mathbf{K}$-module having the set $\{e_S \in E \mid S \in \underline{S}_p \text{ is } \chi\text{-independent}\}$ for basis. We call $C = C(L)$ the *broken-circuit module* of L.

Then $C(L)$ is a submodule of $E(L)$. Recall $I(L)$ and $A(L)$ in (1.4). Let $\varphi = \varphi(L)$ be the natural projection $E(L) \to A(L) =$

$E(L)/I(L)$. Consider its restriction $\psi = \psi(L) = \varphi|_{C(L)} : C(L) \to A(L)$. The main result in this section is

(2.5) THEOREM. *For any geometric lattice L, $\psi(L)$ is an isomorphism of graded* **K**-*modules*:

$$\psi(L) : C(L) \cong A(L).$$

Or equivalently

$$E(L) = C(L) \oplus I(L).$$

(2.6) COROLLARY. *The* **K**-*module $A(L)$ is a free module with a basis $\{e_S + I \in A(L) \mid S \in \underline{S}$ is χ-independent$\}$.* □

The proof of (2.5) completes at the end of this section. Recall $E_X = \Sigma_{\vee S=X}\mathbf{K}e_S$.

(2.7) DEFINITION. For $X \in L$ let $C_X = C_X(L) = C(L) \cap E_X(L)$. Then C_X is a submodule of C.

(2.8) LEMMA. $C = \bigoplus_{X\in L} C_X$.

PROOF. Obvious from the fact that C has a basis $\{e_S \in E \mid S$ is χ-independent$\}$. □

Let $X \in L$. Recall $L_X = \{Y \in L \mid Y \leqslant X\}$. Since $A(L_X) \subseteq A(L)$, the linear order on $A(L)$ induces a linear order on $A(L_X)$. Then we can consider $C(L_X)$ and we have

(2.9) LEMMA. *If $Y \leqslant X$, then $C_Y(L_X) = C_Y(L)$.*

PROOF. Let $S \in \underline{S}(L_X) \subseteq \underline{S}(L)$. It is enough to show that S is a broken-circuit of L_X if and only if it is a broken-circuit of L. If S is a broken-circuit of L_X, then S is obtained by taking off the maximal element of a circuit of L_X, which is also a circuit of L. Thus S is a broken-circuit of L. Conversely if S is a broken-circuit of L, there exists an atom $H \in A(L)$ such that (S,H) is a circuit of L and max $S \prec H$. Since S is independent and (S,H) is dependent, $\vee S = (\vee S) \vee H$. Thus $X \geq \vee S \geq H$ and $(S,H) \in \underline{S}(L_X)$ is a circuit of L_X. This implies that S is a broken-circuit of L_X. □

(2.10) LEMMA. *Let $e_m = e_{H_m} \in E$ correspond to the maximum atom H_m of L. Then $e_m \cdot C \subseteq C$, that is, C is closed under the multiplication of e_m.*

PROOF. It is obvious from the fact that $(S,H_m) \in \underline{S}$ cannot be a broken-circuit. □

(2.11) PROPOSITION. *Let $1 = 1_L$ and $0 = 0_L$ be the top element and the bottom element of L. Put $r = r(L) = r(1)$. Then the* **K***-linear map $\partial : E \to E$ induces an exact sequence*

$$0 \to C_1(L) \xrightarrow{\partial} \bigoplus_{r(Y)=r-1} C_Y(L) \xrightarrow{\partial} \bigoplus_{r(Y)=r-2} C_Y(L) \xrightarrow{\partial} \cdots \xrightarrow{\partial} C_0(L) \to 0.$$

PROOF. Let $S = (H_1, ..., H_p) \in \underline{S}$ be χ-independent. Put $X = \vee S \in L$. Since S is independent, so is each $S_k = (H_1, ..., \hat{H}_k, ..., H_p)$ for $k = 1, ..., p$. Thus $r(\vee S_k) = r(X) - 1$. This shows

$$\partial(C_X) \subseteq \bigoplus_{r(Y)=r(X)-1} C_Y .$$

We shall prove the acyclicity of the complex (C,∂). Assume that $\partial c = 0$, $c \in C$. Then, by (2.10), $e_m c \in C$. We get

$$c = c - e_m(\partial c) = \partial(e_m c) \in \partial C. \quad \square$$

Return to the **K**-linear map $\psi = \psi(L) : C(L) \to A(L)$. Note that $\psi(L)$ induces $\psi_X = \psi_X(L) : C_X(L) \to A_X(L)$.

(2.12) PROPOSITION. *For any geometric lattice L and $X \in L$, the map $\psi_X(L) : C_X(L) \to A_X(L)$ is an isomorphism.*

PROOF. We shall prove by induction on $r(L) = r(1)$. When $r(L) = 0$, L contains no atoms. Thus $C_1(L) = \mathbf{K} = A_1(L)$ and the lemma is obviously true. When $r = r(L) > 0$, let $X \in L$ with $r(X) < r$. Since $r(L_X) < r$, by the induction assumption, $\psi_X(L_X) : C_X(L_X) \xrightarrow{\sim} A_X(L_X)$ is an isomorphism. By (2.9) $C_X(L) = C_X(L_X)$. Moreover, by (1.7), the canonical map $A_X(L_X) \xrightarrow{\sim} A_X(L)$ is also an isomorphism. So we get a commutative diagram

$$\begin{array}{ccc} C_X(L) & \xrightarrow{\psi_X(L)} & A_X(L) \\ \| & & \uparrow\wr \\ C_X(L_X) & \underset{\psi_X(L_X)}{\xrightarrow{\sim}} & A_X(L_X) \end{array}$$

which shows that $\psi_X(L)$ is an isomorphism for $X \in L$ if $r(X) < r$. The remaining part is to prove that $\psi_1 = \psi_1(L)$ is an isomorphism. By connecting the two exact sequences (2.11) and (1.6) by $\psi = \psi(L)$, we obtain a commutative diagram

$$\begin{array}{ccccccccc} 0 \to C_1(L) & \xrightarrow{\partial} & \bigoplus_{r(Y)=r-1} C_Y(L) & \xrightarrow{\partial} & \bigoplus_{r(Y)=r-2} C_Y(L) & \xrightarrow{\partial} & \cdots & \xrightarrow{\partial} & C_0(L) \to 0 \\ \psi_1 \downarrow & & \oplus\psi_Y \downarrow\wr & & \oplus\psi_Y \downarrow\wr & & & & \psi_0 \downarrow\wr \\ 0 \to A_1(L) & \xrightarrow{\partial} & \bigoplus_{r(Y)=r-1} A_Y(L) & \xrightarrow{\partial} & \bigoplus_{r(Y)=r-2} A_Y(L) & \xrightarrow{\partial} & \cdots & \xrightarrow{\partial} & A_0(L) \to 0 \end{array}$$

The both rows are exact and all vertical maps except ψ_1 are known to be isomorphisms. Therefore ψ_1 is also an isomorphism. □

This completes the proof of Theorem (2.5) because

$$\psi = \bigoplus_{X \in L} \psi_X : C = \bigoplus_{X \in L} C_X \to A = \bigoplus_{X \in L} A_X$$

by (2.8) and (1.5).

Let $Y \in L$ and $\mu : L \times L \to \mathbf{Z}$ be the Möbius function.

(2.13) COROLLARY (Rota [8; Prop. 1, p. 358]). *Let m be the cardinality of the set $\{S \in \underline{S} \mid S$ is χ-independent with $\vee S = Y\}$. Then* $\mu(0,Y) = (-1)^{r(Y)} m = (-1)^{r(Y)} \operatorname{rank}_{\mathbf{K}} C_Y(L)$.

PROOF. Easy from (1.8) and (2.12). □

§3. THE EXACT SEQUENCE OF A TRIPLE (A, A', A'')

Let L be a geometric lattice with $\mathcal{A} = \mathcal{A}(L)$ of its atoms. Let H_0 be a fixed element of $\mathcal{A}$ and $e_0 = e_{H_0} \in E(L)$ throughout this section.

(3.1) DEFINITION. The *deletion lattice* $L' = L\backslash H_0$ is defined to be

$$\{H_1 \vee \cdots \vee H_p \in L \mid H_1, \ldots, H_p \in \mathcal{A} \setminus \{H_0\}\}.$$

The partial order on L' is induced from L. Then the rank function on L' is the restriction of the rank function on L. The set of its atoms is $\mathcal{A}(L') = \mathcal{A}\setminus \{H_0\}$ and denoted by $\mathcal{A}'$.

(3.2) DEFINITION. The *contraction lattice* $L'' = L/H_0$ is defined to be $\{X \in L \mid X \geqslant H_0\}$. The partial order on L'' is induced from L. Then the rank function r'' on L'' satisfies $r''(X) = r(X) - 1$ for any $X \in L''$. The set $\mathcal{A}(L'')$ of its atoms is denoted by $\mathcal{A}''$.

Our aim in this section is to construct **K**-linear maps i_A: $A(L') \to A(L)$ and j_A: $A(L) \to A(L'')$ so that the sequence

$$0 \to A(L') \xrightarrow{i_A} A(L) \xrightarrow{j_A} A(L'') \to 0 \tag{3.8}$$

may be exact. This has been proved in [7; 3.13] when **K** is a field. The proof there is still effective over the commutative ring with only one exception, which is the injectivity of i_A. In [7] the dimension count and Brylawski's theorem (3.10) were used to argue the injectivity. In this note we replace those arguments, which cannot be used over a commutative ring, by an argument using the broken-circuit modules.

For simplicity we write $\underline{S} = \underline{S}(L)$, $\underline{S}' = \underline{S}(L')$ and $\underline{S}'' = \underline{S}(L'')$. We use notations like E, E', E'', I, I', I'' or A, A', A'' inherited from the previous sections without comment. We shall define the maps i_A: $A' \to A$ and j_A: $A \to A''$. Define a map

$$\lambda : \mathcal{A}' \to \mathcal{A}'' \tag{3.3}$$

by $\lambda(H) = H_0 \vee H$ for $H \in A'$. Then this maps induces $\lambda : \underline{S}' \to \underline{S}''$ by $\lambda(\ ) = (\ )$ and $\lambda S' = \lambda(H_1, ..., H_p) = (\lambda H_1, ..., \lambda H_p)$ for $S' = (H_1, ..., H_p) \in \underline{S}'$. A surjective **K**-linear map $j : E \to E''$ is defined by

$$j(e_{S'}) = 0, \qquad j(e_0, e_{S'}) = e_{\lambda S'} \tag{3.4}$$

for any $S' \in \underline{S}'$. This is well-defined because $E = E' \oplus e_0 E'$. (This map is called θ in [7; 3.6].) Then we have

(3.5) LEMMA. *There exist* **K**-*linear maps* $i_A : A' \to A$ *and* j_A: $A \to A''$ *induced from the inclusion map* $i : E' \to E$ *and the* **K**-*linear map* j: $E \to E''$ *in* (3.4) *respectively.*

PROOF. Same as the proof [7; 2.35, 3.6] when **K** is a field. □

(3.6) LEMMA. *The sequence of* **K**-*linear maps*

$$A' \xrightarrow{i_A} A \xrightarrow{j_A} A'' \to 0$$

is exact.

PROOF. Same as the proof [7; 3.11] when **K** is a field. □

In order to prove the injectivity of i_A we utilize the broken-circuit module defined in the previous section. Fix a linear order on A such that H_0 is the minimum element. This induces a linear order on the subset A'. Fix these linear orders. Then we can consider the concept of broken-circuit on L and L'. For simplicity we write $C = C(L)$, $C' = C(L')$.

(3.7) LEMMA. $C' \subseteq C$.

PROOF. Let $S' \in \underline{S}'$ be a broken-circuit of L. Then S' is non-void. It is enough to prove that S' is a broken-circuit of L'. Note that an element of $\underline{S}'$ is a circuit of L if and only if it is a circuit of L'. There exists an atom $H \in A$ such that (S',H) is a circuit of L. Since H_0 is the minimum atom, H is different from H_0. This implies that the circuit (S',H) lies in $\underline{S}'$. Therefore S' is a broken-circuit of L'. □

(3.8) THEOREM. *The sequence of* **K**-*linear maps*

$$0 \to A' \xrightarrow{i_A} A \xrightarrow{j_A} A'' \to 0$$

is exact.

PROOF. Thanks to (3.6) the only remaining part is the injectivity of i_A, which is an immediate result from (2.5) and (3.7). □

Note that i_A is a map of degree zero and j_A is of degree -1. Thus we have the following relation among the Poincare series:

(3.9) COROLLARY. $P(A,t) = P(A',t) + tP(A'',t)$. □

REMARK. Recall the characteristic polynomial of a lattice in (1.10). By using (1.11) and (3.9) we have

$$t^{-r(L)}\chi(L,t) = t^{-r(L')}\chi(L',t) - t^{-r(L'')-1}\chi(L'',t), \tag{3.10}$$

which is the so-called Tutte-Grothendieck invariance of $(-1)^{r(L)}\chi(L,t)$ by T. Brylawski [3; Thm. 4.2].

§4. THE ALGEBRA $R(\mathcal{A})$ AND COHOMOLOGY

Let F be a field. In this section L is the intersection lattice (1.2) of an arrangement $\mathcal{A}$ in the ℓ-dimensional vector space V over F: $L = L(\mathcal{A})$. Given an arbitrary linear order on $\mathcal{A}$. Consider the graded F-algebra $\Omega(V) = \oplus_{p=0}^{\ell}\Omega^p(V)$ of rational differential forms on V. If $H \in \mathcal{A}$, choose a linear form $\alpha_H \in V^*$ such that $\ker(\alpha_H) = H$. Define

$$\omega_H = \text{dlog}\ \alpha_H = d\alpha_H/\alpha_H \in \Omega^1(V).$$

Note that although the linear form α_H is only determined up to a constant multiple, the differential form ω_H is uniquely determined by H. Let $\mathbf{K}$ be a subring of F with $1 \in \mathbf{K}$. Then $\Omega(V)$ is also an algebra over $\mathbf{K}$.

(4.1) DEFINITION. Let $R_{\mathbf{K}} = R(\mathcal{A})_{\mathbf{K}}$ be the graded $\mathbf{K}$-subalgebra of $\Omega(V)$ generated by all ω_H, $H \in \mathcal{A}$, and the identity 1.

Let $\underline{S}_p$ be the set of p-tuples of $\mathcal{A}$ and $\underline{S} = \cup_{p \geq 0} \underline{S}_p$. For $S = (H_1, \ldots, H_p) \in \underline{S}$ define $\omega_S = \omega_{H_1} \wedge \cdots \wedge \omega_{H_p} \in R_{\mathbf{K}}$. Then the set of all ω_S, $S \in \underline{S}$, generates $R_{\mathbf{K}}$ as $\mathbf{K}$-module. To specify the coefficient ring $\mathbf{K}$ we let $A(\mathcal{A})_{\mathbf{K}}$ and $C(\mathcal{A})_{\mathbf{K}}$ denote the $\mathbf{K}$-algebras $A(L(\mathcal{A}))$ and $C(L(\mathcal{A}))$ defined in (1.4) and (2.4). We sometimes simply write $A_{\mathbf{K}}$ or $C_{\mathbf{K}}$ instead of $A(\mathcal{A})_{\mathbf{K}}$ or $C(\mathcal{A})_{\mathbf{K}}$. By (2.5) there is an isomorphism $\psi : C_{\mathbf{K}} \xrightarrow{\sim} A_{\mathbf{K}}$. The following theorem was proved in [7; 4.13] when $\mathbf{K}$ is a field.

(4.2) THEOREM. *There exists a* $\mathbf{K}$-*algebra isomorphism* $\gamma_{\mathbf{K}} : A_{\mathbf{K}} \xrightarrow{\sim} R_{\mathbf{K}}$ *satisfying* $\gamma_{\mathbf{K}}(e_S + I) = \omega_S$ *for all* $S \in \underline{S}$.

PROOF. For the existence of a surjective $\mathbf{K}$-algebra homomorphism $\gamma_{\mathbf{K}}: A_{\mathbf{K}} \to R_{\mathbf{K}}$, it is sufficient to verify the statements as follows: $\Sigma_{k=1}^{p}(-1)^k \omega_{S_k} = 0$ if $S = (H_1, \ldots, H_p) \in \underline{S}$ is dependent and $S_k = (H_1, \ldots, \hat{H}_k, \ldots, H_p)$ $(k = 1, \ldots, p)$. This has nothing to do with the coefficient ring. So we can apply the argument in [7; 4.5]. In order to prove the injectivity of $\gamma_{\mathbf{K}}$, consider the F-linear map γ_F: $A_F \to R_F$ which is an isomorphism because of [7; 4.13]. We have a commutative diagram

$$\begin{array}{ccccc} C_{\mathbf{K}} & \xrightarrow[\sim]{\psi} & A_{\mathbf{K}} & \xrightarrow{\gamma_{\mathbf{K}}} & R_{\mathbf{K}} \\ \downarrow & & \downarrow & & \downarrow \\ C_F & \xrightarrow[\psi]{\sim} & A_F & \xrightarrow[\gamma_F]{\sim} & R_F \end{array},$$

where the vertical maps are the canonical ones. By definition of $C_{\mathbf{K}}$ and C_F, the left vertical map is injective. By chasing the diagram we know that $\gamma_{\mathbf{K}}$ is also injective. □

By (2.6) and (4.2) we get

(4.3) COROLLARY. *The* **K**-*module* $R_{\mathbf{K}}$ *is a free module with a basis* $\{\omega_S \in R_{\mathbf{K}} \mid S \in \underline{S}$ *is* χ-*independent in* $L(\mathcal{A})\}$. □

Let $\mathbf{K} = \mathbf{Z} \subseteq F = \mathbf{C}$. Assume $V = \mathbf{C}^{\ell}$. Let $M(\mathcal{A})$ be the complement of an arrangement $\mathcal{A}$ in $\mathbf{C}^{\ell}$: $M(\mathcal{A}) = \mathbf{C}^{\ell} \setminus \cup_{H \in \mathcal{A}} H$.

(4.4) THEOREM (Brieskorn [2; Lemma 5]). *There exists a natural graded* **Z**-*algebra isomorphism* $R_{\mathbf{Z}} \to H^*(M(\mathcal{A}),\mathbf{Z})$ *such that* $\omega_H \mapsto [\omega_H/2\pi i]$ *for all* $H \in \mathcal{A}$, *where* [ ] *implies the cohomology class by the de Rham correspondence.* □

By (4.3) and (4.4) we easily have

(4.5) COROLLARY. *A set*

$$\{[\omega_S/(2\pi i)^p] \in H^p(M(\mathcal{A}),\mathbf{Z}) \mid S \in \underline{S}_p \text{ is } \chi\text{-independent in } L(\mathcal{A})\}$$

is a **Z**-*basis for* $H^p(M(\mathcal{A}),\mathbf{Z})$. □

REFERENCES

1. Aigner, M.: Combinatorial Theory. Grundlehren der Math. Wiss. Band 234, Berlin: Springer 1979.

2. Brieskorn, E.: Sur les groupes de tresses (d'après V.I. Arnold). Séminaire Bourbaki 24ᵉ année, 1971/72, no. 401, Lecture Notes in Math. 317, Berlin-Heidelberg-NewYork: Springer 1973.

3. Brylawski, T.: A decomposition for combinatorial geometries. Trans. Amer. Math. Soc. 171, 235-282 (1972).

4. Brylawski, T.: The broken circuit algebra. Trans. Amer. Math. Soc. 234, 417-433 (1972).

5. Jambu, M. and Leborgne, T.: C.R. Acad. Sci. Paris, 303, Ser. 1, No. 7, (1986).

6. Orlik, P. and Solomon, L.: Combinatorics and topology of complements of hyperplanes. Inventiones Math. 56, 167-189 (1980).

7. Orlik, P., Solomon, L. and Terao, H.: Arrangements of hyperplanes and differential forms. Combinatorics and Algebra. Contemporary Math., 34, 29-65, Providence, R.I.: AMS 1984.

8. Rota, G.-C.: On the Foundations of Combinatorial Theory I. Theory of Möbius Functions. Z. Wahrscheinlichkeitsrechnung 2, 340-368 (1964).

MICHEL JAMBU
INSTITUT DE MATHÉMATIQUES
UNIVERSITÉ DE NANTES
NANTES, FRANCE

and

MATHEMATICS DEPARTMENT
UNIVERSITY OF WUHAN
WUHAN, PEOPLE'S REPUBLIC OF CHINA

HIROAKI TERAO
DEPARTMENT OF MATHEMATICS
INTERNATIONAL CHRISTIAN UNIVERSITY
MITAKA, TOKYO
181, JAPAN

Contemporary Mathematics
Volume **90**, 1989

SIMPLY CONNECTED FOUR-MANIFOLDS OBTAINED FROM WEIGHTED HOMOGENEOUS POLYNOMIALS

M. Kannowski

ABSTRACT. Given f: $(\mathbf{C}^4,0) \to (\mathbf{C},0)$ a weighted homogeneous polynomial (whp) with isolated singularity at the origin, we will examine a four-dimensional manifold which may be obtained from this polynomial. Sometimes the variety we obtain from the whp will be nonsingular and a four-manifold. When we have a singular variety, we will resolve its singularities to obtain this four-manifold. For the manifolds which are odd manifolds, we will find invariants for that manifold which determine it up to homeomorphism.

DEFINITION 1. Let f: $\mathbf{C}^{n+1} \to \mathbf{C}$ be a polynomial of the form $f = \Sigma_{a \in I} c_a z^a$ where c_a is a constant and $z^a = z_0^{a,0} z_1^{a,1} \cdots z_n^{a,n}$. We say that such a polynomial is weighted homogeneous if there is an $(n+1)$-tuple of positive rational numbers $w = (w_0, w_1, \ldots, w_n)$ such that $\Sigma_{i=1}^{n}(a,i)/w_i = 1$.

Let f be a weighted homogeneous polynomial (whp) in four complex variables with an isolated singular point at the origin and $V = \{z \in \mathbf{C}^4\colon f(\underline{z}) = 0\}$. It is known by [5], that the link of the singularity, $K = V \cap \mathbf{S}^7$, is a smooth, simply connected five-manifold. Given f with weights $\underline{w} = (w_0, w_1, w_2, w_3)$ where $w_i = u_i/v_i$ and g.c.d. $(u_i, v_i) = 1$, we define $N = \text{l.c.m.}[u_0, u_1, u_2, u_3]$ and $q,i = N/w_i$ for $i = 0, \ldots, 3$. Now we can define an $\mathbf{S}^1$-action on K via

0271-4132/89 $1.00 + $.25 per page

$\sigma(t;x,y,z,w) = (t^{q,0}x,t^{q,1}y,t^{q,2}z,t^{q,3}w)$. Before we proceed, we recall the following definitions.

DEFINITION 2. If G is a group acting on a set X, then the isotropy subgroup at $x \in X$ is $Gx = \{g \in G: gx = x\}$. The orbit of x is $G(x) = \{y \in X: y = gx \text{ for some } g \in G\}$.

DEFINITION 3. An orbit in S is called exceptional if the isotropy of any point, hence of all points, in the orbit is nontrivial.

DEFINITION 4 [3]. Let X be a G-space with G compact and $S \subseteq X$ be an orbit. By a tube about S we mean a G-equivariant embedding ϕ: $(G \times_H A) \to X$ onto an open neighborhood of S in X, where G/H is the isotropy group for S and A is some space on which H acts. $(G \times_H A)$ denotes the cross product of G with A after the identifications have been made on A under H (i.e. After we identify a and b in A where $a = hb$ for some $h \in H$). For more on slices, we refer the reader to [5] and [9].

DEFINITION 5 [3]. Let $x \in X$, a G-space, and $x \in W_x \subseteq X$ be such that $Gx(W_x) = W_x$. Then W_x is called a slice at x if the map $(G \times_{Gx} W_x) \to X$ taking $[g,s] \to gs$ is a tube about $G(x)$.

We will denote our slice at $z \in K$ by W_z and its slice representation by $[G_q;p_1,p_2]$ where G_q is the isotropy group of the slice and p_1 and p_2 denote the nontrivial coordinate exponents of the restriction of the action to W_z. These will be related to some q,i and q,j modulo a divisor. We know by a theorem of Neumann [6], that for $z \in K$, the quotient space W_z/G_q is a manifold if and only if $q = \text{g.c.d.}(q,p_1)\text{g.c.d.}(q,p_2)$.

Now we define $K^* = K/\sigma$ and note that Randell [8], showed that K^* will be a manifold if and only if W_z/G_q is a manifold for all $z \in K$. This reduces to considering only those W_z with G_q nontrivial.

EXAMPLE 1. $f(x,y,z,w) = x^3 + xy^5 + z^5 + zw^6$ which has

weights $w = (3, 15/2, 5, 15/2)$. The only exceptional orbits are:

(i) $x = 0$, $y = 0$, $z = 0$; which has a slice representation of

$$[\mathbf{Z}_2; 1, 0] \text{ and}$$

(ii) $x = 0$, $z = 0$, $w = 0$; which has a slice representation of

$$[\mathbf{Z}_2; 1, 0].$$

Clearly these satisfy Neumann's criterion and it follows that K^* is a four-manifold.

EXAMPLE 2. $f(x,y,z,w) = x^2 + y^7 + yz^3 + yw^4 + z^2w^2$ which has weights $w = (2, 7, 7/2, 14/3)$. The smooth five-manifold K has the following exceptional orbits:

(i) $x = 0$, $y = 0$, $z = 0$; which has a slice representation of

$$[\mathbf{Z}_3; 1, 1],$$

(ii) $x = 0$, $y = 0$, $w = 0$; which has a slice representation of

$$[\mathbf{Z}_4; 3, 3] \text{ and}$$

(iii) three exceptional orbits arising from $x = 0$, $w = 0$, $y^6 + z^3 = 0$. Each of these orbits has a slice representation of $[\mathbf{Z}_2; 1, 1]$.

It is clear that in this example we have obtained a K^* which is a singular variety. The singular points of K^* arise from these exceptional orbits which we will resolve to obtain a four-manifold K'.

By the nature of the slice and the fact that our $\mathbf{S}^1$-action acts coordinate-wise, we can see that in a neighborhood of a singular point, K^* is a cone over a lens space. Following the methods described in [2], we will resolve the singular point by replacing this neighborhood, whose boundary is a lens space, with a neighborhood of a collection of $\mathbf{S}^2$'s. The boundary of this new neighborhood is the same lens space, however the singular point has been replaced by a finite collection of S^2's. Each of the S^2's has negative definite self-intersection and is arranged so that adjacent S^2's intersect in a single point and so that we can not introduce an S^1 contained in the point set union of the S^2's which is not homotopic to a point in that point set union.

For the K^* which we described in example 2, we resolve as follows, see [2]. For the exceptional orbit with slice representation $[Z_3; 1, 1]$ we define d_j = g.c.d.(p_j,q) for $j = 1,2$. Thus $d_1 = d_2 = 1$. Next we take q_j such that $q = q_j d_j$ for $j = 1,2$, so we have $q_1 = q_2 = 3$. With m = g.c.d.$(q_1,q_2) = 3$, we calculate r_1, r_2 so that $p_j = r_j d_j$. In our example we have $r_1 = r_2 = 1$. Finally we define r_j' such that $r_j r_j' \equiv 1\ (m)$, $0 < r_j < m$ and $n \equiv r_1 r_2'(m)$, $0 < n < m$. When we construct the continued fraction q/n, the number of terms in the expansion tells us the number of S^2's we are introducing while the integers in the expansion are the absolute values of the self-intersection numbers of the S^2's. In the case of $[Z_3; 1, 1]$, we obtain $n = 1$ so that $q/n = 3$ and we introduce one S^2 with self-intersection -3. For $[Z_4; 3, 3]$, $q/n = 4$ and we replace a neighborhood of the singular point in K^* with a neighborhood of one S^2 with self-intersection -4. Finally, for each of the points corresponding to an exceptional orbit with slice representation $[Z_2; 1, 1]$, we introduce one S^2 with self-intersection -2. Using the above methods, we have proven the following.

THEOREM 1. *Every* whp *in four complex variables and with an isolated singularity at the origin, gives rise to a four-manifold.*

Our next task is to attempt to classify these four-manifolds. We will see that in many cases, we can classify K^* or K' up to homeomorphism. We begin with the following result.

THEOREM 2. *The four-manifold described in theorem* 1 *is simply connected.*

PROOF. We recall that K is a smooth, simply connected manifold, [5], and we note that $\mathbf{S}^1$ is a compact, connected Lie group. A theorem of Bredon, [3, II, 6.3] tells us that the fundamental group of K maps onto the fundamental group of K^*. Thus when K^* is a four-manifold, it is simply connected.

Now we consider the case where K^* has one singular point, z, arising from an exceptional orbit in K. We denote by E the collection of $\mathbf{S}^2$'s obtained in resolving z and a neighborhood of E in K' by $N(E)$. Because of the fashion in which the $\mathbf{S}^2$'s intersect, it is clear that $N(E)$ is simply connected. If we consider the orbit, call it Σ, which projects onto z, we can see that $(K\backslash\Sigma)$ is simply connected since K is simply connected and is a real codimension four subset. Again by [3, II, 6.3] we see that $(K^*\backslash z)$ is simply connected and, by the definition of a resolution, $(K'\backslash E)$ is homeomorphic to $(K^*\backslash z)$ so it follows that $(K'\backslash E)$ is simply connected. Now write $K' = U \cup V$, where $U = N(E)$ and $V = (K'\backslash E)$, and use Van Kampen's Theorem to conclude that K' is simply connected.

We note that there are at most a finite number of exceptional orbits in K which correspond to the singular points of K^*. If we call the corresponding collections of $\mathbf{S}^2$'s, E_j for $j = 1$ to n, then the corresponding exceptional orbits Σ_j in K may be joined by an arc in K. Let Σ' denote the union of the Σ_j and the arc joining them. Then $(K\backslash\Sigma')$ is still simply connected as before and the argument above may be repeated taking U to be a neighborhood of the image of Σ' in K' and V to be the complement of the image of Σ' in K'. Therefore K' is simply connected. Q.E.D.

We now turn our attention to the invariants for the manifold, K^* or K', which may be calculated from the whp. We begin with the generating function of a whp which is described by Arnol'd, [1]. This generating function is defined for semiquasihomogeneous functions, however all whp with an isolated singular point at the origin are semiquasihomogeneous. Let $A,i = N/w_i$ and $D,i = N - (A,i)$. Then the generating function $\chi_f(z) = [(z^{D,i} - 1)/(z^{A,i} - 1)]$. It is a result of Arnol'd [1, 4.13], that if f is a whp with an isolated singularity, then χ_f will be a polynomial. Thus we will write $\chi_f(z) = \Sigma\mu_j z^j$. Now set $P = \Sigma(A,i)$, $\mu_+ = \Sigma\mu_j$, where $P + j \equiv N\ (2N)$, and $\mu_- = \Sigma\mu_j$, where $P + j \equiv 0\ (2N)$. Then Dolgachev [4] has shown that for K^* we have $b_2 = 1 + \mu_+ + \mu_-$ and $\sigma = 1 + \mu_+ - \mu_-$. For K' we may calculate these same invariants by keeping track of the number of $\mathbf{S}^2$'s which we introduce and increasing b_2 by one for each $\mathbf{S}^2$ and decreasing by one for each $\mathbf{S}^2$. From b_2 and σ for the manifold, we can also determine $c_2 = b_2 + 2$ and $c_1^2 = 3 + 2c_2$.

Before we return to our examples, we recall that a manifold is said to be even if all the diagonal entries in its intersection form are even, otherwise it is odd. For our purposes we may determine if our nonsingular manifold is even or odd in the following manner. Rochlin's Theorem tells us that for an even manifold, the signature is divisible by sixteen. Since ours are complex manifolds, it follows from the Riemann-Roch Theorem that the expression $(c_1^2 + c_2)$ is divisible by twelve. Thus for an even manifold, we may write $\sigma = 16m$ and $c_2 = 12n - c_1^2$. Using these in the equation $c_1^2 = 3 + 2c_2$, we can see that if we are to have an even manifold, we must also have c_1^2 divisible by eight. If c_1^2 is not divisible by eight, then our manifold is odd and we know that K^* (or K') is homeomorphic to the connect sum of r copies of $\mathbf{CP}^2$ and q copies of $\mathbf{CP}^2$ conjugate, where $r + q = b_2$ and $r - q = \sigma$. Note that when we resolve, as we did in example 2, we can no longer take $r = (1 + \mu_+)$ and $q = \mu_-$.

Let us return to example 1 where $f = x^3 + xy^5 + z^5 + zw^6$. Here we have $\chi_f = (z^4-z^3+z^2-z+1)(z^{12}+\cdots+z+1)^2(z^8-z^7+z^6+z^2-z+1)$. With $P = 12$ and $N = 15$, we can see that $\mu_+ = 2$ and $\mu_- = 20$. Therefore we have $b_2 = 23$, $\sigma = -17$, $c_2 = 25$ and $c_1^2 = -1$. It follows from above that K^* is odd and thus homeomorphic to the connect sum of three copies of $\mathbf{CP}^2$ and twenty copies of $\mathbf{CP}^2$ conjugate.

In example 2 with $f = x^2 + y^7 + yz^3 + yw^4 + z^2w^2$, we have $\chi_f = (z^6-z^5+z^3-z+1)\ (z^8+\cdots+z^2+1)(z^{10}+\cdots+z+1)$. Now we have $P = 16$ and $N = 14$ so that $\mu_+ = 0$ and $\mu_- = 5$. When we resolved the singular variety K^*, we introduced five $\mathbf{S}^2$'s. Thus we now have $b_2 = 11$, $\sigma = -9$, $c_2 = 13$ and $c_1^2 = -1$ for K'. Again we see that our manifold K' is odd and so it is homeomorphic to one copy of $\mathbf{CP}^2$ connect sum with ten copies of $\mathbf{CP}^2$ conjugate.

All of our manifolds have been obtained from whp with an isolated singularity at the origin. With four complex variables, it can be shown, see [7] for example, that there are at most nineteen different forms of whp with an isolated singular point at the origin. Sixteen of these forms will have an isolated singular point at the origin whenever we have a whp. The remaining three forms contain examples of whp with the origin as an isolated singularity and of whp where the origin is not an isolated singularity. The study of these forms may lead to further insights of our four-manifolds.

REFERENCES

[1] Arnol'd, V. I., "Normal Forms of Functions in Neighborhoods of Degenerate Critical Points," *Russian Mathematical Surveys*, 29, pp. 10-50, 1974.

[2] Barth, W., Peters, C. and Van de Ven, A., *Compact Complex Surfaces*, Springer-Verlag, New York, 1984.

[3] Bredon, G., *Introduction to Compact Transformation Groups*, Academic Press, Pure and Applied Mathematics, Vol. 46, New York, 1972.

[4] Dolgachev, I., "Weighted Projective Varieties," *Group Actions and Vector Fields*, Proceedings of a Polish-North American Seminar, Springer-Verlag, New York, 1982.

[5] Milnor, J., *Singular Points of Complex Hypersurfaces*, Princeton University Press, Annals of Mathematics Studies, No. 61, Princeton, N.J., 1968.

[6] Neumann, W. D., *S^1-actions and the -invariants of Their Involutions*, Bonner Mathematische Schriften, No. 44, Bonn, 1970.

[7] Orlik, P. and Randell, R., "The Structure of Weighted Homogeneous Polynomials," *Proceedings of Symposia in Pure Mathematics*, Vol. 30, 1977.

[8] Randell, R., "Quotients of Complete Intersections by C*-actions," *Pacific Journal of Mathematics*, Vol. 74, No. 1, 1978.

[9] Wagreich, P., "The Structure of Quasihomogeneous Singularities," *Proceedings of Symposia in Pure Mathematics*, Vol. 40, part 2, 1983.

DEPARTMENT OF MATHEMATICS
DEPAUW UNIVERSITY
GREENCASTLE, IN 46135

Contemporary Mathematics
Volume **90**, 1989

HOLONOMY LIE ALGEBRAS, LOGARITHMIC CONNECTIONS AND THE LOWER CENTRAL SERIES OF FUNDAMENTAL GROUPS

Toshitake Kohno

Let X be the complement of a complex hypersurface in $\mathbf{C}^n$. We discuss the monodromy representation of the fundamental group of X arising from a connection of the form $\Omega = \Sigma_{i=1}^{s} P_i \, d\log f_i$, with constant matrices P_i. Here we denote by f_i a defining equation of an irreducible component of the hypersurface. As an integrability condition of the above logarithmic connection we obtain a family of relations $\Sigma_{i,j} a_{ij}^k [P_i, P_j] = 0$, $a_{ij}^k \in \mathbf{C}$, $k = 1,2, \dots$. In this note we report on some relations between the Lie algebra defined by the above relations and the fundamental group of X. We also present several algebraic properties of this Lie algebra, which gives information on the lower central series of the fundamental group.

1. DEFINITION OF HOLONOMY LIE ALGEBRAS AND ITS INTERPRETATION

First we give an intrinsic definition of the holonomy Lie algebra for a simplicial complex X. The cup product is considered to be a linear map $\cup : \Lambda^2 H^1(X;\mathbf{Q}) \to H^2(X;\mathbf{Q})$, which induces by duality the homomorphism $\eta : H_2(X;\mathbf{Q}) \to \Lambda^2 H_1(X;\mathbf{Q})$. Let $LH_1(X;\mathbf{Q})$ be the

0271-4132/89 $1.00 + $.25 per page

free Lie algebra generated by $H_1(X;\mathbf{Q})$ over $\mathbf{Q}$. Let J be the homogeneous ideal of $LH_1(X;\mathbf{Q})$ generated by image η. Here we identify $\Lambda^2 H_1(X;\mathbf{Q})$ with the degree two part of $LH_1(X;\mathbf{Q})$. The quotient Lie algebra $LH_1(X;\mathbf{Q})/J$ is called the *holonomy Lie algebra* of X over $\mathbf{Q}$, which we shall denote by $\mathbf{G}_X$.

Let us discuss the meaning of the holonomy Lie algebra in the case of the complement of a hypersurface in $\mathbf{C}^n$. Let D_i, $1 \leqslant i \leqslant s$, be a finitely many family of irreducible hypersurfaces defined by f_i. We consider a matrix valued 1-form $\Omega = \Sigma_i P_i d\log f_i$ with some constant matrices P_i. Let X be the complementary space of the union $\cup D_i$. The 1-form Ω is considered to be a connection of a trivial vector bundle over X in a natural way.

This connection is integrable if and only if the condition

$$\Omega \wedge \Omega = 0 \tag{1.1}$$

is satisfied. Let $e : H_1(X;\mathbf{C}) \to gl(m,\mathbf{C})$ be the linear map defined by $e(X_i) = P_i$, $1 \leqslant i \leqslant s$, where X_i denotes the homology class of the hypersurface D_i. Then we observe that the integrability condition (1.1) is satisfied if and only if e can be lifted to a Lie algebra homomorphism of $\mathbf{G}_X \otimes \mathbf{C}$.

We have also another interpretation of the holonomy Lie algebra due to K. Aomoto [A] from the viewpoint of the Schlesinger equation (monodromy preserving condition).

In the case of the complement of the union of finitely many complex hyperplanes H_j, $1 \leqslant j \leqslant s$, in $\mathbf{C}^n$, the holonomy Lie algebra is generated by X_j, $1 \leqslant j \leqslant s$, with relations:

$$\left[X_{j_\nu}, X_{j_1} + \dots + X_{j_p}\right] = 0, \quad 1 \leqslant \nu \leqslant p, \tag{1.2}$$

for any maximal family $\{H_{j_1}, \dots, H_{j_p}\}$ such that $\operatorname{codim}_{\mathbf{C}}(H_{j_1} \cap \dots \cap H_{j_p}) = 2$ (see [K1]).

To describe basic results on a relation between holonomy Lie algebras and fundamental groups, we recall several notations. Let

$\Gamma_1\pi_1(X) \supset \dots \supset \Gamma_m\pi_1(X) \supset \dots$ denote the *lower central series* of $\pi_1(X)$ defined inductively by $\Gamma_1\pi_1(X) = \pi_1(X)$, $\Gamma_{m+1}\pi_1(X) = [\pi_1(X), \Gamma_m\pi_1(X)]$, $m \geq 1$. The lower central series of $\mathbf{G}_X$ is defined in the same way by using the bracket product. The graded Lie algebra $[\mathrm{gr}\ \pi_1(X)] \otimes \mathbf{Q}$ is defined to be

$$\oplus_j\ [\Gamma_j\pi_1(X)/\Gamma_{j+1}\pi_1(X)] \otimes \mathbf{Q}$$

with the bracket product induced from the commutator product (see [MKS]). The graded Lie algebra gr $\mathbf{G}_X$ is defined to be $\oplus_j\ [\Gamma_j\mathbf{G}_X/\Gamma_{j+1}\mathbf{G}_X]$.

Let us denote by $[\pi_1(X)/\Gamma_j\pi_1(X)] \otimes \mathbf{Q}$ the *Malcev Lie algebra* associated with the nilpotent group $\pi_1(X)/\Gamma_j\pi_1(X)$, which is a nilpotent Lie algebra over $\mathbf{Q}$ defined inductively by means of the central extensions:

$$\Gamma_j\pi_1(X)/\Gamma_{j+1}\pi_1(X) \to \pi_1(X)/\Gamma_{j+1}\pi_1(X) \to \pi_1(X)/\Gamma_j\pi_1(X)$$

(see [M] for a precise definition).

THEOREM 1.3 ([K1]). *Let X be the complement of a complex hypersurface in $\mathbf{C}^n$. Then we have the following isomorphism of nilpotent Lie algebras over $\mathbf{Q}$:*

$$[\pi_1(X)/\Gamma_j\pi_1(X)] \otimes \mathbf{Q} \simeq \mathbf{G}_X/\Gamma_j\mathbf{G}_X, \quad j \geq 1. \tag{1.3.1}$$

In particular, we have an isomorphism:

$$[\mathrm{gr}\ \pi_1(X)] \otimes \mathbf{Q} \simeq \mathrm{gr}\ \mathbf{G}_X. \tag{1.3.2}$$

The proof is based on Sullivan's result [S] which describes a relation between the nilpotent completion of fundamental groups and 1-minimal models together with the Morgan's mixed Hodge structure on the 1-minimal model [M].

Let us go back to the situation of an integrable connection $\Omega = \Sigma\ P_i d\log f_i$ on the trivial bundle E on the complement of a hypersurface. This defines a foliation transversal to the fibers on the total space of E, which gives a linear representation of the fundamental group:

$$\theta : \pi_1(X,x) \to \mathrm{Aut}(E_x)$$

as the holonomy. Here E_x denotes the fiber over the base point x.

Let us now recall the notion of the *iterated integrals* due to K. T. Chen [C], which gives an expression of our monodromy representation θ.

(1.4) DEFINITION. Let X be a smooth manifold and let ω_i, $1 \leqslant i \leqslant r$, be matrix valued 1-forms on X. For a path $\gamma : [0,1] \to X$ we put $\gamma^*\omega_i = A_i(t)dt$ and we define the *iterated integral* by

$$\int_\gamma \omega_1\omega_2 \dots \omega_r = \int_0^1 A_1(t_1)\int_0^{t_1} A_2(t_2) \dots \int_0^{t_{r-1}} A_r(t_r)dt_r \dots dt_1.$$

LEMMA 1.5 (K. T. Chen [C]). *Let us suppose that the connection Ω is integrable. Then the monodromy representation θ is given by*

$$\theta(\gamma) = 1 + \int_\gamma \Omega + \int_\gamma \Omega\Omega + \int_\gamma \Omega\Omega\Omega + \dots .$$

We see that for each γ there exists $M > 0$ such that

$$\| \int_\gamma \underbrace{\Omega \dots \Omega}_{r} \| = 0(M^r/r!),$$

since the expression appearing in the definition of the iterated integral may be considered as an integral over the standard simplex. Hence the above series is convergent.

As a universal expression of (1.5), we obtain a homomorphism

$$\theta^\sim : \pi_1(X,x) \to U(\mathbf{G}_X \otimes \mathbf{C})^\wedge$$

by the iterated integral of the Chen's formal connection

$$\Omega^{\sim} = \Sigma\, X_i \otimes d\log f_i.$$

Here $U(\mathbf{G}_X \otimes \mathbf{C})^{\wedge}$ denotes the completion of the universal enveloping algebra of $\mathbf{G}_X \otimes \mathbf{C}$ with respect to the natural filtration by degrees and X_i is the homology class corresponding to the hypersurface D_i. The algebra $U(\mathbf{G}_X \otimes \mathbf{C})^{\wedge}$ may also be described in the following way. Let $C\langle\langle X_i\rangle\rangle$ denote the ring of the non-commutative power series with indeterminates X_i, $1 \leqslant i \leqslant s$ and let J be the two-sided ideal of $\mathbf{C}\langle\langle X_i\rangle\rangle$ generated by the quadratic relations among X_i appearing in the definition of the holonomy Lie algebra. Then we have an isomorphism

$$U(\mathbf{G}_X \otimes \mathbf{C})^{\wedge} \simeq \mathbf{C}\langle\langle X_i\rangle\rangle / J$$

The monodromy representation θ is obtained from the universal expression $\theta^{\sim}$ by the substitution $X_i = P_i$.

For a group G we denote by $\mathbf{C}[G]$ its group algebra over $\mathbf{C}$. Let $\epsilon : \mathbf{C}[G] \to \mathbf{C}$ denote the augmentation homomorphism and we put $IG = \operatorname{Ker} \epsilon$. Let $\mathbf{C}[G]^{\wedge}$ be the completion of $\mathbf{C}[G]$ with respect to the topology defined by $\{I^k G\}_{k\geqslant 1}$, where $I^k G$ signifies the k-th power of IG. Let $j : G \to \mathbf{C}[G]^{\wedge}$ be the natural homomorphism.

THEOREM 1.6. ([A], [H]). *We have an isomorphism of complete Hopf algebras*

$$\mathbf{C}[\pi_1(X)]^{\wedge} \simeq U(\mathbf{G}_X \otimes \mathbf{C})^{\wedge}$$

such that the following diagram is commutative.

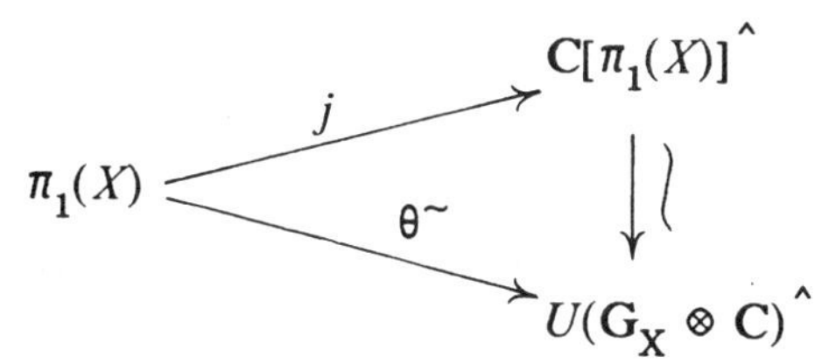

(1.7) REMARK. By taking the primitive part P of $A = U(\mathbf{G}_X \otimes \mathbf{C})^{\wedge}$, i.e., $P = \{x \in A;\ \Delta x = x \mathbin{\hat{\otimes}} 1 + 1 \mathbin{\hat{\otimes}} x\}$, we obtain the

Malcev Lie algebra of the nilpotent completion of $\pi_1(X)$ over **C**. This describes a relation between Theorem 1.3 and Theorem 1.6.

2. POINCARÉ SERIES ASSOCIATED WITH FUNDAMENTAL GROUPS

In the case $X = \{(z_1, \ldots, z_n) \in \mathbf{C}^n;\ z_i \neq z_j \text{ if } i \neq j\}$, the fundamental group of X is the *pure braid group on n strings.* The holonomy Lie algebra $\mathbf{G}_X$ is generated by X_{ij}, $1 \leqslant i < j \leqslant n$, with the relations:

$$[X_{ij}, X_{ik} + X_{jk}] = [X_{ij} + X_{ik}, X_{jk}] = 0, \quad i < j < k$$

$$[X_{ij}, X_{k\ell}] = 0, \quad i,j,k,\ell \text{ distinct.}$$

The above relations relevant to the rational solutions of the classical Yang-Baxter equations are of particular interest (see [K4] for details).

In this section, we first report on some algebraic properties of $\mathbf{G}_X$ proved in [K2]. In general for a simplicial complex X, we introduce the following complex $(R_., \partial_.)$. Let $\{\omega^i\}_{1 \leqslant i \leqslant s}$ be a basis of $H^1(X;\mathbf{Q})$ and let $\{X_i\}_{1 \leqslant i \leqslant s}$ be its dual basis of $H_1(X;\mathbf{Q})$. We put

$$R^k = U(\mathbf{G}_X) \otimes_{\mathbf{Q}} H^k(X;\mathbf{Q})$$

$$R_k = \mathrm{Hom}_{U(\mathbf{G}_X)}(R^k, U(\mathbf{G}_X)), \quad k \geqslant 0.$$

Let $\delta^k : R^{k-1} \to R^k$ be the left $U(\mathbf{G}_x)$-module homomorphism defined by

$$\delta^k(1 \otimes \varphi) = \sum_{i=1}^{s} X_i \otimes \omega^i \cup \varphi$$

for $\varphi \in H^{k-1}(X;\mathbf{Q})$. Let $\partial_k : R_k \to R_{k-1}$ be the dual of δ^k, which is a right $U(\mathbf{G}_X)$-module homomorphism.

THEOREM 2.1 ([K2]). *If* $X = \{(z_1, \ldots, z_n) \in \mathbf{C}^n ;\ z_i \neq z_j$ *if* $i \neq j\}$, *then the associated complex* $(R_., \partial_.)$ *defined above is acyclic. This determines a minimal free resolution of* $\mathbf{Q}$ *as a* $U(\mathbf{G}_X)$*-module.*

COROLLARY 2.2. *In the above situation, we have canonical isomorphisms:*

$$H^{\cdot}(X;\mathbf{Q}) \simeq H^{\cdot}(\pi_1(X);\mathbf{Q}) \simeq H^{\cdot}(\mathbf{G}_X;\mathbf{Q}).$$

COROLLARY 2.3. *Let* $P_X(t)$ *be the Poincaré polynomial of the above* X. *Then we have*

$$\prod_{j=1}^{\infty} (1 - t^j)^{-\varphi_j} = P_X(-t)^{-1} \text{ in } \mathbf{Z}[[t]].$$

Here $\varphi_j = \text{rank}[\Gamma_j\pi_1(X)/\Gamma_{j+1}\pi_1(X)]$.

PROOF OF (2.3): Let $H(X;t)$ denote the Hilbert series of $U(\mathbf{G}_X)$ defined by

$$H(X;t) = \sum_p \{\dim U(\mathbf{G}_X)^p\} t^p.$$

Here $U(\mathbf{G}_X)^p$ denotes the degree p part of $U(\mathbf{G}_X)$. The minimal free resolution $R. \to \mathbf{Q} \to 0$ induces the following exact sequence of $\mathbf{Q}$ vector spaces:

$$0 \to R_n^p \to R_{n-1}^{p+1} \to \cdots .$$

Since $\dim R_j^p = b_j(X). \dim U(\mathbf{G}_X)^p$, we obtain the formula

$$H(X;t).\ P_X(-t) = 1.$$

By means of the Poincaré-Birkhoff-Witt theorem and the isomorphism (1.3.2), we obtain

$$H(X;t) = \prod_{j=1}^{\infty} (1 - t^j)^{-\varphi_j(X)} .$$

This completes the proof.

The formula (2.4) may be considered as a generalization of

Witt formula for free groups (see [MKS]). The same type of formula is known for:

i) Riemann surface (Labute [L])

ii) complement of a fiber type arrangement (Falk-Randell [FR])

iii) complement of an arrangement of type D ([K3]).

In general, we have a cohomological criteria for this type of formula.

THEOREM 2.5. *Let X be the complement of a hypersurface in* $\mathbf{C}^n$ *(or any space satisfying the condition* 1.3.2 *(see Remarque* 2.16 *of* [K2])). *If we have an isomorphism*

$$H^{\cdot}(\mathbf{G}_X;\mathbf{Q}) \cong H^{\cdot}(X;\mathbf{Q}).$$

then we have the formula

$$\prod_{j=1}^{\infty} (1 - t^j)^{-\varphi_j(X)} = P_X(-t)^{-1}.$$

with $\varphi_j(X) = \operatorname{rank}[\Gamma_j\pi_1(X)/\Gamma_{j+1}\pi_1(X)]$.

PROOF. We put $K_0 = U(\mathbf{G}_X)$. Let K_1 be a free $U(\mathbf{G}_X)$-module with basis $\{u_1, \ldots, u_s\}$, where $s = b_1(X)$ and $\deg u_j = 0$. We define a right $U(\mathbf{G}_X)$-module homomorphism $\partial_1 : K_1 \to K_0$ by $\partial_1(u_j) = X_j$, $1 \leqslant j \leqslant s$. Since $\mathbf{G}_X$ is defined by the relations of degree two, we see that $\operatorname{Ker} \partial_1$ is a $U(\mathbf{G}_X)$-module of finite type generated by elements of degree one (cf. [K2], Lemma 4.10). Let $\{Y_j\}_{1\leqslant j\leqslant p}$ be a $\mathbf{Q}$ basis of the degree one part of $\operatorname{Ker} \partial_1$ and we define K_2 to be the free $U(\mathbf{G}_X)$-module with basis $\{v_j\}_{1\leqslant j\leqslant p}$, $\deg v_j = 0$. Let $\partial_2 : K_2 \to K_1$ be the right $U(\mathbf{G}_X)$-module homomorphism defined by $\partial_2(v_j) = Y_j$, $1 \leqslant j \leqslant p$. Repeating this argument, we obtain a free resolution of $\mathbf{Q}$ as a $U(\mathbf{G}_X)$-module:

$$\longrightarrow K_j \xrightarrow{\partial_j} K_{j-1} \longrightarrow \cdots \longrightarrow K_0 \xrightarrow{\epsilon} \mathbf{Q} \longrightarrow 0$$

such that

i) $\deg \partial_j = 1$

ii) K_j is of finite rank, $j \geq 0$.

It follows that rank $K_j = \dim H^j(G_X;Q)$. Combining this with the hypothesis, we have rank $K_j = b_j(X)$. By means of the argument of the proof of 2.3, we deduce that $H(X;t) = P_X(-t)^{-1}$. The desired formula follows from this by using (1.3.2) and Poincaré-Birkhoff-Witt theorem.

(2.6) REMARK. In particular, if the Sullivan's minimal model of X is generated by elements of degree 1, we have the formula 2.4. This result was also obtained by M. Falk [F] in the case of the complement of hyperplanes by an explicit computation of the minimal model.

Now we are going to discuss a relation between the lower central series of fundamental groups and the Poincaré series of certain Noetherian local rings. For a local ring R with its maximal ideal m and $k = R/m$, let us recall that the Poincaré series of R is defined to be

$$P(R;t) = \sum_{n\geq 0} [\dim \mathrm{Tor}_n^R(k,k)]t^n.$$

It was pointed out by E. Roos [R] that this series is relevant to the series

$$\sum_{n\geq 0} [\dim H^n(\Omega X; Q)]t^n$$

for X a simply connected CW complex and ΩX its loop space. We shall show another interpretation of the Poincaré series.

Let X be a compact Kähler manifold or the complement of a hypersurface in $\mathbf{C}^n$ (any space satisfying the condition 1.3.2 will do). We consider as before the series

$$\prod_{j=1}^{\infty} (1 - t^j)^{-\varphi_j(X)}$$

defined by means of the lower central series of the fundamental group of X. As we have shown in the proof of Corollary 2.3, this series is equal to the Hilbert series of the universal enveloping algebra of the holonomy Lie algebra of X. We associate to X a Noetherian local ring R_X in the following way. Let $\{\omega^j\}_{1 \leq j \leq s}$ be a basis of $H^1(X;\mathbf{C})$ and let K be the kernel of the natural homomorphism $\wedge H^1(X;\mathbf{C}) \to H^*(X;\mathbf{C})$. Let $\{\Sigma_{i<j} c^k_{ij}\omega^i \wedge \omega^j\}_k$ be a basis of $K^{(2)}$ (degree two part of K). The commutative ring R_X is defined to be

$$\mathbf{C}[X_1, \ldots, X_s]/\left(\sum_{i<j} c^k_{ij}X_iX_j,\ m^3\right).$$

Here m is the maximal ideal of $\mathbf{C}[X_1, \ldots, X_s]$.

THEOREM 2.7. *Let X be a simplicial complex satisfying the condition 1.3.2. Then the Poincaré series of the associated Noetherian local ring R_X defined above is given by*

$$P(R_X;t) = \frac{Z(t)}{1 + t - Z(t)(1 - b_1(X)t + pt^2)}$$

where

$$Z(t) = \prod_{j=1}^{\infty}(1 - t^j)^{-\varphi_j(X)}, \quad p = \binom{b_1(X)}{2} - \dim K^{(2)}.$$

PROOF. We put $R = R_X$, $\mathbf{G} = \mathbf{G}_X \otimes \mathbf{C}$. Let $\varphi : R/R^2 \otimes R/R^2 \to R^2$ be the $\mathbf{C}$ linear map induced by the multiplication. We put $A = T((R/R^2)^*)/(\text{image } \varphi^*)$. Here T stands for the tensor algebra. Let $T(X_1, \ldots, X_s)/(\Sigma_{i<j} a^k_{ij}[X_i,X_j])$ be a presentation of $U(\mathbf{G})$. Putting $[X,Y]_+ = XY + YX$ (anti-commutator), we define the associative algebra $U(\mathbf{G})_+$ by

$$U(\mathbf{G})_+ = T(X_1, \ldots, X_s)/\left(\sum_{i<j} a^k_{ij}[X_i,X_j]_+\right).$$

We see that $U(\mathbf{G})_+$ is isomorphic to the algebra A defined above. It follows from a work of Löfwall [L] that the Poincaré series of R is given by

$$\frac{t\, H(A;t)}{1 + t - H(A;t)(1 - \dim (R/R^2)t + \dim (R^2)t^2)}$$

where $H(A;t)$ is the Hilbert series of A. It is clear that the Hilbert series of A is equal to that of $U(\mathbf{G})$, which is equal to $Z(t)$ by means of the proof of Corollary 2.3. This completes the proof.

REFERENCES

[A] K. Aomoto, Fonctions hyperlogarithmiques et groupes de monodromie unipotents, J. Fac. Sci. Tokyo, 25 (1978), 149-156.

[C] K. T. Chen, Extensions of C^∞ function algebra by integrals and Malcev completion of π_1, Advances in Math. 23 (1977), 181-210.

[F] M. Falk, The minimal model of the complement of an arrangement of hyperplanes, preprint, The Univ. of Iowa.

[FR] M. Falk and R. Randell, The lower central series of a fiber-type arrangement, Invent. Math. 82 (1985), 77-88.

[H] R. Hain, The de Rham homotopy theory of complex algebraic varieties, preprint, University of Washington.

[K1] T. Kohno, On the holonomy Lie algebra and the nilpotent completion of the fundamental group of the complement of hypersurfaces, Nagoya Math. J. 92 (1983), 21-37.

[K2] ________, Série de Poincaré-Koszul associée aux groupes de tresses pures, Invent. Math. 82 (1985), 57-75.

[K3] ________, Poincaré series of the Malcev completion of generalized pure braid groups, preprint, IHES, 1985.

[K4] ________, Linear representation of braid groups and classical Yang-Baxter equations, to appear in Contemp. Math. "Artin's Braid Groups," Santa Cruz.

[L] J. P. Labute, On the descending central series of groups with a single defining relation, J. of Algebra 14 (1970), 16-23.

[Lö] C. Löfwall, Une algèbre nilpotente dont la série de Poincaré-Betti est non rationelle, C. R. Acad. Sc. Paris, 288 (1979), 327-330.

[M] J. Morgan, The algebraic topology of smooth algebraic varieties, Publ. Math. IHES, 48 (1978), 137-204.

[MKS] W. Magnus, A. Karrass, and D. Solitar, Combinatorial group theory, John Wiley, 1966.

[R] E. Roos, Relations between the Poincaré-Betti series of loop spaces and of local rings, Lect. Note in Math. 740 (1979), 285-322.

[S] D. Sullivan, Infinitesimal computations in topology, Publ. Math. 47 (1977), 269-331.

DEPARTMENT OF MATHEMATICS
NAGOYA UNIVERSITY
CHIKUSA-KU, NAGOYA, 464
JAPAN

Contemporary Mathematics
Volume **90**, 1989

TANGENT CONES FOR DEFORMATIONS OF TWO-DIMENSIONAL QUASI-HOMOGENEOUS SINGULARITIES

Henry B. Laufer*

SECTION 1. INTRODUCTION

In [15], Zariski asked if the topology of a hypersurface singularity (V,p) determines its multiplicity. Using results of Varchenko [11], O'Shea [8] and Greuel [4] answered Zariski's question in the affirmative for families of singularities whose initial fiber was an isolated quasi-homogeneous singularity. A'Campo's work [1], [7] shows how the tangent cone to V at p can often be used to give positive results towards Zariski's question.

We consider a deformation $\lambda: \mathcal{V} \to T$ of the two-dimensional isolated quasi-homogeneous singularity (V,p). Assume that all of the fibers are topologically equivalent, so as to be in the area of Zariski's question. Then μ_t is constant and constant multiplicity is known. There is also a weak simultaneous resolution [5] $\pi: \mathcal{M} \to \mathcal{V}$. This paper has two main results. Constant multiplicity is re-proved (Theorem 2.5) using the pull-backs to M_t of the maximal ideals m_t of the singularities (V_t,p_t). The multiplicity is computed using Proposition 2.1, a variation of Wagreich's result [12]. While Theorem 2.5 is not new and is only proved in dimension two, it

*Research partial supported by NSF Grant No. DMS8601871.

does explain constant multiplicity from a different viewpoint than does the work of O'Shea and Greuel. As examples of Briancon and Speder [3] and of the author [6] already show, the pull-backs of the m_t need not behave well (not even flatly) under deformation. Nonetheless, in the proof of Theorem 2.5 one can see how multiplicity is preserved.

The second main result, Theorem 3.1, is new and is philosophically related to the observations of Le and Teissier [7] on A'Campo's work. Namely, the abstract topology of the (reduced) tangent cone to (V_t,p_t) is constant for all small t. An example, (3.1), shows that the embedded tangent cone need not deform in an equisingular manner.

SECTION II. CONSTANT MULTIPLICITY

Let π: $M \to V$ be a resolution of the normal two-dimensional singularity (V,p). Let $A = \pi^{-1}(p)$ be the (reduced) exceptional set in M. Let $A = \cup A_i$ be the decomposition of A into irreducible components. We shall use the following notation. Let f be a holomorphic function on M which vanishes on A. Then we shall decompose the divisor (f) as f as $(f) = D_f + W_f$, where D_f is a divisor whose support is A and W_f is a divisor on M whose support meets A only at a finite number of points. Let m denote the maximal ideal at p. Let $\pi^{-1}(m)$ denote the sheaf on M which is the pull-back of m. Let ν denote the multiplicity of V at p. We then have the following variation of a theorem of Wagreich [12].

PROPOSITION 2.1. *Let f and g be holomorphic functions on M which generate $\pi^{-1}(m)$ and have $W_f \cap W_g = \emptyset$. Then $\nu = -D_f \cdot D_g$.*

PROOF. We shall reduce Proposition 2.1 to Wagreich's result as follows. If additionally $\pi^{-1}(m)$ is locally principal, we let D denote the divisor on A associated to m, i.e., Yau's [14] maximal ideal cycle. Then, Wagreich [12], $\nu = -D_m \cdot D_m$. There are cycles

$E_f \geqslant 0$ and $E_g \geqslant 0$ such that $D_f = D_{m} + E_f$ and $D_g = D_{m} + E_g$. Since f and g generate $\pi^{-1}(m)$, $E_f \cap E_g = \emptyset$. Also, if A_i is in the support of E_f, then $A_i \cdot D_g = 0$. Then $D_f \cdot D_g = D_{m} \cdot D_{m}$ and Proposition 2.1 holds.

If $\pi^{-1}(m)$ is not locally principal, we let $\sigma: M' \to M$ be the result of quadratic transformations which make the pullback of m to M' locally principal. Let $f' = \sigma^*(f)$ and $g' = \sigma^*(g)$. Let $\sigma^{-1}(D)$ denote the total transform of a divisor D. Then $D_{f'} = \sigma^{-1}(D_f) + F_f$ for a suitable divisor $F_f \geqslant 0$. Similarly for g. Since $W_f \cap W_g = \emptyset$, $F_f \cap F_g = \emptyset$ and also $F_f \cdot \sigma^{-1}(D_g) = 0$. Then $D_{f'} \cdot D_{g'} = \sigma^{-1}(D_f) \cdot \sigma^{-1}(D_g) = D_f \cdot D_g$. This reduces the general case to the already proved special case.

Let $f(x,y,z) = 0$ be the defining equation for a quasi-homogeneous hypersurface V with an isolated singularity at the origin 0. Let x, y, and z have respective weights w_x, w_y, and w_z. Let $\pi: M \to V$ be the minimal resolution of V. With a minor abuse of the previous notation, we shall write $(\pi^*(x)) = D_x + W_x$, rather than writing $(\pi^*(x)) = D_{\pi^*(x)} + W_{\pi^*(x)}$. Let $A = \pi^{-1}(0)$ be the exceptional set in M. Since [10, Theorem 2.1] [9] the isolated quasi-homogeneous singularity (V,p) is determined by suitable information on M, we may use the following description of (M,A): Let $A = \cup A_i$ be the decomposition of A into irreducible components, with A_0 denoting the central curve. Each orbit in $M - A$ under the $\mathbf{C}^*$ action has exactly one point of A in its closure. We shall call such a closed set in M a fiber and the point in A is the base of the fiber. Every point in A_0 which is not on an arm is the base of a fiber. Each arm, as is (2.1) below, contains exactly one fiber base, which we shall call ∞ (for that arm). The point ∞ lies in that curve A_ℓ in the arm which meets only one other A_i.

Let a, b, and c denote the respective orders of the zeros on A_0 of $\pi^*(x)$, $\pi^*(y)$, and $\pi^*(z)$, i.e., the coefficients of A_0 in D_x, D_y, and D_z respectively. Observe that the ratios $a{:}b{:}c$ and $w_x{:}w_y{:}w_z$ are equal. Order x, y, and z so that $a \leqslant b \leqslant c$.

The following lemma (stated only for x, but also true for y, z, or any function on (V,p) which is weighted homogeneous) shows that if W_x is non-zero or has a simple zero at ∞, then D_x near the arm is determined by the order of vanishing along the central curve.

$$\underset{A_0}{*} - - - - \underset{A_1}{*} - - - - \cdots - - - - \underset{A_\ell}{*} \tag{2.1}$$

LEMMA 2.2. *Consider an arm plus central curve, as in* (2.1). *Let the integer a be given. Consider cycles* $C = \Sigma\, c_i A_i$, $0 \leqslant i \leqslant \ell$ *such that*

(i) $c_0 = a$

(ii) $0 \geqslant A_i \cdot C$ *for* $0 \leqslant i \leqslant \ell$.

Suppose that $W_x \cdot A_\ell \leqslant 1$. *Suppose that a is the order of vanishing of x on the central curve* A_0, *i.e. the coefficient of* A_0 *in* D_x. *Then*

$D := D_x$ *restricted to the arm plus central curve is the minimal cycle with the properties* (i) *and* (ii).

PROOF. D satisfies (i) trivially and also (ii) since it comes from the divisor of a holomorphic function. Moreover, $A_i \cdot D = 0$ for $0 < i < \ell$. By taking minima for the coefficients, as in the proof of the existence of the fundamental cycle [2], a unique minimal cycle Y satisfying (i) and (ii) can be seen to exist. Write

$$D = Y + E, \quad \text{with} \quad E \geqslant 0 \tag{2.2}$$

To prove the Lemma, it suffices to show that $E = 0$. From (2.2), for $0 < i < \ell$, $A_i \cdot E \geqslant 0$. Since $0 = A_\ell \cdot (D_x + W_x) \leqslant A_\ell \cdot D_x + 1$, $A_\ell \cdot E \geqslant -1$. Write $E = \Sigma\, e_i A_i$, $k \leqslant i \leqslant \ell$, $k > 0$. $e_k \neq 0$. Then, since there are no -1 curves on the arm, an easy induction shows that $e_{i+1} > e_i$, $k \leqslant i < \ell$. But then $A_\ell \cdot E \geqslant -1$ is impossible.

When we use the word "generic," we shall mean as usual "for values off of an analytic subvariety of positive codimension."

PROPOSITION 2.3. *For the quasi-homogeneous singularity* (V,p)

make a generic choice of the coordinate function y. *Let* α *be a complex number. Let* $f = \pi^*(x + \alpha z)$ *and* $g = \pi^*(y)$. *Then for a generic choice of* α, f *and* g *satisfy the hypotheses of Proposition* 2.1.

PROOF. The homogeneous case is clear. So we shall only consider the case $a < c$.

Consider the conditions of Proposition 2.1 near a point q which is in A_0 but not in an arm. It suffices to show that $W_x \cap W_y \cap A_0 = \emptyset$. For then if q is not in W_x, since $c > a$, $D_f + W_f = D_x = D_m$ near q. While if q is not in W_y, the other possibility, since $c \geqslant b$, $\pi^*(z)$ is in the ideal of g near q and hence f and g generate $\pi^*(m)$. But if some common component C of W_x and W_y should meet A_0 at q, then, since $c > 1$, the functions x, y, and z cannot separate points on $\pi(C)$.

Now let $L = A_1 + \cdots + A_\ell$ be an arm, with $A_1 \cdot A_0 = 1$. There are various possibilities along L, which we shall also need in later proofs.

CASE 1: $W_x \cdot L = 0$. By Lemma 2.3, $\pi^*(x)$ generates $\pi^*(m)$ near L. Moreover, since $c > a$ and D_x is minimal along L, $D_f + W_f = D_x = D_m$ near L.

CASE 2: $W_y \cdot L = 0$. By Lemma 2.2, the ideal $\pi^*(y)$ includes $\pi^*(z)$ near all points of L. Then f and g generate $\pi^{-1}(m)$ near L.

CASE 3: We are in neither Case 1 nor in Case 2. Then the fiber C which is above the point ∞ in A_ℓ lies in both W_x and W_y. Since the functions x, y, and z separate points on V, ∞ is not a point of W_z and moreover, A_ℓ has multiplicity 1 in D_z. Since $W_z \cdot A_\ell = 0$ and we are not in Case 2, our generic choice of y implies that $b < c$. It is impossible to satisfy simultaneously both $W_x \cdot A_\ell \geqslant 2$ and $W_y \cdot A_\ell \geqslant 2$, for then the Jacobian of the map (x,y,z) could not have rank two along $\pi(C)$. So either $W_x \cdot A_\ell = 1$ or $W_y \cdot A_1 = 1$. By Lemma 2.2, D_z is minimal along L with respect to c and either D_x or D_y is similarly minimal. Recall that A_ℓ has multiplicity 1 in D_z. Then the minimal one of D_x and D_y also has A_ℓ

with multiplicity 1. Then $a < b < c$, D_x minimal, and $W_x \cdot A_\ell = 1$ is impossible. Then either $a = b$ or D_y is minimal with $W_y \cdot A_\ell = 1$. For $a = b$, our generic choice of y insures that D_y is minimal with $W_y \cdot A_\ell = 1$.

In summary so far for Case 3: $b < c$, $W_x \cdot A_\ell > 1$, $W_y \cdot A_\ell = 1$, and $W_z \cdot A_\ell = 0$. D_y and D_z are minimal along L with respect to b and c respectively. A_ℓ has multiplicity 1 in both D_y and D_z. Then the ideal of $\pi^*(y)$ contains $\pi^*(z)$ at all points of L except ∞. At ∞, $\pi^*(z)$ generates $\pi^*(m)$. In all, f and g do satisfy the hypotheses of Proposition 2.1.

PROPOSITION 2.4. *Let* $\lambda\colon \mathcal{V} \to T$ *be a deformation of* (V,p) *which has a weak simultaneous resolution* $\pi\colon \mathcal{M} \to \mathcal{V}$. *The functions* x, y, *and* z *on* $(V,p) = (V_0,p_0)$ *may be lifted to* $\mathcal{V}$ *so that their order of vanishing remains constant on the central curves* $A_{0,t}$ *of the* M_t.

PROOF. Look at the initial fiber. The $\mathbf{C}^*$ action on (V,p) gives a direct sum decomposition, by weight, on $H^1(M, \mathcal{O})$. More precisely, for all $k \geq 0$, let C^k be the least cycle on A such that $C^k \geq 0$, A_0 has coefficient k in C^k, and $A_i \cdot C^k \leq 0$ for $i > 0$. Restricting $-C^k$ to A_0 yields the divisor $D^{(k)}$ of Pinkham [10]. For all $k \geq 0$, consider the exact sheaf sequence on M:

$$0 \to \mathcal{O}(-C^{k+1}) \to \mathcal{O}(-C^k) \to \mathcal{O}(-C^k)/\mathcal{O}(-C^{k+1}) \to 0 \qquad (2.3)$$

There is a corresponding long exact cohomology sequence for (2.3). Take a Leray cover $\{U_j\}$ of M for the sheaves in (2.3) given by: U_j is a neighborhood of an arm for $j > 0$ and U_0 meets no arm. The $\mathbf{C}^*$ action gives a grading on each $\Gamma(U_j, \mathcal{O}(-C^k))$. Then

$$H^0(M, \mathcal{O}(-C^k)/\mathcal{O}(-C^{k+1})) \cong H^0(A_0, \mathcal{O}_{A_0}(D^{(k)}))$$

and

$$H^1(M, \mathcal{O}(-C^k)/\mathcal{O}(-C^{k+1})) \cong H^1(A_0, \mathcal{O}_{A_0}(D^{(k)})).$$

From [10, Theorem 5.7], for all $k \geq 0$,

$$\dim H^1(M, \mathcal{O}(-C^k)) = \dim H^1(M, \mathcal{O}(-C^{k+1})) + \dim H^1(M, \mathcal{O}(-C^k)/\mathcal{O}(-C^{k+1})) \quad (2.4)$$

Now look at (2.3) and (2.4) for nearby fibers M_t. From the long exact cohomology sequence for (2.3), (2.4) becomes

$$\dim H^1(M_t, \mathcal{O}(-C^k)) \leq \dim H^1(M_t, \mathcal{O}(-C^{k+1})) + \dim H^1(M_t, \mathcal{O}(-C^k)/\mathcal{O}(-C^{k+1})) \quad (2.5)$$

Since there is a simultaneous resolution and $C^0 = 0$, $\dim H^1(M_t, \mathcal{O}(-C^0))$ is a constant function of t. Additionally, $\dim H^1(M_t, \mathcal{O}(-C^k))$ and $\dim H^1(M_t, \mathcal{O}(-C^k)/\mathcal{O}(-C^{k+1}))$ are upper-semi-continuous functions of t. Then there is equality in (2.5) for all k and $\dim H^1(M_t, \mathcal{O}(-C^k))$ is in fact a constant function of t, for all k. Then, for each k, elements of $\Gamma(M, \mathcal{O}(-C^k))$ lift to elements of $\Gamma(M_t, \mathcal{O}(-C^k))$. This proves the Proposition.

THEOREM 2.5 [O'Shea, Greuel, Varchenko]. *ν_t is a constant function of t.*

Let us examine Proposition 2.3 and its proof, under the deformation λ. We again only need to consider the inhomogeneous case. First look along the central curves away from the arms. Since $b \leq c$, for all t, for q in A_0 and q not in $W_{y,t}$, the ideal of $\pi^*(y)$ contains $\pi^*(z)$. Since a_t and b_t is constant, for all small t, $W_x \cap W_y = \emptyset$. Since $a < c$, near any q in $A_0 \cap W_{y,t}$, $\pi^*(f)$ generates $\pi^*(m)$.

The minimality conditions in the Cases in the proof of Proposition 2.3 insure that for all t, f_t and g_t satisfy the hypotheses of Proposition 2.1. Thus to complete the proof of Theorem 2.5, we need to examine the values of $D_{f,t} \cdot D_{g,t}$. We now shall abuse notation and not distinguish between a cycle on the exceptional set A_t, $t \neq 0$ and its homotopically equivalent cycle on A. Since the coefficients in the divisor of a function cannot increase if the function lifts under a deformation, there are cycles $E_f \geq 0$ and $E_g \geq 0$ such that for small $t \neq 0$,

$$D_{f,t} + E_f = D_{f,0} \qquad D_{g,t} + E_g = D_{g,0}$$

We need to show that $D_{f,0} \cdot D_{g,0} = D_{f,t} \cdot D_{g,t}$. Since, by Proposition 2.4, the order of vanishing on the central curves is constant for x, y, and z, E_f and E_g are supported on the arms. So we may consider each arm L separately.

Consider the cases in the proof of Lemma 2.3. Observe that any minimal cycle on the initial fiber lifts to an equal minimal cycle on the nearby fibers. Case 1: Near L, $E_f = 0$. Also, $A_i \cdot D_{f,0} = 0$ for A_i in L. Case 2: Near L, $E_g = 0$. Also, $A_i \cdot D_{g,0} = 0$ for A_i in L. Case 3: Near L, $E_g = 0$. For A_i in L, $i \neq \ell$, $A_i \cdot D_g = 0$. Finally, A_ℓ is not in the support of E_f.

Looking at all of the cases in the previous paragraph, we see that $E_f \cap E_g = \emptyset$, $E_f \cdot D_g = 0$ and $E_g \cdot D_f = 0$. This completes the proof of Theorem 2.5.

SECTION III. THE REDUCED TANGENT CONE.

Recall the observations of Le and Teissier [7], which are based on A'Campo's work [1]. Let (V,p) be a dimension two, isolated hypersurface singularity. Let $C_{V,p}$ be the reduced tangent cone, i.e. the Whitney [13] tangent cone $C(V,p)$. Let $PC_{V,p}$ denote the hypersurface in P^2 over which $C_{V,p}$ is a cone. Then, work of A'Campo shows that the multiplicity of (V,p) is determined by the embedded topology of (V,p) in case the topological Euler number $\chi(PC_{V,p})$ is non-zero. The same arguments also show that, for isolated hypersurface two-dimensional singularities, the embedded topology and the multiplicity determine $\chi(PC_{V,p})$.

Let $PT_{V,p}$ be the non-reduced curve in P^2 which is associated with the non-reduced tangent cone to (V,p), i.e. the defining equation for $PT_{V,p}$ is given in homogeneous coordinates by the lowest order (non-zero) terms in the defining equation for (V,p). $PC_{V,p}$ is the reduction of $PT_{V,p}$. If there is constant multiplicity ν for the family $\{V_t,0\}$ given by $f(x,y,z,t) = 0$, then there is a deformation of

$PT_{V,p}$ given by the terms of order ν in x, y, and z in the power series expansion of $f(x,y,z,t)$ in x, y, and z.

We shall now give an example of a deformation of zero weight of a quasi-homogeneous singularity $(V_0,0)$ such that for the initial fiber, $PC_{V_0,0}$ is two transversely intersecting spheres, while for the general fiber, $PC_{V_t,0}$ is two spheres which meet tangentially. Observe that for every fiber, $\chi(PC_{V_t,0}) = 0$. Namely, with (x,y,z) ambient coordinates and t the parameter for the deformation, consider the family

$$(xz + ty^2)z^2 + y^5 + x^{10} = 0 \tag{3.1}$$

In (3.1), for all t, $\mu = 84$. Of course, for all t, the multiplicity is 4.

It has been suggested by Wahl, in a private conversation, that perhaps a stronger form of the Zariski multiplicity questions hold, at least for deformations of two-dimensional singularities. Namely, under topologically trivial deformations, is there an induced equisingular deformation of the $PT_{V,p}$? Example (3.1) shows that Wahl's suggestion is not true exactly as stated. However, Theorem 3.1 below does hold. Perhaps one may omit the condition in Theorem 3.1 that the initial fiber be quasi-homogeneous.

THEOREM 3.1. *Let $\{V_t,p_t\}$ be a deformation of singularities whose initial fiber, at $t = 0$, is a two-dimensional isolated quasi-homogeneous singularity and such that the (V_t,p_t) are all homeomorphic to the initial (V_0,p_0). Then for all small t, CP_{V_t,p_t} is homeomorphic to the initial CP_{V_0,p_0}.*

PROOF. Start by considering just one two-dimensional singularity (V,p). $PT_{V,p}$ is also given as the proper transform of p via a single blow-up $\sigma'\colon V' \to V$ at p (or equivalently, at the maximal ideal m). Let $\sigma''\colon V'' \to V$ be obtained from σ' by normalizing V' to get V''. Let $\pi: M \to V$ be a resolution of V with $\pi^{-1}(m)$ locally principal. Let D be the maximal ideal cycle. The (reduction of) $\sigma''^{-1}(p)$ is obtained from the exceptional set A in M by identifying to a point those curves A_i such that $A_i \cdot D = 0$. $PC_{V,0}$ is obtained from (the set underlying) $\sigma''^{-1}(0)$ by a finite-to-one map, i.e. by undoing the normalization.

Let us examine $PC_{V,p}$ from the viewpoint of the resolution p and its deformations. Recall that the map σ from an exceptional set A with a locally principal $\pi^{-1}(m)$ onto $PC_{V,p}$ in P^2 is given by: for q in A, $\sigma(q) = (\pi^*(x) : \pi^*(y) : \pi^*(z))$.

By Theorem 2.5, the fibers (V_t,p_t) in the deformation have constant multiplicity. Hence the lowest order terms in the defining equation give a deformation of PT_{V_0,p_0}.

Recall that D denotes the maximal ideal cycle. Whenever there is no ambiguity, we shall drop the subscripts and denote (V_0,p_0) by (V,p). There are now various cases.

CASE 1: $a = b = c$. This is the homogeneous case. Since (V,p) has an isolated singularity, $PT_{(V,p)}$ is a nonsingular plane curve. The induced deformation of $PT_{(v,p)}$ is topologically trivial.

CASE 2: $a = b < c$. By the beginning of the proof of Proposition 2.3, on the minimal resolution, $\pi^{-1}(m)$ has no embedded points on the central curve A_0. In $PC_{V,p}$, A_0 is mapped to $\{z = 0\}$. By Proposition 2.4, for all t, A_0 is mapped to $\{z = 0\}$.

Now look near an arm $L = \Sigma A_i$, $1 \leqslant i \leqslant \ell$, as in (2.1). As in the proof of Case 3 of Proposition 2.3, $W_x \cdot A_\ell \geqslant 2$ and $W_y \cdot A_\ell \geqslant 2$ is impossible. Take $W_x \cdot A_\ell \leqslant 1$. If $W_x \cdot A_\ell = 0$, then x generates $\pi^{-1}(m)$ near L, and $A_i \cdot D = 0$ for all A_i in L. So the entire arm is mapped to a point in $PC_{V,p}$. Since the order of vanishing of x is minimal along the arm, for all t, the entire arm is mapped to a point in $PC_{V,p}$. On the other hand, if $W_x \cdot A_\ell = 1$, then $W_z \cdot A_\ell = 0$. In order for the blow-down map π to separate points in the fiber above ∞, z must vanish to order 1 on A_ℓ. Since x and z vanish to minimal order on the arm and $a < c$, x vanishes to order at most that of z on A_ℓ, i.e. x also vanishes to order 1 on A_ℓ. So σ maps all A_i in the arm other than A_ℓ to a point. $\sigma(A_\ell) = \{\alpha x + \beta y = 0\}$ for suitable constants α and β. Moreover, since A_ℓ is a P^1, the ratio $(x : z)$ gives a biholomorphic map of A_ℓ. Then, under deformation, A_ℓ is mapped to a deformation of $\{\alpha x + \beta y = 0\}$. There can be at most one arm which falls into this sub-case; for otherwise z fails to separate the points which in the fibers above the ∞ in two such arms.

CASE 3: $a < b = c$. σ maps A_0 to a point. Let us observe that on the minimal resolution, $\pi^{-1}(m)$ has no embedded points on A_0 away from an arm. Indeed, were W_x to meet A_0 at a point q not on an arm, then after a linear change between y and z, we could also have W_y meet A_0 at q. Since $b = c > a \geqslant 1$, z would fail to separate points in the fiber above q. Now consider an arm. Some linear combination $\alpha y + \beta z$ of y and z vanishes to order greater than 1 on A_ℓ and also satisfies $W_{\alpha y+\beta z} \cdot A_\ell \geqslant 2$. As in the proof of Case 3 of Proposition 2.3, $W_x \cdot A_\ell \leqslant 1$. If $W_x \cdot A_\ell = 0$, then σ maps the arm to a point. This is preserved under deformation. If $W_x \cdot A_\ell = 1$, then, because of separation of points in the fiber above ∞, z must vanish to order 1 on A_ℓ. By minimality for x and z, also x vanishes to order 1 on A_ℓ. So σ maps all A_i in the arm other than A_ℓ to a point. $\sigma(A_\ell) = \{\alpha y + \beta z = 0\}$. Under deformation, A_ℓ is mapped to a deformation of $\alpha y + \beta z = 0$. All of the $\{\alpha y + \beta z = 0\}$ which can occur in this subcase must be distinct. Otherwise, y and z could not separate the points in the fibers above the ∞ in two of the arms. So deformations preserve the (embedded) topology of $PT_{V,p}$.

CASE 4: $a < b < c$. σ maps A_0 to a point. On the minimal resolution, $\pi^{-1}(m)$ may have embedded points in A_0. Suppose then that W_x meets A_0 at a point q away from the arms. Then neither W_y nor W_z can meet A_0 at q, since y and z must separate points above in the fiber above q. Then σ maps a suitable blow-up of q to $\{z = 0\}$. $\{z = 0\}$ may have high multiplicity in $PT_{V,p}$. Since, under deformation, we may apply Proposition 2.4, this component of $PT_{V,p}$ deforms also to $\{z = 0\}$, with suitable multiplicity.

Now consider an arm L as in (2.1).

CASE 4.1.: $W_x \cdot A_\ell = 0$. Then σ maps the arm to a point. This is preserved under deformation.

CASE 4.2.: $W_x \cdot A_\ell = 1$. Then x generates $\pi^{-1}(m)$ along the arm except at ∞. σ maps all A_i, $i \neq \ell$, in the arm to a point. Let us observe that $W_y \cdot A_\ell = 0$. Namely, if $W_y \cdot A_\ell \geqslant 1$, $W_z \cdot A_\ell = 0$ and, in order to separate points in the fiber above ∞, z must vanish to order 1 on A_ℓ. x and z vanish to minimal order along the arm. Hence also x vanishes to minimal order 1 on A_ℓ. Now compute the minimal orders of vanishing for z. Recall that $c \geqslant a + 2$. To D_x, we first add $A_0 + A_1 + \cdots + A_{\ell-1}$. This yields a divisor D with $D \cdot A_\ell = 0$. To reach D_z, we must add at least one additional A_0 and then increase the divisor until $D_z \cdot A_i \leqslant 0$ for all A_i in the arm. But it is then impossible for A_ℓ to have coefficient 1 in D_z. So $W_y \cdot A_\ell = 0$. Blow up any embedded points in the arm. σ now maps the arm onto $\{z = 0\}$. This will remain true under deformation.

CASE 4.3: $W_x \cdot A_\ell \geqslant 2$.

CASE 4.3.1: $W_y \cdot A_\ell = 0$. By Lemma 2.2, D_y is minimal along the arm. Since $a < b$, σ cannot map the arm to just a point. Since $b < c$, after needed blow-ups for embedded points, σ maps the arm onto $\{z = 0\}$. Also under deformation, the arm is mapped to $\{z = 0\}$.

CASE 4.3.2: $W_y \cdot A_\ell > 0$. Separation of points requires that $W_z \cdot A_\ell = 0$ and also that z vanish to order 1 on A_ℓ. Moreover, Case 4.3.2 can occur for at most one arm. The proof of Case 3 of Proposition 2.3 implies that $W_y \cdot A_\ell = 1$. Then D_y and D_z are both minimal near the arm. Then y vanishes to order 1 on A_ℓ. Also $c = b + 1$.

CASE 4.3.2.1: x vanishes to order 1 on A_ℓ. Then D_x is also minimal along the arm, as in the proof of Lemma 2.2 using the truncated "arm" ΣA_i, $1 \leq i \leq \ell - 1$. Then x generates $\pi^{-1}(m)$ along the arm except at ∞. σ maps A_ℓ in a one-to-one manner onto, for suitable constants r and k, $r \neq 0$, $\{ry^k = xz^{k-1}\}$. Since (V,p) is quasi-homogeneous, in fact $k = b - a + 1$. The image of the arm under σ is thus $\{ry^{b-a+1} = xz^{b-a}\}$. Since D_x, D_y, and D_z are all minimal along the arm, this image deforms in an equisingular manner.

CASE 4.3.2.2: x vanishes to order greater than one on A_ℓ. Since $a < b$, there must be an embedded point for $\pi^{-1}(m)$ on the arm. σ maps the blow-up of these embedded points to $\{z = 0\}$. σ maps A_ℓ to $\{x = 0\}$. In all, σ maps the arm onto $\{xz = 0\}$.

In contrast to all of the other cases, Case 4.3.2.2 need not be preserved under deformation, e.g. Example (3.1). If Case 4.3.2.2 does occur (and it can occur only along at most one arm), then $PC_{V,p}$ is precisely $\{xz = 0\}$. Topologically, $PC_{V,p}$ is the wedge of two real 2-spheres. Consider a deformation of (V,p). The central curve and the arms not covered by Case 4.3.2.2 contribute at most $\{z = 0\}$ to $PC_{V,p}$. Look near the arm of Case 4.3.2.2. D_y and D_z are fixed near the arm. In particular, all of the arm (including any blown-up embedded points) except A_ℓ is mapped into $\{z = 0\}$. If, after deformation, x continues to vanish to order greater than 1 on A_ℓ, then $PC_{V,p}$ remains $\{xz = 0\}$. Finally, if, after deformation, x vanishes to order 1 on A_ℓ, then σ still maps A_ℓ in a one-to-one manner and the image $\sigma(A_\ell)$ meets the set $\{z = 0\}$ in just one point. So to conclude the proof of Theorem 3.1, we just need to show that for $t \neq 0$, $\{z = 0\}$ is still part of PC_{V_t,p_t}. There are two last cases for $t \neq 0$. First, if for some A_i, $1 \leq i \leq \ell - 1$, $D_{x,t} \cdot A_{i,t} < 0$, then x_t cannot generate $\pi^{-1}(m_t)$ near $A_{i,t}$ and this part of the arm must be mapped by σ_t onto $\{z = 0\}$. Second, if the first case does not hold, then $D_{x,t}$ is minimal on the truncated arm $\Sigma A_{i,t}$, $1 \leq i \leq \ell - 1$. Then the coefficient of A_1 in $D_{x,0}$ is strictly greater than the coefficient of A_1 in $D_{x,t}$. Then $D_{x,t}$ meets $A_{0,t}$ in at least one point

q_t near, but not on, the arm. This point q_t is an embedded point for $\pi^{-1}(m_t)$ whose blow-up is mapped onto $\{z = 0\}$ by σ_t.

REFERENCES

1. A'Campo, N., La fonction zeta d'une monodromie, Comment. Math. Helv. 50 (1975), 233-248.

2. Artin, M., On isolated rational singularities of surfaces, Amer. J. Math. 88 (1966), 129-136.

3. Briancon, J. and Speder, J., La trivialité topologique n'implique pas les conditions de Whitney, C. R. Acad. Sci. Paris, Set A, 280 (1970), 365.

4. Greuel, G.-M., Constant Milnor number implies constant multiplicity for quasihomogeneous singularities, Preprint.

5. Laufer, H., Weak simultaneous resolution for deformations of Gorenstein surface singularities, Proc. Symp. Pure Math. 40 Part 2 (1983), 1-29.

6. Laufer, H., Strong simultaneous resolution for surface singularities, Adv. Studies in Pure Math. 8 (1986), 207-214.

7. Le, D. T. and Teissier, B., Report on the problem session, Proc. Symp. Pure Math. 40 Part 2 (1983), 105-116.

8. O'Shea, D., Topologically trivial deformations of isolated quasihomogeneous hypersurface singularities are equimultiple, Preprint.

9. Orlik, P. and Wagreich, P., Isolated singularities of algebraic surfaces with **C***-action, Ann. Math. 93 (1971), 205-228.

10. Pinkham, H., Normal surface singularities with **C***-action, Math. Ann. 227 (1977), 183-193.

11. Varchenko, A., A lower bound for the codimension of the stratum μ-constant in terms of the mixed Hodge structure, Vestnik Mos. Univ. Seria Matem. 37, No. 6 (1982), 28-31.

12. Wagreich, P., Elliptic singularities of surfaces, Amer. J. Math. 92 (1970), 419-454.

13. Whitney, H., Tangents to an analytic variety, Ann. Math. 81 (1965), 496-549.

14. Yau, S.S.-T., Normal singularities of surfaces, Proc. Symp. Pure Math., Vol. 32 (1978), 195-198.

15. Zariski, O., Some open questions in the theory of singularities, Bull. A.M.S. 77 (1971), 481-491.

STATE UNIVERSITY OF NEW YORK AT STONY BROOK
LONG ISLAND, NY 11794

Contemporary Mathematics
Volume **90**, 1989

ON THE WEAK SIMULTANEOUS RESOLUTION OF A NEGLIGIBLE TRUNCATION OF THE NEWTON BOUNDARY

Mutsuo Oka

Dedicated to Professor N. Iwahori for his 60th *Birthday*

§1. INTRODUCTION

Let $f(z_1, \ldots, z_n, t)$ be an analytic function defined in an open set $U \times D$ where U is a neighborhood of $\vec{0} \in \mathbf{C}$ and D is an open disk in $\mathbf{C}$ containing the unit interval I and let $V = \{(z,t) \in U \times D : f(z,t) = 0\}$ and let $\pi : V \to D$ be the projection. We also use the notation $f_t(z) = f(z,t)$ when t is fixed. We assume that for each $t \in D$, the hypersurface $V_t = \pi^{-1}(t) = f_t^{-1}(0)$ has an isolated singularity at the origin and that the Milnor number $\mu(f_t)$ is constant $(= \mu(f_0))$. The question which we are interested in this paper is: *Are* V_t $(t \in D)$ *topologically equivalent to* V_0? The assertion is true for $n \neq 3$ by Lê and Ramanujam [10]. Thus the question is open only for $n = 3$. The paper [6] seems to have some gaps in the proof. If $\mu^*(f_t)$ $(t \in D)$ are constant, the assertion is true for Teissier [8]. In particular, if the Newton boundary $\Gamma(f_t)$ is non-degenerate in the sense of Kouchnirenko [1] and if $\Gamma(f_t) = \Gamma(f_0)$, the assertion is true by [8] and [4]. If there is a uniform stable radius for the Milnor fibration of f_t at the origin, the assertion is also true. However there is an example of a μ-constant family which has no uniform stable radius. See Appendix.

0271-4132/89 $1.00 + $.25 per page

$\Psi : \widetilde{V} \to V$ is called a *weak simultaneous resolution* of $\pi : V \to D$ if the following conditions are satisfied.

(i) Ψ is a proper modification.

(ii) $\pi\circ\Psi : \widetilde{V} \to D$ is a flat map.

(iii) $\Psi : \widetilde{V}_t \to V_t$ is a resolution of V_t.

(iv) Let $E = \Psi^{-1}(\vec{0} \times D)$. Then $\pi\circ\Psi : E \to D$ is simple.

See Teissier [9] and Laufer [2] for further detail. In our case, the existence of a weak resolution is equivalent to the topological stability of $\{V_t\}$ by Theorem 6.4 of [2].

Briancon and Speder gave the following example of μ-constant family which is not a μ^*-constant family.

$$z_1^5 + tz_1z_3^6 + z_2z_3^7 + z_2^{15} = 0. \tag{1.1}$$

This has a weak simultaneous resolution by [7,12]. The purpose of this paper is to generalize this in the following case.

Let $f_1(z) = \Sigma_\nu b_\nu z^\nu$ be an analytic function defined in a neighbourhood of the origin. We assume that f_1 has a non-degenerate Newton boundary which is convenient and let A be a vertex of $\Gamma(f_1)$. Let $f_t(z) = f_1(z) - (1-t)\, b_A z^A$. We say that $f_t(z)$ is a *negligible truncation* if the following conditions are satisfied.

(i) There exists an open disk D in $\mathbf{C}$ containing the unit interval [0,1] such that $f_t(z)$ has a non-degenerate Newton boundary $\Gamma(f_t)$ for each $t \in D$.

(ii) f_0 is convenient and $\Gamma_-(f_1)$ is a proper subset of $\Gamma_-(f_0)$.

(iii) $\nu(\Gamma_-(f_1)) = \nu(\Gamma_-(f_0))$.

Here $\nu(W)$ is the Newton number of the polyhedron W. See §2 for the definition. f is called *convenient* if the intersection of $\Gamma(f)$ and each coordinate ax is non-empty.

The example of Briancon-Speder does not satisfy the convenience condition in (ii), but this can be modified by adding y^N for a sufficiently large N for which the isomorphism class of f_t

does not change.

Let $f_t(z)$ be a negligible truncation and let $\pi : V \to D$ be as above. Then the following is the result.

THEOREM (1.2). *$\pi : V \to D$ has a weak simultaneous resolution.*

In general, the family of a negligible truncation is not μ^*-constant.

§2. POSITIVITY OF THE NEWTON NUMBERS IN $\mathbf{R}^3$

Let W be a polyhedron in $\mathbf{R}^n_+ = \{(x_i) \in \mathbf{R}^n;\ x_i \geqslant 0\}$. Recall that *the Newton number* $\nu(W)$ is defined by

$$\sum_I (-1)^{n-|I|}\, |I|!\ |I|\text{-dim.volume } (W^I)$$

where the sum is taken for every subset I of $\{1, \ldots, n\}$ and $W_I = \{(x_i) : x_i = 0 \text{ for } i \notin I\}$. The corresponding term for $I = \emptyset$, is $(-1)^n$ for 0 according to $\vec{0} \in W$ or $\vec{0} \notin W$. Let $W = W_1 \cup W_2$ be a polyhedral decomposition of W. Then we have

$$\nu(W) = \nu(W_1) + \nu(W_2) - \nu(W_1 \cap W_2).$$

Now we consider the case that $n = 3$ and W is a three dimensional simplex with integral vertices A, B, C and D. Let $A = (a_1,a_2,a_3)$, ..., $D = (d_1,d_2,d_3)$. We assume that $\vec{0} \notin W$. Let h be the number of I's such that $|I| = 2$ and $\dim W^I = 2$. Let ℓ be the number of i's such that $\dim W^{\{i\}} = 1$.

LEMMA (2.2). *Assume that $\vec{0}$ is not in W. Then $\nu(W) \geqslant 0$. The equality holds if and only if one of the following conditions is satisfied (up to a permutation of the coordinates).*

(i) *$h = 2$, $\ell = 1$. We assume that A, C are in $\mathbf{R}^{\{3\}}$, B is in $\mathbf{R}^{\{1,3\}}$ and D are in $\mathbf{R}^{\{2,3\}}$. Then either $b_1 = 1$ or $d_2 = 1$.*

(ii) $h = 1$ *and* $\ell = 0$. *Assume that* A, B *and* C *are in* $\mathbf{R}^{\{1,3\}}$. *Then* $d_2 = 1$.

PROOF. If $h = 0$, it is obvious that $\nu(W) > 0$. Assume that $h = 1$ and A, B and C be in $\mathbb{R}^{\{1,3\}}$. Then we have

$$\nu(W) \geq sd_2 - s = s(d_2 - 1) \geq 0$$

where $s = 2$ volume $W^{\{1,3\}}$. The first equality holds if and only if $\ell = 0$. Thus $\nu(W) = 0$ if and only if $d_2 = 1$ and $\ell = 0$. Assume that $h = 2$ and $b_2 = d_1 = 0$ and $a_i = c_i = 0$ for $i = 1,2$ and that $c_3 > a_3$. Then we have

$$\nu(W) = (c_3 - a_3)(d_2 b_1 - b_1 - d_2 + 1) = (c_3 - a_3)(d_2 - 1)(b_1 - 1).$$

Thus $\nu(W) = 0$ if and only if either b_1 or d_2 is 1. The case $h = 3$ is eliminated by the assumption on W.

REMARK (2.3). The analogous assertion in Lemma (2.2) is not true for the higher dimensions. For example, let $n = 4$ and let W be the simplex which is spanned by $A = (1+t,0,0,0)$, $B = (1,0,0,0)$, $C = (1,2,3,0)$, $D = (1,3,2,0)$ and $E = (1,1,1,1)$. Then $\nu(W)$ is $-t$.

§3. NEGLIGIBLE TRUNCATIONS

Let $f_1(z_1,z_2,z_3) = \Sigma_\nu b_\nu z^\nu$ be an analytic function defined in a neighbourhood of the origin and assume that $f_1(z)$ has a non-degenerate Newton boundary $\Gamma(f_1)$. We also assume that f_1 is convenient. Namely $\Gamma(f_1)^{\{i\}}$ is non-empty for each i. Let $\Gamma_-(f_1)$ be the cone of $\Gamma(f_1)$ with the origin i.e., $\Gamma_-(f_1) = \{rx;\ x \in \Gamma(f_1),\ 0 \leq r \leq 1\}$. Recall that $\nu(\Gamma_-(f_1)) = \mu(f_1)$ by Kouchnirenko [1] where $\mu(f_1)$ is the Milnor number of f_1 at $\vec{0}$. Let $A = (a_1,a_2,a_3)$ be a vertex of $\Gamma(f)$ and let $f_t(z) = f(z) - (1-t)b_A z^A$ be a negligible truncation. Let W be the closure of $\Gamma(f_0) - \Gamma(f_1)$ and let $\hat{W}$ be the cone of W with vertex A. Then it is easy to see that $\Gamma_-(f_0) = \Gamma_-(f_1) \cup \hat{W}$. Thus the assumption that $\nu(\Gamma_-(f_1)) = \nu(\Gamma_-(f_0))$ and (2.1) implies $\nu(\hat{W}) = 0$. We can triangulate W into finite two-simplices so that any vertex of W which is on a coordinate axis is contained in a unique two-simplex. Then by the equality $\nu(\hat{W}) = 0$ and (2.1) and Lemma (2.2), we conclude that W is a simplex spun by three integral vertices B, C, D and $\hat{W}$ satisfies one of the conditions in Lemma (2.2). We say that $f_t(z)$ is *a negligible truncation of type* (i) or *of type* (ii) when $\hat{W}$ is of type (i) or of type (ii) of Lemma (2.2) respectively. Before we proceed to prove Theorem (1.2), we give some examples.

EXAMPLE (3.1) (E_7, type (i)). Let $f_t(x,y,z) = x^2 + y^3 + yz^3 + tz^5 + z^k$ for $k > 5$. Then f_t is a μ^*-constant family.

More generally, assume that $f_t(z)$ be a negligible truncation of type (i). Then one can show that $f_t(z)$ is a μ^*-constant family.

EXAMPLE (3.2) (Type (ii), Briancon-Speder). Let

$$f_t(x,y,z) = z^5 + ty^6z + y^7x + y^k + x^{15} \quad (k \geqslant 8)$$

This is not a μ^*-constant family.

EXAMPLE (3.3) (Type (ii)). Let

$$f_t(x,y,z) = x^8 + y^\ell + z^\ell + tx^5z^2 + x^3yz^3$$

where $\ell \geqslant 16$. Then μ^* is not constant. In fact, $\mu(f_t) = 2\ell^2 + 18\ell + 7$ and the Milnor numbers of the generic hyperplane sections of f_1 and f_0 are $2\ell + 23$ and $2\ell + 24$ respectively. These examples show that negligible truncations of type (ii) are not generally μ^*-constant.

§4. RESOLUTION BY TOROIDAL EMBEDDING THEORY

In this section, we briefly recall the resolution of $V_1 = f_1^{-1}(0)$. The dual vector space of $\mathbf{R}^3$ can be identified canonically with itself. To distinguish vectors in $\mathbf{R}^3$ and in the dual space, we write dual vectors by column vectors. Let N^+ be the subset of the dual vectors which are nonnegative. Let P be a vector in N^+. We denote by $\Delta(P)$ the face of $\Gamma_+(f_1)$ where P takes its minimal value $d(P)$ as a function on $\Gamma_+(f)$. Here $\Gamma_+(f_1)$ is equal to $\{rx\ ;\ x \in \Gamma(f_1),\ r \geqslant 1\}$ by the convenience of f_1. We introduce an equivalence relation $\sim$ in N^+ by $P \sim Q$ if and only if $\Delta(P) = \Delta(Q)$. This gives a conical polyhedral subdivision of N^+ which we call *the dual Newton diagram* and denote it by $\Gamma^*(f_1)$. We can subdivide $\Gamma^*(f_1)$ into a cone over a simplicial complex Σ^* such that each two simplex $\sigma = (P_1,P_2,P_3)$ of Σ^* is a unimodular matrix. We call Σ^* a *unimodular simplicial subdivision* of $\Gamma^*(f_1)$. See Varchenko [11] and Oka [5] for details. Let Σ^* be a unimodular simplicial subdivision. For each two simplex $\sigma = (P_1,P_2,P_3) = (p_{ij})$ of Σ^*, we associate a three-dimensional affine space $\mathbf{C}^3_\sigma$ with coordinates $\mathbf{y}_\sigma = (y_{\sigma 1},y_{\sigma 2},y_{\sigma 3})$ and the birational morphism $\pi_\sigma : \mathbf{C}^3_\sigma \to \mathbf{C}^3$ which is defined by $\pi_\sigma(\mathbf{y}_\sigma) = (z_1,z_2,z_3)$ where

$$z_i = y_{\sigma 1}^{p_{i1}}\, y_{\sigma 2}^{p_{i2}}\, y_{\sigma 3}^{p_{i3}} \quad (i = 1, 2, 3).$$

Then we glue $\mathbf{C}^3_\sigma$'s in a canonical way to get a smooth complex manifold X and proper birational morphism $\hat{\pi} : X \to \mathbf{C}^3$. Let $\tilde{V}_1$ be the proper transform of V_1 and let $\pi : \tilde{V}_1 \to V_1$. Then $\pi : \tilde{V}_1 \to V_1$ gives a resolution of the singularity $\vec{0}$ of V_1 if $\Gamma(f_1)$ is nondegenerate. In $\mathbf{C}^3_\sigma$, $\tilde{V}_1$ is defined by $f_{1,\sigma}(\mathbf{y}_\sigma) = 0$ where

$$f_{1,\sigma}(\mathbf{y}_\sigma) = f_1(\pi_\sigma(\mathbf{y}_\sigma)) \Big/ \prod_{i=1}^{3} y_{\sigma i}^{d(P_i)} .$$

For each strictly positive vector P such that $\dim \Delta(P) \geqslant 1$, there is a corresponding exceptional divisor $E(P)$. Suppose that $P = P_1$. Then $E(P)$ is defined in $\mathbf{C}^3_\sigma$ by

$$h_\sigma(\mathbf{y}_\sigma) = f_{1,\Delta(P)}(\mathbf{y}_\sigma) \Big/ \prod_{i=1}^{3} y_{\sigma i}^{d(P_i)} .$$

See §10 of [11] or §4 of [5].

§5. PROOF OF THEOREM (1.2)

Let $f_1(z) = \Sigma_\nu b_\nu z^\nu$ and let $f(z,t) = f_1(z) - (1-t)b_A z^A$ be a negligible truncation as in Theorem (1.2) of §1. Let $V = \{(z,t) \in \mathbf{C}^3 \times D \; ; \; f_t(z) = 0\}$. Let Σ^* be a unimodular simplicial subdivision. We apply the mapping $\pi : X \to \mathbf{C}^3$ simultaneously to f_t to resolve the singularities of V. Namely let $\hat{\Pi} : X \times D \to \mathbf{C}^3 \times D$ be the projection defined by $\hat{\Pi}(\mathbf{y},t) = (\hat{\pi}(\mathbf{y}),t)$ and let $\widetilde{V}$ be the proper transform of V. Let $\omega : V \to D$ be the projection into D. We denote the restriction of $\hat{\Pi}$ to $\widetilde{V}$ by Π. Let $\widetilde{V}_t$ be the inverse image of $V_t = \omega^{-1}(t)$ by Π. We are going to show that $\Pi : \widetilde{V} \to V$ is a simultaneous resolution of V if we choose a suitable unimodular simplicial subdivision Σ^*.

Let $\hat{W}$ and A, B, C and D be as in §3. We will prove Theorem (1.2) assuming $\hat{W}$ is of type (ii) of Lemma (2.2). The case of (i) can be proved in a similar way. Let $A = (a_1,0,a_3)$, $B = (b_1,0,b_3)$, $C = (c_1,0,c_3)$ and $D = (d_1,1,d_3)$. Let Δ_1 and Δ_2 be the faces of $\Gamma(f_1)$ which are spanned by A, B, D and A, C, D respectively. Let $P = {}^t(p_1,p_2,p_3)$ and $Q = {}^t(q_1,q_2,q_3)$ be the respective weight vectors of Δ_1 and Δ_2. By the definition, $\Delta(P) = \Delta_1$ and $\Delta(Q) = \Delta_2$. (Strictly speaking, Δ_1 may be a subset of $\Delta(P)$ and Δ_2 may be a subset of $\Delta(Q)$.) By the definition, P and Q must satisfy the following.

(5.1) $$p_1a_1 + p_3a_3 = p_1b_1 + p_3b_3 = p_1d_1 + p_2 + p_3d_3 < p_1c_1 + p_3c_3$$

(5.2) $$q_1a_1 + q_3a_3 = q_1c_1 + q_3c_3 = q_1d_1 + q_2 + q_3d_3 < q_1b_1 + q_3b_3.$$

From (5.1) and (5.2), we have

(5.3) $$p_2 = p_1(a_1 - d_1) + p_3(a_3 - d_3) \text{ and}$$

(5.4) $$q_2 = q_1(a_1 - d_1) + q_3(a_3 - d_3).$$

As P and Q are primitive vectors, (5.3) and (5.4) implies

(5.6) $$\text{G.C.D.}(p_1,p_3) = \text{G.C.D.}(q_1,q_3) = 1$$

By the geometric consideration of $\Gamma(f_1)$, we may assume that $b_1 > a_1$

$> c_1$, or equivalently $b_3 < a_3 < c_3$. From (5.1), (5.2) and (5.6), we have

$$(5.7) \qquad (b_1 - a_1) = rp_3, \quad (a_3 - b_3) = rp_1$$

$$(5.8) \qquad (a_1 - c_1) = sq_3, \quad (c_3 - a_3) = sq_1$$

for some positive integers r and s. Thus by the inequality of (5.1), we have

$$(5.9) \qquad q_1p_3 - q_3p_1 > 0$$

Let $R = {}^t(0,1,0)$. Then the interior $\mathrm{Int}(T)$ of the triangle T with vertices P, Q and R is an equivalence class in $\Gamma^*(f_1)$. Namely for a dual vector S, $S \in \mathrm{Int}(T)$ if and only if $\Delta(S) = \{A\}$. By (5.6), we have $\det(P,R) = \det(Q,R) = 1$. Thus we do not need any other vertex on the line segment $\overline{PR}$ and $\overline{QR}$. On the other hand, using (5.3) and (5.4), we can see that

$$(5.10) \qquad \det(P,Q) = \det(P,Q,R) = q_1\ p_3 - q_3\ p_1.$$

This implies the following. Let $T_1, \ldots, T_k$ be any unimodular subdivision of $\overline{PQ}$ with $T_0 = P$ and $T_{k+1} = Q$. Then two-simplices (T_i,T_{i+1},R) $(i = 0, \ldots, k)$ are already unimodular. Thus we do not need any other vertices in T to subdivide $\Gamma^*(f_1)$. We take a unimodular simplicial subdivision Σ^* whose restriction on the two-simplex T is as described above. Now we consider $\Pi : \widetilde{V} \to V$ which is associated to Σ^*. It is easy to see that $\widetilde{V} - \widetilde{V}_0$ and $\widetilde{V}_t$ $(t \neq 0)$ are non-singular. Let $\sigma = (P_1,P_2,P_3)$. Then in the coordinate chart $\mathbf{C}^3_\sigma \times D$, $\widetilde{V}$ is defined by

$$f_\sigma(\mathbf{y}_\sigma,t) = f(\pi_\sigma(\mathbf{y}_\sigma),t)/\prod_{i=1}^{3} \mathbf{y}_{\sigma i}^{d(P_i)} = 0.$$

and $E(P_1)$ is defined by

$$h_\sigma(\mathbf{y}_\sigma,t) = f_{\Delta(P_1)}(\pi_\sigma(\mathbf{y}_\sigma),t)/\prod_{i=1}^{3} \mathbf{y}_{\sigma i}^{d(P_i)} = 0.$$

This is a polynomial of $y_{\sigma 2}$, $y_{\sigma 3}$ and t and

$$f_\sigma(\mathbf{y}_\sigma,t) \equiv h_\sigma(\mathbf{y}_\sigma,t) \quad \text{modulo } (y_{\sigma 1})\mathbf{C}[y_{\sigma 1},y_{\sigma 2},y_{\sigma 3}].$$

Let $\xi(\sigma) = \cap_{i=1}^{3}\Delta(P_i)$. Then the constant term of $h_\sigma(\mathbf{y}_\sigma,t)$ with respect to $\mathbf{y}_\sigma$ is $b_{\xi(\sigma)}$ if $\xi(\sigma) \neq A$. Thus in this case, $\tilde{V}$ and $\tilde{V}_0$ is non-singular and $\Pi : E \to D$ is simple in this chart where $E = \Pi^{-1}(\bar{0} \times D)$. Assume that $\sigma = (T_i,T_{i+1},R)$ ($T_0 = P$ and $T_{k+1} = Q$). (This is the chart to be studied carefully.) Then $E(T_i)$ is defined by

$$\begin{aligned} h_\sigma(y_{\sigma 2},y_{\sigma 3},t) &= b_A\, t + b_D\, y_{\sigma 3} = 0, \quad (0 < i < k-1) \\ &= b_A\, t + b_D\, y_{\sigma 3} + b_B\, y_{\sigma 2}^r = 0 \quad (i = 0 \ \text{ or } \ i = k-1) \end{aligned}$$

This can be easily seen by a direct calculation from the formula for h_σ above. Thus $E(T_i)$ is non-singular by the existence of the linear term $b_D y_{\sigma 3}$. By the same reason, $\tilde{V}_0$ and V are non-singular and $\Pi : E \to D$ is simple over D. The smoothness of $E(Q)$ is proved in a similar way. Note that the vanishing order of the ideal $\underline{I}(E_t)$ of the non- reduced exceptional divisor E_t along $E(T_i)$ jumps up at $t = 0$. This implies that the above resolution is not strongly simultaneous.

Now we consider the case of a negligible truncation of type (i). Then by the same notation as above, we have to replace $A = (0,0,a_3)$, $B = (b_1,0,b_3)$, $C = (0,0,c_3)$ and $D = (0,1,d_3)$. Also the weight vector Q is simply ${}^t(1,0,0)$. The rest of the argument is completely parallel to the above argument. This completes the proof of Theorem (1.2).

APPENDIX

In this appendix, we give an example of μ-constant family which has no uniform stable radius. This example is due to J. Briancon. Let

$$f(x,y,z,t) = z^5 + tzy^6 + (x + y + z)y^7 + (x + y + z)^{15}$$

be a modified family of (1.1). In this family, $\{x = 0\}$ gives a generic plane section and $\mu_t^{(2)} = 26$ and $\mu_0^{(2)} = 28$. Thus using the Curve Selection Lemma ([3]), we can find an analytic curve $p(s) = (0,y(s),z(s),t(s))$ $(0 \leqslant s \leqslant 1)$ such that

(i) $t(s) = s^c$ $(c > 0)$,

(ii) $\dfrac{\partial f}{\partial y}(p(s)) = \lambda(s)\overline{y(s)}$, $\dfrac{\partial f}{\partial z}(p(s)) = \lambda(s)\overline{z(s)}$ and

(iii) $f(p(s)) = 0$, $p(0) = 0$ and $p(s) \neq 0$ for $s \neq 0$.
Here $\lambda(s)$ is a suitable Laurent series.

Let $y(s) = \alpha s^a +$ (higher) and $z(s) = \beta s^b +$ (higher). We use the notations

$$f_x(s) = \frac{\partial f}{\partial x}(p(s)), \quad f_y(s) = \frac{\partial f}{\partial y}(p(s)) \quad \text{and} \quad f_z(s) = \frac{\partial f}{\partial z}(p(s)).$$

ASSERTION. $\mathrm{ord}_s f_x(s) \geqslant \min(\mathrm{ord}_s(f_y(s)), \mathrm{ord}_s(f_z(s)))$.

Assume the assertion for a while. Assume that $\mathrm{ord}_s f_x(s) \geqslant \mathrm{ord}_s f_y(s)$ for instance. We consider the μ-constant family $h(x,y,z,s) = f(x,y - \epsilon(s)x,z,s^c)$ where $\epsilon(s) = f_x(s)/f_y(s)$. It is easy to see that $h(p(s)) = 0$ and

$$\frac{\partial h}{\partial x}(p(s)) = 0, \quad \frac{\partial h}{\partial y}(p(s)) = \lambda(s)\overline{y(s)} \quad \text{and} \quad \frac{\partial h}{\partial z}(p(s)) = \lambda(s)\overline{z(s)}.$$

This says that the family $h(x,y,z,s)$ has no uniform stable radius.

PROOF OF ASSERTION. Note that $y(s) \neq 0$ and $z(s) \neq 0$. Let $g(y,z,t) = f(0,y,z,t)$. The Newton boundary $\Gamma(g)$ of g as a function of three variables is non-degenerate and consists of a 2-simplex with vertices $P = (8,0,0)$, $Q = (0,5,0)$ and $R = (6,1,1)$. Let Δ be the face of $\Gamma(g)$ where the linear function "$ay + bz + ct$" on $\Gamma_+(g)$ takes its minimal value d. As $\partial f/\partial x = y^7 + 15(x+y+z)^{14}$, $\mathrm{ord}_s f_x(s) \geqslant 7a$. On the other hand, considering the initial terms of $f(p(s)) \equiv 0$ and $(df/ds)(p(s)) = 0$, we have

(i) $g_\Delta(u) = 0$ and

(ii) $$\frac{\partial g_\Delta}{\partial y}(u)a\alpha^{a-1} + \frac{\partial g_\Delta}{\partial z}(u)b\beta^{b-1} + \frac{\partial g_\Delta}{\partial t}(u)c = 0 \quad \text{where } u = (\alpha,\beta,1).$$

(i) implies that $\dim \Delta > 0$.

CASE 1. $\Delta = \{P,Q,R\}$. Then $8a = 5b = 6a + b + c = d$ and $d - a$, $d - b \leqslant 7a$. Thus the assertion is immediate from the above equation (ii) and the non-degeneracy of $\Gamma(g)$.

CASE 2. $\Delta = \{Q,R\}$. Then $\text{ord}_s f_y(s) = d - a \leqslant 7a$.

CASE 3. $\Delta = \{P,R\}$. Then $d = 8a = 6a + b + c < 5b$. If $\text{ord}_s f_y(s) > d - a = 7a$, then $\text{ord}_s f_z(s) = d - b < 7a$ as $b > a$.

CASE 4. $\Delta = \{P,Q\}$. Similarly we get either $\text{ord}_s f_y(s) = d - a = 7a$ or $\text{ord}_s f_z(s) = d - b < 7a$. This completes the proof of the assertion.

REFERENCES

[1] A. G. Kouchnirenko, Polyèdres de Newton et Nombres de Milnor, Inventiones Math., **32** (1976), 1-32.

[2] H. B. Laufer, Weak simultaneous resolution for deformations of Gorenstein surface singularities, Proceedings of Symposia in Pure Math., **40 part 2** (1983), 1-29.

[3] J. Milnor, Singular Points of Complex Hypersurface, **61**, Princeton Univ. Press, Princeton, 1968.

[4] M. Oka, On the topology of the Newton boundary II, J. Math. Soc. Japan, **32** (1980), 65-92.

[5] M. Oka, On the Resolution of Hypersurface Singularities, Advanced Study of Pure Mathematics, **8** (1986), 405-436.

[6] B. Perron, <<μ-constant>> implique <<type topologique constant>> en dimension trois, C.R. Acad. Sci. Paris, **295** (1982), 735-738.

[7] H. Pinkham, Deformations of normal surface singularities with **C*** action, Math. Ann., **232** (1978), 65-84.

[8] B. Teissier, Cycles évanescents, sections planes et conditions de Whitney, Asterisque, **7-8** (1973), 285-362.

[9] B. Teissier, Resolution simultanée I, II, Seminaire sur les Singularités des Surfaces, **777**, Springer-Verlag, Berlin-Heidelberg-New York, 1980, 71-146.

[10] Lê Dũng Trang and C. P. Ramanujam, The Invariance of Milnor's Number Implies the Invariance of the Topological Type, Amer. J. Math., **98** (1976), 67-78.

[11] A. N. Varchenko, Zeta-Function of Monodromy and Newton's Diagram, Inventiones Math., **37** (1976), 253-262.

[12] J. Wahl, Equisingular deformations of normal surface singularities, Ann. of Math., **104 (2)** (1976), 325-356.

DEPARTMENT OF MATHEMATICS
FACULTY OF SCIENCES
TOKYO INSTITUTE OF TECHNOLOGY
OH-OKAYAMA, MEGURO-KU, TOKYO 152

Contemporary Mathematics
Volume **90**, 1989

BASIC DERIVATIONS FOR UNITARY REFLECTION GROUPS

Peter Orlik*

§1. INTRODUCTION

Let V be a complex vector space of dimension ℓ. A hyperplane in V is a vector subspace of codimension 1. An *arrangement* $A = (A,V)$ in V is a finite set of hyperplanes. Let S be the C-algebra of polynomial functions on V. Let Der_S be the S-module of derivations of S. A non-zero element $\theta \in \mathrm{Der}_S$ is homogeneous of degree b if $\theta(S_p) \subset S_{p+b}$. This makes Der_S a graded S-module. For each $H \in A$ choose a linear form $\alpha_H \in V^*$ with $\ker(\alpha_H) = H$. Define $Q \in S$ by

$$Q = \Pi\, \alpha_H. \tag{1.1}$$

The polynomial Q is uniquely determined by A up to a constant multiple. Define

$$D_S(A) = \{\theta \in \mathrm{Der}_S \mid \theta Q \in QS\}. \tag{1.2}$$

Then $D_S(A)$ is a graded submodule of Der_S.

The arrangement A is *free* if $D_S(A)$ is a free S-module. If A is free then there exists a basis for $D_S(A)$ as S-module consisting of ℓ homogeneous elements $\theta_1, \ldots, \theta_\ell$.

*This work was supported in part by the National Science Foundation.

Let $G \subset \mathrm{GL}(V)$ be a finite group generated by reflections [9]. Let A be the set of reflecting hyperplanes. Then A is a *reflection arrangement.* Although it follows from [5] and work of Terao [10] that every reflection arrangement is free, an explicit basis is not known for every reflection arrangement.

Let $R = S^G$ be the algebra of G-invariant polynomials. By Chevalley's theorem there exists a set $B = \{f_1, ..., f_\ell\}$ of homogeneous polynomials such that $R = \mathbb{C}[f_1, ..., f_\ell]$. We call B a *basic set* for G and f_i *basic invariants* for G. The degrees $d_i = \deg f_i$ are uniquely determined by G. Let $\mathrm{Der}_S{}^G$ be the R-module of G-invariant derivations. We showed [5] that $\mathrm{Der}_S{}^G$ is a free R-module of rank ℓ. We call a basis $D = \{\theta_1, ..., \theta_\ell\}$ of homogeneous elements a set of *basic derivations.* It follows from Terao's work that

$$D_S(A) \simeq \mathrm{Der}_S{}^G \otimes_R S. \tag{1.3}$$

Thus basic derivations $D = \{\theta_1, ..., \theta_\ell\}$ may be viewed as a basis for $D_S(A)$.

Shephard and Todd [9] classified finite irreducible unitary reflection groups. Basic invariants for these groups are found in the literature [1,2,3,4,12]. Basic derivations were defined recently [10,6]. They are not known for all the groups. Our aim in this paper is to construct explicit basic derivations using the classification and the known basic invariants.

In Section 2 we describe the Hessian map and the matrix equation used to construct basic derivations from basic invariants. In Section 3 we list the groups with known basic derivations. It follows from the classification [9] that the groups numbered 24, 27, 29, 31, 33, 34 by Shephard and Todd remain. The rest of the paper is concerned with a construction of basic derivations for five of these six groups. In Section 4 we establish some facts for these groups and prove the existence of a unique matrix of invariants $\underline{C}$ associated to $B = \{f_1, ..., f_\ell\}$ with the property that knowledge of $\underline{C}$ is sufficient to find a (unique) $D = \{\theta_1, ..., \theta_\ell\}$ associated to $B = \{f_1, ..., f_\ell\}$. We also give an algorithm for computing $\underline{C}$. In Section 5 we give an example to illustrate the use of this algorithm and present the numerical results of the paper. The method of Section 4 also works for the last group. The calculation was carried out by H. Terao and Y. Enta. Their results appear in the appendix.

I would like to thank L. Solomon and H. Terao for many helpful conversations on this subject.

§2. THE HESSIAN MAP AND THE MATRIX EQUATION

In this section we state some of our results from [6]. Let $\Omega_S = \mathrm{Hom}_S(\mathrm{Der}_S, S)$ be the S-module of differential 1-forms dual to Der_S. Let $\{e_i\}$ be a basis for V and let $\{x_i\}$ be the dual basis for V^*. Then $\{D_i = \partial/\partial x_i\}$ and $\{dx_i\}$ are bases for Der_S and Ω_S as S-modules. In the natural grading of Ω_S $\deg(dx_i) = 1$. Let $\Omega_S{}^G$ be the R-module of G-invariant 1-forms. Then $\Omega_S{}^G$ is also a free R-module of rank ℓ with a homogeneous basis $df_1, \ldots, df_\ell$. We have $\deg(df_i) = d_i = m_i+1$ where the m_i are the *exponents* of G. It is customary to order the exponents so that $m_1 \leqslant \ldots \leqslant m_\ell$. Similarly, if $D = \{\theta_1, \ldots, \theta_\ell\}$ is a set of basic derivations then $\deg(\theta_i) = n_i-1$ where the n_i are the *coexponents* of G. Although it is customary to order the n_i in a similar fashion, it is more convenient here not to do so.

If $f \in S$ define the map $\mathrm{Hess}(f)\colon \mathrm{Der}_S \to \Omega_S$ by

$$\mathrm{Hess}(f)\theta = \Sigma\, \theta(D_i f) dx_i. \tag{2.1}$$

Then $\mathrm{Hess}(f)$ is an S-module homomorphism which does not depend on the choice of basis for V. Let $\underline{H}(f)$ denote the matrix of the map $\mathrm{Hess}(f)$ with respect to the pair of bases $\{D_i\}$, $\{dx_i\}$. Then $\underline{H}(f)$ is the usual Hessian matrix of second partial derivatives of f. Both Der_S and Ω_S are graded G-modules as follows:

$$(g\theta)(f) = g(\theta(g^{-1}f)) \quad \text{and} \quad (g\omega)(\theta) = g(\omega(g^{-1}\theta)) \tag{2.2}$$

where $\theta \in \mathrm{Der}_S$ and $\omega \in \Omega_S$. We showed in [6, 2.18] that if $f \in R$ is a form of degree p then $\mathrm{Hess}(f)\colon \mathrm{Der}_S{}^G \to \Omega_S{}^G$ is a graded R-module map of degree p. We choose bases in $\mathrm{Der}_S{}^G$ and in $\Omega_S{}^G$ to derive a matrix equation. If $f_1, ..., f_\ell \in S$ define their Jacobian matrix $\underline{J}(f_1,..., f_\ell)$ by $\underline{J}_{ij} = D_i f_j$. If $B = \{f_1, ..., f_\ell\}$ is a basic set we write $\underline{J}(B) = \underline{J}(f_1, ..., f_\ell)$. If $\theta_1, ..., \theta_\ell \in \mathrm{Der}_S$ define a matrix $\underline{Q}(\theta_1, ..., \theta_\ell)$ by $\underline{Q}_{ij} = \theta_j(x_i)$. Thus $\theta_j = \Sigma\, \underline{Q}_{ij} D_i$. If $D = \{\theta_1, ..., \theta_\ell\}$ is a set of basic derivations we write $\underline{Q}(D) = \underline{Q}(\theta_1, ..., \theta_\ell)$. We proved [6, 4.5]:

(2.3) PROPOSITION. *Let $G \subseteq \mathrm{GL}(V)$ be a unitary reflection group. Let $B = \{f_1, ..., f_\ell\}$ be a basic set and let $D = \{\theta_1, ..., \theta_\ell\}$ be a set of basic derivations. If $f \in R$ then there exists $\underline{C} \in M_\ell(R)$ such that*

$$\underline{H}(f)\ \underline{Q}(D) = m_1 \underline{J}(B)\ \underline{C}. \tag{2.4}$$

If f is homogeneous then the nonzero entries of $\underline{C}$ are homogeneous.

The factor m_1 in (2.4) is included so that $\theta_1 = \Sigma\, x_i D_i$ is the usual Euler derivation, see [6, (1.6)]. Recall that each reflection in G fixes a hyperplane in V and $A = A(G)$ is the set of these hyperplanes. The subgroup of G fixing H is cyclic of order e_H. It is known [6, 2.26 and 2.27] that:

(2.5.i) If $B = \{f_1, ..., f_\ell\}$ is a set of basic invariants then

$$\det \underline{J}(B) \simeq \prod_{H \in A} \alpha_H^{\,e_H - 1}.$$

(2.5.ii) If $D = \{\theta_1, ..., \theta_\ell\}$ is a set of basic derivations then

$$\det \underline{Q}(D) \simeq \prod_{H \in A} \alpha_H .$$

Here and in what follows we write $a \simeq b$ if a and b are equal up to a nonzero constant multiple.

(2.6) PROPOSITION. *Let $D = \{\theta_1, ..., \theta_\ell\}$ and $D' = \{\theta_1', ..., \theta_\ell'\}$ be basic derivations for G. There exists an invertible matrix of invariant polynomials $\underline{U} \in M_\ell(R)$ such that $\underline{Q}(D') = \underline{Q}(D)\ \underline{U}$.*

PROOF. This follows from (2.4) and the fact that a change of basis in $\mathrm{Der}_S{}^G$ corresponds to elementary column operations on $\underline{Q}(D)$.

Our aim is to use (2.4) to construct basic derivations. Suppose we know basic invariants $B = \{f_1, ..., f_\ell\}$ and an invariant f such that $\det \underline{H}(f) \neq 0$. Then we may view (2.4) as an equation which contains two known matrices $\underline{J}(B)$, $\underline{H}(f)$ and two unknown matrices $\underline{Q}(D)$, $\underline{C}$. Clearly, either one of $\underline{Q}(D)$ and $\underline{C}$ determines the other by (2.4). We show next how to reduce the problem of finding $\underline{Q}(D)$ to an algorithm for finding a suitable $\underline{C}$.

(2.7) DEFINITION. Given basic invariants $B = \{f_1, ..., f_\ell\}$ and $f \in R$ such that $\det \underline{H}(f) \neq 0$ we call $\underline{C} \in M_\ell(R)$ *admissible* if there exist basic derivations $D = \{\theta_1, ..., \theta_\ell\}$ such that (2.4) holds. Write $c_{ij} = \underline{C}_{ij}$.

(2.8) PROPOSITION. *If $\underline{C}$ is admissible then*

$$\det \underline{C} \doteq \det \underline{H}(f)/\prod \alpha_H{}^{e_H-2}.$$

(2.9) REMARK. The matrix $\underline{J}(B)$ has homogeneous columns of degree m_i. The matrix $\underline{Q}(D)$ has homogeneous columns of degree n_j. If f is homogeneous of degree d then it follows from (2.4) that

$$\deg(c_{ij}) = d - 2 + n_j - m_i\ .$$

In particular $c_{ij} = 0$ if $d - 2 + n_j - m_i$ is not an invariant degree.

(2.10) PROPOSITION. *Suppose $\underline{C}$ is admissible. If for some p, q we have $c_{p,q} \in \mathbf{C}^*$ then there exists an admissible matrix $\underline{C}'$ with $c'_{p,q} = 1$ and $c'_{p,r} = 0$ for $r \neq q$.*

PROOF. This is a consequence of (2.6).

§3. KNOWN BASIC DERIVATIONS

In this section we recall the basic derivations found in the literature. Let G be an irreducible unitary reflection group. Let $A = A(G)$ be its set of reflecting hyperplanes and recall Q from (1.1). The next lemma [6, (2.28)] is an analog of a result of Saito [8, (1.8.ii)].

(3.1) LEMMA. *A set* $\eta_1, \ldots, \eta_\ell$ *of homogeneous elements of* $\mathrm{Der}_S{}^G$ *is a set of basic derivations if and only if* $\det \underline{Q}(\eta_1, \ldots, \eta_\ell) \doteq Q$.

This may be used to check that the derivations given in (3.2) - (3.4) are in fact basic derivations.

(3.2) For the monomial groups $G(r,p,\ell)$ with $1 \leqslant p < r$ we have $Q \doteq \prod_{i=1}^{\ell} x_i \prod_{1 \leqslant i<j \leqslant \ell} (x_i^r - x_j^r)$. We may choose

$$\theta_i = \sum_{j=1}^{\ell} x_j^{(i-1)r+1} D_j \qquad 1 \leqslant i \leqslant \ell.$$

(3.3) For the monomial groups $G(r,r,\ell)$ we have $Q \doteq \prod_{1 \leqslant i<j \leqslant \ell} (x_i^r - x_j^r)$. We may choose θ_i as in (3.2) for $1 \leqslant i \leqslant \ell-1$ together with

$$\theta_\ell = \sum_{j=1}^{\ell} (x_1 \ldots x_{j-1} x_{j+1} \ldots x_\ell)^{r-1} D_j \,.$$

(3.4) For groups with $\ell = 2$ we may choose

$$\theta_1 = x_1 D_1 + x_2 D_2,$$
$$\theta_2 = (- D_2 Q) D_1 + (D_1 Q) D_2.$$

(3.5) In [6] we defined a class of unitary reflection groups which we called Shephard groups. We proved in [6, (1.5)] that if G is a Shephard group then for any basic invariants $B = \{f_1, \ldots, f_\ell\}$ there exist basic derivations $D = \{\theta_1, \ldots, \theta_\ell\}$ such that

$$\underline{H}(f_1)\underline{Q}(D) = m_1\ \underline{J}(B).$$

In light of (2.4) this amounts to the assertion that for a Shephard group the identity matrix $\underline{C} = \underline{I}$ is admissible. We listed explicit basic derivations for all Shephard groups in [7].

(3.6) REMARK. It follows from the classification [9] that the groups numbered 24,27,29,31,33,34 by Shephard and Todd remain. The rest of this paper is concerned with a construction of basic derivations for five of these six groups. The method of Section 4 also works for the group numbered 34. The calculation was carried out by H. Terao and Y. Enta. Their results appear in the appendix.

§4. THE ALGORITHM FOR C

In this section we outline a construction of basic derivations for the groups $\Gamma = \{G_{24},G_{27},G_{29},G_{31},G_{33}\}$. Shephard and Todd [9] give references where basic invariants may be found. The next assertion is checked case by case.

(4.1) PROPOSITION. *Let $G \in \Gamma$. Then the following hold:*

(i) $d_1 < d_2$ *so up to constant there is a unique invariant* f_1 *of minimal positive degree,*

(ii) det $\underline{H}(f_1)$ *may be taken as one of the basic invariants for* G,

(iii) G *is generated by reflections of order* 2. *Thus* $e_H = 2$ *for all* $H \in A(G)$. *If we choose* $f = f_1$ *in* (2.4) *then it follows from* (2.8) *that* $\det \underline{C} \simeq \det \underline{H}(f_1)$.

(4.2) DEFINITION. Let $G \in \Gamma$. We call a basic set $B = \{f_1, \ldots, f_\ell\}$ *proper* if for some k we have $f_k \simeq \det \underline{H}(f_1)$.

(4.3) THEOREM. *Let $G \in \Gamma$ and suppose that $B = \{f_1, \ldots, f_\ell\}$ is proper. There is an ordering of the n_i such that there exists a unique admissible matrix $\underline{C}$ with the following properties:*

(i) $\underline{C}$ *is upper triangular,*

(ii) *the diagonal elements are:* $c_{q,q} = f_k \simeq \det \underline{H}(f_1)$ *and* $c_{i,i} = 1$,

(iii) $c_{i,j} = 0$ *for* $i \neq q$,

(iv) *the elements* $c_{q,j}$ *for* $j > q$ *are polynomials in* $\mathbb{C}[f_1, \ldots, f_{k-1}, f_{k+1}, \ldots, f_\ell]$.

PROOF. Let $\underline{C}$ be an admissible matrix. By (4.1.iii) and the assumption that B is proper we have $\det \underline{C} \simeq f_k$. Since the basic invariants are algebraically independent, one of the summands of $\det \underline{C}$ is a constant multiple of f_k. This proves (i) and (ii). We use (2.10) to obtain (iii). Finally, (iv) is obtained by elementary column transformations. It remains to show uniqueness. Suppose there exist basic derivations $D = \{\theta_1, \ldots, \theta_\ell\}$ and $D' = \{\theta_1', \ldots, \theta_\ell'\}$ and admissible matrices $\underline{C}$ and $\underline{C}'$ such that $\underline{H}(f_1)\underline{Q}(D) = m_1\underline{J}(B)\underline{C}$ and $\underline{H}(f_1)\underline{Q}(D') = m_1\underline{J}(B)\underline{C}'$. It follows from (2.6) that $\underline{C}' = \underline{C}\,\underline{U}$ for an invertible $\underline{U} \in M_\ell(R)$. Since both $\underline{C}$ and $\underline{C}'$ satisfy (i) - (iv) above we may assume without loss of generality that $q = 1$ so we have:

$$c_{ij} = c_{ij}' = \delta_{ij} \quad \text{for } 2 \leqslant i \leqslant \ell \text{ and } 1 \leqslant j \leqslant \ell,$$
$$c_{11} = c_{11}' = f_k, \text{ and}$$
$$c_{1j} \text{ and } c_{1j}' \in \mathbb{C}[f_1, \ldots, f_{k-1}, f_{k+1}, \ldots, f_\ell] \quad \text{for } 2 \leqslant j \leqslant \ell.$$

Write $u_{ij} = \underline{U}_{ij}$. If we check the equation $\underline{C}' = \underline{C}\,\underline{U}$ term for term, it follows that $u_{ij} = \delta_{ij}$ for $2 \leqslant i \leqslant \ell$ and $1 \leqslant j \leqslant \ell$. For $i = 1$ we get $u_{11}f_k = f_k$ and $u_{1j}f_k + c_{1j} = c_{1j}'$. Thus $u_{11} = 1$ and by the assumptions on c_{1j} and c_{1j}' and the algebraic independence of the basic invariants we get $u_{1j} = 0$ for $2 \leqslant j \leqslant \ell$. Thus $\underline{C}' = \underline{C}$.

(4.4) DEFINITION. Let $G \in \Gamma$ and let $B = \{f_1, \ldots, f_\ell\}$ be a proper basic set for G. We call the unique admissible matrix $\underline{C}$ which satisfies (4.3) the *standard* $\underline{C}$ associated to B. We call the nonzero entries $c_{q,j}$ for $j > q$ in $\mathbb{C}[f_1, \ldots, f_{k-1}, f_{k+1}, \ldots, f_\ell]$ the *nontrivial* elements of $\underline{C}$.

(4.5) DEFINITION. Let $G \in \Gamma$ and let $B = \{f_1, \ldots, f_\ell\}$ be a proper basic set for G. Let $\underline{C}$ be the standard matrix associated to

B. We call the unique basic derivations $D = \{\theta_1, ..., \theta_\ell\}$ which satisfy $\underline{H}(f_1)\underline{Q}(D) = m_1\underline{J}(B)\underline{C}$ the basic derivations *associated* to B.

(4.6) COROLLARY. *Let $G \in \Gamma$ and let $B = \{f_1, ..., f_\ell\}$ be a proper basic set for G. In order to find the basic derivations $D = \{\theta_1, ..., \theta_\ell\}$ associated to B it suffices to find the standard $\underline{C}$ associated to B.*

In view of (1.3) and (2.4) the problem of finding a basis for the free arrangement A is reduced to an algorithm for finding the entries of the standard $\underline{C}$. The steps are as follows:

(4.7) Construct a proper basic set $B = \{f_1, ..., f_\ell\}$.

(4.8) Compute the matrix of degrees $d_1 - 2 + n_j - m_i$ of the entries of an admissible $\underline{C}$.

(4.9) Determine the ordering of the n_i which makes $\underline{C}$ upper triangular and determine all trivial entries of the standard $\underline{C}$.

(4.10) Knowledge of the degrees of the nontrivial entries permits determination of these entries up to constants. Since det $\underline{H}(f_1) \simeq f_k$ we may define a matrix $\underline{A}$ by $\underline{A}\underline{H}(f_1) = f_k\underline{I}$. Up to constant $\underline{A}$ is just the adjoint of $\underline{H}(f_1)$ and hence it may be computed. Left multiplication of (2.4) by $\underline{A}$ gives

$$f_k\underline{Q}(D) = m_1\underline{A}\underline{J}(B)\underline{C}. \tag{4.11}$$

Compute $\underline{P} = m_1\underline{A}\underline{J}(B)$ and write $p_{ij} = \underline{P}_{ij}$. The nontrivial entries of $\underline{C}$ are computed from the congruences

$$\Sigma\, p_{is}c_{sj} \equiv 0 \pmod{f_k}. \tag{4.12}$$

§5. NEW BASIC DERIVATIONS

We demonstrate the algorithm for finding the standard $\underline{C}$ on G_{29}. Shephard and Todd [9, p. 287] indicate which of the forms found by Maschke [4] may be taken as basic invariants for G.

Using Maschke's notation, they are Φ_1, F_8, F_{12} and F_{20} defined in [4, (11) and (12)]. These do not satisfy (4.7). The Hessian determinant of Φ_1 has degree 8. It is convenient to choose the following proper basic set B:

$$(5.1)\quad \begin{aligned} f_1 = \Phi_1 &= z_1^4 + z_2^4 + z_3^4 + z_4^4 - 6(z_1^2z_2^2 + z_1^2z_3^2 + z_1^2z_4^2 \\ &\quad + z_2^2z_3^2 + z_2^2z_4^2 + z_3^2z_4^2). \\ f_2 &= (-1/20736)\det \underline{H}(f_1), \\ f_3 &= F_{12}, \\ f_4 &= F_{20}\,. \end{aligned}$$

For (4.8) we find the exponents and coexponents of G_{29} in [5]. They are $m_i = 3,7,11,19$ and $n_i = 1,9,13,17$. The matrix of the degrees $d_1 - 2 + n_j - m_i$ of the entries of $\underline{C}$ is

$$(5.2)\qquad \begin{bmatrix} 0 & 8 & 12 & 16 \\ -4 & 4 & 8 & 12 \\ -8 & 0 & 4 & 8 \\ -16 & -8 & -4 & 0 \end{bmatrix}$$

Thus when the n_i are ordered 1,13,9,17, $\underline{C}$ is upper triangular. For (4.9) it follows from (2.10), (4.3) and (5.2) that the standard $\underline{C}$ associated to B has the form:

$$(5.3)\qquad \underline{C} = \begin{bmatrix} 1 & 0 & 0 & 0 \\ 0 & f_2 & c_{23} & c_{24} \\ 0 & 0 & 1 & 0 \\ 0 & 0 & 0 & 1 \end{bmatrix}$$

For (4.10) it follows from (4.3) and the known degrees (5.2) that

$$c_{23} = a_1f_1 \quad\text{and}\quad c_{24} = a_2f_1^3 + a_3f_3 \quad\text{with}\quad a_1,a_2,a_3 \in \mathbf{C}.$$

Thus the problem of finding the standard $\underline{C}$ associated to B of (5.1) is reduced to the determination of these three coefficients. The relevant congruences from (4.12) are for $j = 3$ and $j = 4$:

$$p_{i2}a_1f_1 + p_{i3} \equiv 0 \pmod{f_2} \tag{5.4}$$

$$p_{i2}(a_2f_1^{\,3} + a_3f_3) + p_{i4} \equiv 0 \pmod{f_2} \tag{5.5}$$

These are solved using MACSYMA. We may choose $i = 1$. Substitute "generic" values, say 1, 2, 3, for three of the four variables. Call the resulting polynomials in one variable $r_{12}, r_{13}, r_{14}, h_1, h_2, h_3$. The congruences (5.4) and (5.5) become

$$r_{12}a_1h_1 + r_{13} \equiv 0 \pmod{h_2} \tag{5.6}$$

$$r_{12}(a_2h_1^{\,3} + a_3h_3) + r_{14} \equiv 0 \pmod{h_2} \tag{5.7}$$

The MACSYMA command *remainder* computes the remainders of the left hand sides after division by h_2. The coefficients of the remainders contain a_1, a_2, a_3 as linear terms. Setting these coefficients equal to zero gives the solution

$$a_1 = -63/80, \qquad a_2 = -57/320, \qquad a_3 = -3/5.$$

Thus the standard $\underline{C}$ associated to B of (5.1) is

$$\underline{C} = \begin{bmatrix} 1 & 0 & 0 & 0 \\ 0 & f_2 & -\frac{63}{80} f_1 & -\frac{57}{320} f_1^{\,3} - \frac{3}{5} f_3 \\ 0 & 0 & 1 & 0 \\ 0 & 0 & 0 & 1 \end{bmatrix}$$

In view of Corollary (4.6) this completes the determination of the basic derivations $D = \{\theta_1, ..., \theta_4\}$ associated to the basic invariants $B = \{f_1, ..., f_4\}$ of (5.1).

For the remaining groups we give only a proper basic set $B = \{f_1, ..., f_\ell\}$, the ordering of the n_i and the standard $\underline{C}$ associated to it. In the description of basic sets for the groups G_{24} and G_{27} it is convenient to have the following definition. Let u and v be polynomials in x,y,z. Define

$$\operatorname{bord}(u,v) = \det \begin{bmatrix} \partial^2 u/\partial^2 x & \partial^2 u/\partial x\partial y & \partial^2 u/\partial x\partial z & \partial v/\partial x \\ \partial^2 u/\partial x\partial y & \partial^2 u/\partial^2 y & \partial^2 u/\partial y\partial z & \partial v/\partial y \\ \partial^2 u/\partial x\partial z & \partial^2 u/\partial y\partial z & \partial^2 u/\partial^2 z & \partial v/\partial z \\ \partial v/\partial x & \partial v/\partial y & \partial v/\partial z & 0 \end{bmatrix}$$

For G_{24}, Shephard and Todd indicate that we may use the invariants of Klein [3]. We choose the proper basic set B:

$$f_1 = x^3y + y^3z + z^3x,$$

$$f_2 = (1/54) \det \underline{H}(f_1),$$

$$f_3 = (1/9) \operatorname{bord}(f_1,f_2).$$

The n_i are ordered 1,9,11. The associated $\underline{C}$ is:

$$\underline{C} = \begin{bmatrix} 1 & 0 & 0 \\ 0 & f_2 & -48f_1^{\,2} \\ 0 & 0 & 1 \end{bmatrix}$$

For G_{27}, Shephard and Todd indicate that we may use the invariants of Wiman [12]. We choose the proper basic set B:

$$f_1 = 10x^3y^3 + 9z(x^5+y^5) - 45x^2y^2z^2 - 135x\ y\ z^4 + 27z^6,$$

$$f_2 = (1/20250) \det \underline{H}(f_1),$$

$$f_3 = (1/24300) \operatorname{bord}(f_1,f_2).$$

The n_i are ordered 1,19,25. The associated $\underline{C}$ is:

$$\underline{C} = \begin{bmatrix} 1 & 0 & 0 \\ 0 & f_2 & \frac{13}{9} f_1^3 \\ 0 & 0 & 1 \end{bmatrix}$$

For G_{31}, Shephard and Todd indicate that we may use the invariants of Maschke [4, p. 505]. We choose the proper basic set B:

$$f_1 = F_8 = \Sigma\, z_i^8 + 14\, \Sigma\, z_i^4 z_k^4 + 168\, z_1^2 z_2^2 z_3^2 z_4^2,$$

$$f_2 = F_{12},$$

$$f_3 = F_{20},$$

$$f_4 = (1/265531392) \det \underline{H}(f_1).$$

The n_i are ordered 1,29,13,17. The associated $\underline{C}$ is:

$$\underline{C} = \begin{bmatrix} 1 & 0 & 0 & 0 \\ 0 & f_4 & -\frac{1}{5} f_1 & -\frac{1}{1620} f_2 \\ 0 & 0 & 1 & 0 \\ 0 & 0 & 0 & 1 \end{bmatrix}$$

For G_{33}, Shephard and Todd indicate that we may use the invariants of Burkhardt [1, p. 208]. We choose the proper basic set B:

$$f_1 = J_4 = Y_0^4 + 8Y_0(Y_1^3 + Y_2^3 + Y_3^3 + Y_4^3) + 48\, Y_1 Y_2 Y_3 Y_4,$$

$$f_2 = J_6,$$

$$f_3 = (1/63700992) \det \underline{H}(f_1),$$

$$f_4 = J_{12},$$

$$f_5 = J_{18}.$$

The n_i are ordered 1,13,7,9,15. The associated $\underline{C}$ is:

$$\underline{C} = \begin{bmatrix} 1 & 0 & 0 & 0 & 0 \\ 0 & f_3 & -\frac{1}{576} f_1 & -\frac{1}{192} f_2 & -\frac{1}{192} f_4 \\ 0 & 0 & 1 & 0 & 0 \\ 0 & 0 & 0 & 1 & 0 \\ 0 & 0 & 0 & 0 & 1 \end{bmatrix}$$

The reader should be warned that Burkhardt's paper contains several misprints. He gives three expressions for each invariant. In order to compare them we give them different names: $J_n = J_n(\Phi, u, \Psi, \ldots)$, $H_n = H_n(Y_0, \ldots, Y_4)$ and $K_n = K_n(Y_0, a_1, \ldots, a_4)$, where $\Phi, u, \Psi, \ldots$ are polynomials defined in the paper, and $a_1, \ldots, a_4$ are certain symmetric polynomials in $Y_1, \ldots, Y_4$. Our calculations indicate that not all of Burkhardt's expressions are equal. In particular:

$$J_6 = H_6 \neq K_6$$

In K_6 the sign of $20a_1Y_0^3$ should be +.

$$J_{12} = H_{12} \neq K_{12}$$

In K_{12} the signs of $33a_1Y_0^5$ and $108a_1Y_0$ should be +.

$$J_{18} \neq H_{18} = K_{18}$$

In H_{18} the coefficient of	is given as	but should be
$Y_0^8\Sigma\, Y_1^4Y_2^4Y_3Y_4$	12	6
$Y_0^7\Sigma\, Y_1^5Y_2^2Y_3^2Y_4^2$	- 36	- 18
$Y_0\Sigma\, Y_1^5Y_2^5Y_3^5Y_4^2$	- 27	- 72
$Y_0\Sigma\, Y_1^{11}Y_2^2Y_3^2Y_4^2$	12	- 12

The symbol $\Sigma\, Y_1^6Y_2^3Y_3^3Y_4^3$ should be $\Sigma\, Y_1^9Y_2^3Y_3^3Y_4^3$ and the symbol $Y_0^7\Sigma\, Y_1^5Y_2^2Y_3^2Y_4^3$ should be $Y_0^7\Sigma\, Y_1^5Y_2^2Y_3^2Y_4^2$.

In K_{18} the coefficient of	is given as	but should be
$a_4^{1/3}a_2Y_0^8$	12	6
$a_4^{2/3}a_1$	$36Y_1^7$	$18Y_0^7$
$a_4^{2/3}Y_0$	$45a_3 + 18a_1a_2 - 12a_1^3$	$162a_3 - 54a_1a_2 + 12a_1^3$.

This completes the computation of the standard $\underline{C}$ associated to B for $G \in \Gamma$. It is sufficient information to compute $\underline{Q}(D)$. The simplest way is to use (4.11). Since we already have $\underline{P} = m_1\underline{AJ}(B)$ we get $\underline{Q}(D)$ by dividing each entry of $m_1\underline{AJ}(B)\underline{C} = \underline{PC}$ by f_k. The actual calculations require computing power. In G_{31} polynomials of degree 41 in four variables must be divided by a polynomial of degree 24. In G_{33} polynomials of degree 25 in five variables must be divided by a polynomial of degree 10.

APPENDIX

BASIC DERIVATIVES FOR G_{34}

Hiroaki Terao and Yoichi Enta

We compute the matrix **C** for the last unitary reflection group G_{34}, sometimes called the Mitchell group. It is known [9] that (4.1)(i) and (iii) hold true. As for (4.1)(ii), we can verify it by MACSYMA computation using the basic invariants obtained (also using MACSYMA) by J. H. Conway and N. J. A. Sloane [2]. Thus we can choose the proper basic set B in the sense of (4.2) using the short vectors in the 3-basis [2; p. 425, Table 2, p. 438, Th. 10]:

$$f_1 = (1/1944)\mu_6 = -\Sigma\, x_i^6 + 10\,\Sigma\, x_i^3 x_k^3 + 180\, x_1x_2x_3x_4x_5x_6\,,$$

$$f_2 = (1/3888)\mu_{12}, \quad f_3 = (1/1944)\mu_{18},$$

$$f_4 = (1/23328000000)\ \det \mathbf{H}(f_1),\ f_5 = (1/1944)\mu_{30},\ f_6 = (1/1944)\mu_{42}.$$

The n_i are ordered 1,31,13,19,25,37. The associated $\mathbf{C}$ is:

$$\mathbf{C} = \begin{bmatrix} 1 & 0 & 0 & 0 & 0 & 0 \\ 0 & f_4 & c_{23} & c_{24} & c_{25} & c_{26} \\ 0 & 0 & 1 & 0 & 0 & 0 \\ 0 & 0 & 0 & 1 & 0 & 0 \\ 0 & 0 & 0 & 0 & 1 & 0 \\ 0 & 0 & 0 & 0 & 0 & 1 \end{bmatrix}$$

Here

$$c_{23} = -\frac{25623}{175}\, f_1, \quad c_{24} = \frac{29}{437500}\, f_1^{\,2} - \frac{2}{984375}\, f_2,$$

$$c_{25} = -\frac{18958101273}{28700}\, f_1^{\,3} + \frac{4383422469}{315700} f_1 f_2 - \frac{1197907}{6314} f_3,$$

$$\begin{aligned} c_{26} = &-\frac{7830732709990364241670851 6}{12141111960078125}\, f_1^{\,5} \\ &+\frac{566404141998190388920389}{1184498727812500}\, f_1^{\,3} f_2 \\ &-\frac{1197786870889112467288053}{2671044631217187 50}\, f_1 f_2^{\,2} \\ &-\frac{457854601208874156113 4}{2428222392015625}\, f_1^{\,2} f_3 \\ &+\frac{1729346695224875667163}{1068417852486875 00}\, f_2 f_3 - \frac{2079266672313}{8239991150} f_5\,. \end{aligned}$$

ACKNOWLEDGMENT. We are grateful to the Integrated Learning Center of the International Christian University for allowing us to use UNIX-MACSYMA.

DEPARTMENT OF MATHEMATICS
INTERNATIONAL CHRISTIAN UNIVERSITY
OSAWA, MITAKA
TOKYO, JAPAN

REFERENCES

1. H. Burkhardt, *Untersuchungen auf dem Gebiete der hyperelliptischen Modulfunctionen (Zweiter Teil)*, Math. Annalen, 38 (1891), 161-224.

2. J. H. Conway and N. J. A. Sloane, *The Coxeter - Todd lattice, the Mitchell group, and related sphere packings*, Math. Proc. Camb. Phil. Soc., 93 (1983), 421-440.

3. F. Klein, *Ueber die transformationen siebenter Ordnung der elliptischen Funktionen*, Math. Annalen, 14 (1879), 428-471.

4. H. Maschke, *Ueber die quarternare, endliche, lineare Substitutionsgruppe der Borchardt'schen Moduln*, Math. Annalen, 30 (1887), 496-515.

5. P. Orlik and L. Solomon, *Unitary reflection groups and cohomology*, Invent. Math., 59 (1980), 77-94.

6. P. Orlik and L. Solomon, *The Hessian map in the invariant theory of reflection groups*, Nagoya Math. J., 109 (1988), 1-21.

7. P. Orlik and L. Solomon, *Discriminants in the invariant theory of reflection groups*, Nagoya Math. J., 109 (1988), 23-45.

8. K. Saito, *Theory of logarithmic differential forms and logarithmic vector fields*, J. Fac. Sci. Univ. Tokyo, Sect. IA, 27 (1980), 265-291.

9. G. C. Shephard and J. A. Todd, *Finite unitary reflection groups*, Can. J. Math., 6 (1954), 274-304.

10. H. Terao, *Free arrangements of hyperplanes and unitary reflection groups*, Proc. Japan Acad. Ser. A, 56 (1980), 389-392.

11. J. A. Todd, *The invariants of a finite collineation group in five dimensions*, Proc. Camb. Phil. Soc., 46 (1950), 73-90.

12. A. Wiman, *Ueber eine einfache Gruppe von 360 ebenen Collineationen*, Math. Annalen, 47 (1896), 531-556.

UNIVERSITY OF WISCONSIN
MADISON, WI 53706

Contemporary Mathematics
Volume **90**, 1989

SPACES OF LIMITING TANGENT SPACES

Donal B. O'Shea

INTRODUCTION

In this paper we introduce a pair of spaces constructed from limiting tangent hyperplanes to the smooth part of singular analytic varieties. When specialized to varieties which can be thought of as smooth families of singular varieties, this pair seems to play an important role in coding some of the equisingularity properties of the family. What follows is a description of this pair of spaces and a list of facts, questions, and conjectures concerning it. Unfortunately, the questions and conjectures outnumber the facts. Hopefully, however, there are enough of the latter to persuade the reader that the former are interesting.

The author is grateful to Lê Dũng Trang for his attentive reading of a preliminary version of this paper and the numerous suggestions that he made, almost all of which have found their way into the present paper.

DEFINITIONS

Let $W^m \subset U \subset \mathbf{C}^{m+1}$ be a reduced analytic hypersurface in an open set U in $\mathbf{C}^{m+1}$.

AMS 1980 Subject Classification: Primary 32B30; Secondary 32G11, 14B07. Research partially supported by the National Science Foundation and Mount Holyoke Faculty Grants.

0271-4132/89 $1.00 + $.25 per page

1. The space $K(W,x)$. Let W^0 denote the set of smooth points of W. If $x \in W^0$, let $T(W,x)$ denote the tangent space to W^0 at x considered as an element of the (dual) projective space $\mathbf{P}^m$ of hyperplanes in $\mathbf{C}^{m+1}$. Let $\mathbf{g}$: $W^0 \to \mathbf{P}^m$ be the morphism which takes x to $T(W,x)$ and let $\hat{W}$ be the closure in $W \times \mathbf{P}^m$ of the graph of $\mathbf{g}$. According to Whitney [15, p. 532], $\hat{W}$ is an analytic set and the projection $\mathbf{p}$: $\hat{W} \to W$ is a proper analytic morphism called the *Nash blow-up* of W. Let $|\mathbf{p}^{-1}(x)|$ denote the reduced analytic space underlying $\mathbf{p}^{-1}(x)$ and define $K(W,x)$ be setting

$$|\mathbf{p}^{-1}(x)| = \{x\} \times \mathrm{K}(W,x).$$

2. The space $K(f,x)$. Let f be a local equation for W (so $W = \{f = 0\}$ in U) and let U^0 be the set of $x \in U$ which are not critical points of f considered as a map f: $U \to \mathbf{C}$. By shrinking U, we may assume that $U - U^0 \subset W$. (This follows from the Bertini-Sard theorem. Alternatively, see Milnor [11, chapters 2, 3] for the algebraic case and Lê [5] for the analytic case). If $x \in U^0$, let $T(f,x)$ denote the tangent space of $f^{-1}(f(x))$ considered as an element of the (dual) projective space $\mathbf{P}^m$ of hyperplanes in $\mathbf{C}^{m+1}$. Let $\mathbf{h}$: $U^0 \to \mathbf{P}^m$ be the morphism defined by $\mathbf{h}(x) = T(f,x)$ and let $\hat{U}$ be the closure in $U \times G$ of the graph of $\mathbf{h}$. By [15], $\hat{U}$ is an analytic set and the projection $\mathbf{q}$: $\hat{U} \to U$ is a proper analytic morphism. As above, we let $|\mathbf{q}^{-1}(x)|$ denote the reduced analytic space underlying $\mathbf{q}^{-1}(x)$. Define $K(f,x)$ by setting

$$|\mathbf{q}^{-1}(x)| = \{x\} \times K(f,x).$$

3. The pair $K^*(W,x)$. If $x \in W$, it is clear that $K(W,x) \subset K(f,x)$. We let $K^*(W,x)$ denote the pair $(K(f,x), K(W,x))$ of reduced analytic spaces.

Although we shall only consider hypersurfaces, it is clear that the pair $K^*(W,x)$ could be defined if W were any complete intersection.

The space $K(W,x)$ is the fiber of the Nash blow-up. The space $K(f,x)$, the fiber of the blow-up of the Jacobian ideal of f, is the fiber of the relative conormal morphism (where the relative morphism is that induced by projection on the installation of f by its graph). In the case of hypersurfaces, the Nash blow-up is the blow-up of an ideal. (See Noble [12].) In fact, it is the fiber of the strict transform of the Jacobian ideal restricted to the hypersurface. (See Lê [6].) Thus, if Z is the singular locus of W, the dimension of $\mathbf{p}^{-1}(Z)$ is $m - 1$. The dimension of $\mathbf{q}^{-1}(Z)$ is less than or equal to m.

EXAMPLES

Here are some examples of the pair $K^*(W,x)$.

1) If $x \in W^0$, then $K(W,x) = K(f,x)$ consists of the single point $T(W,x)$. It can happen, of course, that $K(W,x)$ is a single point, but x is not smooth. On the other hand, $x \in W^0$ if and only if $K(f,x)$ is a single point (see theorem 5 below).

2) If W is a reduced plane curve with singular point x, then $K(W,x)$ is a finite number of points corresponding to the lines in the reduced tangent cone of W at x and $K(f,x)$ is isomorphic to the projective line $\mathbf{P}^1$ (because x is an isolated singularity; see below).

3) If W^m has an isolated singular point at x, then dim $K(W,x) = m - 1$ and $K(f,x)$ is isomorphic to m-dimensional projective space $\mathbf{P}^m$. The first statement follows from the fact cited above that, for hypersurfaces, the Nash blow-up is the blow-up of an ideal. The second statement follows from the interpretation of the Milnor number as the degree of a map (see Milnor [11, chap. 5]). Alternatively, one can proceed as in Lê [6,7] (see also [2]) to show that the generic polar curve is nonvoid. One then observes, following Lê, that $K(f,x) = \mathbf{P}^m$ whenever the generic polar curve is nonvoid. Henaut [3] has obtained a nice expression for the exceptional divisor of the Nash blow-up in the case of an isolated singularity.

4) Let W be a swallowtail; that is, the discriminant locus of the semi-universal deformation of the polynomial $g(w) = w^4$. Then

$$W = \{(x,y,z): f(x,y,z) = 256x^3 - 27y^4 - 128x^2z^2 + \\ + 144xy^2z + 16xz^4 - 4y^2z^3 = 0\}.$$

By direct computation, one finds that $K(W,0)$ = {single point} and $K(f,0) = \mathbf{P}^2$. More generally, if W is a generalized swallowtail (that is, the discriminant locus of the semi-universal unfolding of the polynomial $g(w) = w^k$) or, even more generally, the discriminant locus of a simple singularity, then $K^*(W,0) = (\mathbf{P}^m, \text{point})$ where $m+1$ is the dimension of the parameter space. In the latter case, a theorem due to Teissier [14, theorem 5.5.1] implies that $K(W,0)$ is a single point. On the other hand, the following argument due to Lê shows that $K(f,0) = \mathbf{P}^m$. By the classification of simple singularities (see, for example, Arnold [1]), π_1 of the complement of W is different from π_1 of the complement of a generic plane section through 0. Hence (see [7] or [2]), the generic polar curve is nonvoid and the result follows. Although Teissier's result applies to any versal deformation of an isolated hypersurface singularity, it seems doubtful that $K^*(W,0) = (\mathbf{P}^m, \text{point})$ whenever W is a semi-universal unfolding of an isolated hypersurface singularity.

5) Suppose X is the surface given by the equation $y^2 - z^2x^2 + x^3 = 0$. We think of X as a family of isolated singularities of curves. Let u be the origin and v an arbitrary point $(0, 0, t)$, other than the origin, on the z-axis. Then, $K(X,u) = \mathbf{P}^1$, each point of which corresponds to a plane passing through the z-axis; $K(f,u) = \mathbf{P}^2$, each point of which corresponds to a plane passing through the origin in $\mathbf{C}^3$; $K(X,v)$ consists of two points, corresponding to the tangent planes $\{y - tx = 0, y + tx = 0\}$; $K(f,v) = \mathbf{P}^1$, each point of which corresponds to a plane passing through the z-axis. This can be seen by direct computation or by the recipe in [4].

PROPERTIES OF $K^*(W,x)$. The following result is the first indication that the pair $K^*(W,x)$ is worthy of study.

PROPOSITION 1. *If* $W \subset U \subset \mathbf{C}^m$ *is a hypersurface and* $x \in W$, *the pair* $K^*(W,x)$ *only depends on the germ* (W,x) (*and not on the embedding of* W *in* U).

PROOF. We may assume that W is embedded so that x is the origin $\mathbf{0} \in \mathbf{C}^m$ and that $\mathbf{0}$ is actually a singular point of W. If F: $\mathbf{C}^m, 0 \to \mathbf{C}^m, 0$ represents a holomorphic change of coordinates at the origin, then its derivative dF_0 clearly induces an isomorphism of K^*. Therefore, we need only check that K^* does not depend on the choice of equation $f = 0$ representing W. If $g = 0$ is another equation representing W, then $g = uf$ for some unit $u \in C\{x_1, ..., x_m\}$. Let $\mathbf{a}$ be a unit vector in $\mathbf{C}^m$ representing an element of $K(f,x)$. Since $K(f,x)$ is analytic, there exists a path $\alpha : \mathbf{D}, 0 \to \mathbf{C}, 0$ such that $\mathbf{a}$ is the limit of $df(\alpha(t)/\|df(\alpha(t)\|$ as t tends to zero. We claim that $\mathbf{a}$ is the limit of $dg(\alpha(t))/\|dg(\alpha(t)\|$ as t tends to zero. This will follow if we can show that lim $\|f(\alpha(t))du(\alpha(t))\|/\|df(\alpha(t))\|$ $= 0$. But this holds because f being singular at $\mathbf{0}$ and $f(\mathbf{0}) = 0$ imply that $|f(\alpha(t)| \leq C \sup|\alpha_i(t)\partial f/\partial z_i(\alpha(t))|$ for all sufficiently small t where α_i is the ith component function of $\alpha(t)$ (see, for example, [14, p. 291]). (If $\mathbf{a}$ represents an element of $K(W,x)$, we may actually assume that $f(\alpha(t)) \equiv 0$). Thus, multiplying by the local equation of W by a unit u fixes K^* pointwise. This establishes the proposition.

By a theorem of Hironaka (cited in [8, p. 181]) $K(W,x)$ contains the dual to the reduced tangent cone of W at x. It also contains much more, as can be seen by reflecting on the example above. The question of what comprises the extra elements is delicate. When W is a surface, a partial characterization of the structure of $K(W,x)$ was obtained by Henry and Lê in [4] and by Lê in [8]. A result due to Henaut [3] leads to a nice expression for $K^*(W,x)$ in the case when x is an isolated singularity of W. Lê and Teissier [10] have recently obtained a far more complete characterization valid in any dimension. As a result of this work, there is a reasonable understanding in principle of the extra elements in $K(W,x)$. It is still not easy to compute $K(W,x)$ or $K(f,x)$. Proposiiton 1 gives rise to the hope that cruder, and hopefully more computable, objects associated to the pair $K^*(W,x)$, will suffice for many purposes.

In general, there is much one would want to know about the pair $K^*(W,x)$. For example, one might ask what possible pairs of integer values the dimensions ($\dim K(f,x)$, $\dim K(W,x)$) can take. This is interesting because examples suggest that $\dim K(f,x)$ plays the role of an embedding dimension, while the difference $d(x) \equiv \dim K(f,x) - \dim K(W,x)$ is the minimal codimension of a stratum containing x along which W is equisingular. It would be useful to know to what extent this holds for general varieties. Phrased somewhat differently, if we stratify W into pieces along which $d(x)$ is locally constant, what properties does the resulting stratification have? Ideally, we might hope to recover more delicate numerical invariants from the pair $K^*(x)$.

We consider here just the simplest question raised by the tentative interpretation of $d(x)$ above. Namely, when is $d(x)$ as small or as large as possible? That is, when is $d(x) = 0$? When is it m? (The latter case occurs only when $\dim (W,x) = 0$ and $\dim K(f,x) = m$ which means that the Nash blowup of W is finite and the local polar curve of W at x is nonvoid.)

FAMILIES OF ISOLATED SINGULARITIES

We provide an example here of when $d(x) = 0$. (In an earlier paper, the author had unwisely conjectured that such examples could not exist.) This example will also indicate that variations in $K^*(W,x)$ as x ranges over the singular locus indicate breakdowns in equisingularity at x.

We suppose that $W \subset U \subset \mathbf{C}^{m+1}$ is a hypersurface which has a one-dimensional singular locus which is smooth (as a point set). We allow ourselves to shrink U without changing notation and we let $\{f = 0\}$ be a local equation for the hypersurface in U. Fix a projection of $\mathbf{C}^{m+1}$ onto $\mathbf{C}$ and let $\pi\colon W \to D$ denote the induced morphism onto a neighbourhood D of the origin in $\mathbf{C}$. Do this in such a way that π admits a section σ such that $W - \sigma(D)$ is smooth. Set $W_t = \pi^{-1}(t)$ and $f_t = f|W_t$ for $t \in D$. Choose coordinates $x_1, \ldots, x_{m+1}$ on U such that $\sigma(D)$ is given by $x_1 = \ldots = x_m = 0$ and the map π by the x_{m+1} coordinate. We identify $\sigma(D)$ and D. In particular, we shall refer to the singular locus of W as D and the letter t will serve to denote a point $t \in D$, the value of the x_{m+1} coordinate, and the point $\sigma(t) = (0, \ldots, 0, t)$. In accord with this convention, if x is the point with coordinates $x_1 = \ldots = x_m = 0$, $x_{m+1} = t$, we let $K(X,t)$ and $K(f,t)$ denotes the spaces $K(W,x)$ and $K(f,x)$, respectively, and $K^*(t)$ the pair $K^*(W,x)$. Similarly, we let $\mu_t = \mu(f_t)$ be the Milnor number, $\mu_t^* = \mu^*(f_t)$ be Teissier's multiplicity sequence [14], and $m_t = m(W_t)$ be the multiplicity of f_t at the same point x. We say that the family is *μ-constant* (respectively, *μ^*-constant, equimultiple*) if $\mu_0 = \mu_t$ (resp., $\mu_0^* = \mu_t^*$, $m_0 = m_t$) for all small $t \in D - 0$.

Teissier [14] and Lê and Saito [9] established a necessary and sufficient condition for a family to be μ-constant. In terms of our notation, their result can be stated as follows.

THEOREM 2. *$\mu_0 = \mu_t$ for all $t \in D$ if and only if $K(f,0) = K(f,t) = \mathbf{P}^{m-1}$ for all $t \in D$. Moreover, if $\mu_0 > \mu_t$ for nonzero $t \in D$, then $K(f,0) = \mathbf{P}^m$.*

Thus, in a μ-constant family, one has $K(f,t) = \mathbf{P}^{m-1}$ for all t. It is natural to ask whether one can have $K(W,t) = \mathbf{P}^m$. The following gives a partial answer to this question.

THEOREM 3. *Suppose that f is a μ-constant family, but that $\mu_t^{(m-1)} \neq \mu_0^{(m-1)}$ for $t \in D - 0$. Then $K(f,0) = K(W,0) = \mathbf{P}^{m-1}$ is the subset of all planes containing the x_{m+1}-axis.*

PROOF. We know that $K(f,0) = \mathbf{P}^{m-1}$ by the theorem cited above. We show that $K(W,0) = \mathbf{P}^{m-1}$. If $H \in \mathbf{P}^m$ is a fixed hyperplane, then recall that the polar curve C_H with respect to the hyperplane H of a function g: $\mathbf{C}^{m+1} \to \mathbf{C}$ with a singularity at the origin is the closure of the set of points $\{x \in \mathbf{C}^{m+1}: [T(g^{-1}(g(x)), x)] = H\}$. Let $C_H{}^t$ denote the polar curve with respect to the hyperplane $H_t = H \cap W_t$ of the function f_t. If H_t is generic for f_t, then a formula due to Teissier [14] and Lê [6] states that $f_t \cdot C_H{}^t = \mu_t + \mu_t{}^{(m-1)}$ where the left hand side is the intersection index at the origin of the polar curve $C_H{}^t$ and the hypersurface $f_t{}^{-1}(0)$. But it is easy to see that if H_0 generic for f_0 and if H contains the x_{m+1}-axis then H_t generic for all t sufficiently close to 0. Since $\mu_t{}^{(m-1)}$ jumps at $t = 0$, we have $f_0 \cdot C_H{}^0 > f_t \cdot C_H{}^t$ for all small $t \neq 0$. However, by a standard argument (the curve selection lemma or the principle of continuity) this implies that there is a smooth curve $\alpha : [0,\delta) \to \mathbf{C}^{m+1}$ such that $\alpha(t) \in f_t{}^{-1}(0) \cap C_H{}^t$ and $\|\alpha(t)\| \geqslant 0$ with equality only when $t = 0$. Since the pair $(W - \{x_{m+1}\text{-axis}\}, \{x_{m+1}\text{-axis}\})$ satisfies Whitney condition a) at the origin, it is easy to check that the limit of the tangent hyperplanes $T(W,\alpha(t))$ as $t \to 0$ exists and is just the hyperplane H. Since we carry out this construction for an open dense set of the subset of hyperplanes in $\mathbf{C}^{m+1}$ containing the x_{m+1}-axis and since $K(W,x)$ is an analytic set, we must have $K(W,x) = \mathbf{P}^{m-1}$.

QUESTION. Is the converse of the above true?

By elaborating the above method, one can show the following.

THEOREM. *Suppose* $\mu_t^* \neq \mu_0^*$ *for* $t \in D - 0$. *Then* $K(W,0) = \mathbf{P}^{m-1}$.

The above also follows from Lê and Teissier's characterization in [10] of $K(W,x)$ in terms of tangent cones to polar varieties.

I am quite optimistic that studying variations in the pair $K^*(t)$ will shed light on the ways in which μ_t^* can vary in a family of isolated singularities. Rather than pursue this theme here, we restrict ourselves to considering other situations in which $d(x)$ is maximal or minimal.

WHEN IS dim $K(f,x)$ AS SMALL (AS LARGE) AS POSSIBLE?

Fix m and suppose that $W = f^{-1}(0)$ where $f\colon \mathbf{C}^{m+1}, \mathbf{0} \to \mathbf{C}, 0$ is reduced.

We first ask under what conditions $K(f,\mathbf{0})$ is as small as possible (that is, a single point). The following result, conjectured in the preliminary version of this paper, is due to Lê.

THEOREM 5 (Lê). *Let f be as above. Then $K(f,\mathbf{0})$ is a single point if and only if $\mathbf{0}$ is a nonsingular point of f.*

PROOF. The 'only if' part of the statement is trivial. To prove the 'if' part, note that $K(f,x)$ is the fiber of the blow up of the Jacobian ideal. Since the fiber has dimension 0, so do all nearby fibers over points on which the ideal vanishes. However, the codimension in $\mathbf{C}^{m+1}$ of the zero set of the Jacobian ideal is at most two (by the Bertini-Sard theorem). Thus, the dimension of the exceptional set is at most $m - 1$. But if x were singular, this dimension would have to be m (since we are blowing up an ideal). Thus, x is simple.

Equally interesting is the question of characterizing those singularities for which $K(f,0)$ is as large as possible, namely $\mathbf{P}^m$. In other words, we would like to characterize singularities for which the polar curve is nonvoid [6]. This is always the case for isolated singularities, so that we are interested in finding classes of nonisolated singularities with this property. Here's a partial result in that direction

PROPOSITION 6. *Let f be homogeneous, reduced and singular at the origin. Suppose that no partial derivative of f vanishes identically in a neighbourhood of the origin $\mathbf{0}$. Then $K(f,\mathbf{0}) = \mathbf{P}^m$.*

PROOF. Let $l \in \mathbf{P}^m$ be given in homogeneous coordinates by $l = (l_0,l_1, ..., l_m)$. If $l_0 \neq 0$, say, then any nontrivial solution of the m homogeneous equations $l_0(\partial f/\partial x_i) = l_i(\partial f/\partial x_0)$, $1 \leqslant i \leqslant n$, gives a line of points x in $\mathbf{C}^{m+1}$ through the origin (and, in particular, points x arbitrarily close to $\mathbf{0}$) along which $[df(x)] = l$. Bezout's theorem states that nontrivial solutions always exist. Thus, $l \in \mathbf{P}^m$.

COROLLARY 7. *If* (W,x) *is a cone, then* $K^*(W,x) = (\mathbf{P}^n, \mathbf{P}(C(W,x))$ *where* ~ *indicates the dual.*

Proposition 6 could also be proved in the style of [2, 6, 7] without appeal to Bezout's theorem by showing that the generic polar curve is nonvoid. I conjecture that proposition 6 holds if f is quasihomogeneous. One thing for sure, the above proof does not hold, as it is possible for some polar curves to be empty. This conjecture would be interesting in that, if true, it would form the basis of an elementary proof (using the Lê-Saito valuative criterion for μ-constancy) that the μ-constant stratum of an isolated quasihomogeneous hypersurface singularity is smooth. Currently, the only available proof (due to Varchenko [16]) uses mixed Hodge theory.

REFERENCES

[1] V. I. Arnold, Normal forms for functions near degenerate critical points, the Weyl groups A_k, D_k, E_k and Lagrangian singularities, *Functional Analysis and its Applications*, **6** (1972) 254-272.

[2] H. Hamm and Lê Dũng Tráng, Un théorème de Zariski du type de Lefschetz, *Ann. Sc. de L'Ecole Norm. Sup.*, **6** (1973) 317-366.

[3] A. Henaut, Cycle exceptionel de l'éclatement de Nash d'une hypersurface analytique complexe à singularités isolée, *Bull. Soc. Math. France*, **109** (1981) 475-481.

[4] J.-P. Henry and Lê Dũng Trang, Limites d'éspaces tangent, Sem. F. Norguet, *Lecture Notes in Mathematics*, **482**, Springer-Verlag, 1975.

[5] Lê Dũng Trang, Topologie des singularités isolées des hypersurfaces complexes, *Asterisque*, **7-8** (1973) 171-181.

[6] Lê Dũng Tráng, Calcul du nombre de cycles évanouissants d'une hypersurface complexe, *Ann. Inst. Fourier, Grenoble*, **23** (1973) 261-270.

[7] Lê Dũng Tráng, Topological use of polar curves, *Proc. Symp. Pure Math.*, **29** (1974) 507-512.

[8] Lê Dũng Tráng, Limites d'éspaces tangents sur les surfaces, in Proc. Symp. on Algebraic Singularities, Reinhartsbrunn 1978, *Nova Acta Leopoldina*, **52** (1981) 119-137.

[9] Lê Dũng Tráng and K. Saito, La constance de nombre Milnor donne des bonnes stratifications, *C.R. Acad. Sci. Paris*, **277** (1973) 793-795.

[10] Lê Dũng Tráng and B. Teissier, Limites d'éspaces tangents en géometrie analytique, preprint, 1986.

[11] J. Milnor, *Singular Points of Complex Hypersurfaces*, Ann. of Math. Studies, No. 61, Princeton Univ. Press, 1968.

[12] A. Nobile, Some properties of the Nash blowing-up, *Pac. J. Math.*, **60** (1975) 297-305.

[13] B. Teissier, The hunting of invariants in the geometry of discriminants, in Proc. Nordic Summer School (Oslo, 1967): *Real and Complex Singularities*, (ed. P. Holm), Sijthoff and Noordhoff, 1977.

[14] B. Teissier, Cycles évanescents, sections planes et conditions de Whitney, *Asterisque*, **7-8** (1973) 285-362.

[15] H. Whitney, Tangents to an analytic variety, *Ann. Math.*, **81** (1965) 496-549.

[16] A. N. Varchenko, A lower bound for the codimension of the stratum μ-const. in terms of the mixed Hodge structure (Russian), *Vestnik MGU*, *Seria Matem.*, 1083.

MATHEMATICS DEPARTMENT
MOUNT HOLYOKE COLLEGE
SOUTH HADLEY, MA 01075

Contemporary Mathematics
Volume **90**, 1989

SERIES OF ISOLATED SINGULARITIES

Ruud Pellikaan

INTRODUCTION. We start with quoting two remarks of V.I. Arnol'd: "Although the series undoubtedly exist, it is not all clear what a series of singularities is" [A] page 153.

"However a general definition of series of singularities is not known. It is only clear that the series are associated with singularities of infinite multiplicity (for example $D \sim x^2y$, $T \sim xyz$), so that the hierarchy of series reflects the hierarchy of non-isolated singularities" [A] page 154.

We do not give a definition of series of singularities either, but start to look at functions with a one dimensional singular locus Σ. We characterize those functions which have constant transversal R-type at every point of $\Sigma \setminus \{0\}$. We show that hypersurfaces with prescribed transversal singularities exist. We look at $f + x^{k+1}$, where x is a coordinate function such that $\Sigma \cap V(x) = \{0\}$. For $k >> 0$, $f + x^{k+1}$ has an isolated singularity. We give a formula for the Milnor number μ_k of $f + x^{k+1}$ in algebraic terms. Yomdin [Y] and Lê [Lê] proved a formula which relates the Euler characteristic of f and $f + x^{k+1}$, in greater generality. We combine these two results to get a formula for the Euler characteristic of the Milnor fibre of f.

$\mathcal{O}$ denotes the local ring of germs of analytic functions f: $(\mathbb{C}^{n+1},0) \to \mathbb{C}$, and $\mathcal{M}$ denotes the maximal ideal of $\mathcal{O}$.

Sometimes in proofs we take representatives of the germs on an open neighbourhood U of 0 in $\mathbb{C}^{n+1}$ and then $\mathcal{O}$ is the sheaf of analytic functions on U.

If local coordinates $z_0, z_1, \ldots, z_n$ are chosen of $(\mathbb{C}^{n+1},0)$ then $f_0, f_1, \ldots, f_n$ denote the partial derivatives of f with respect to $z_0, z_1, \ldots, z_n$. We let $J_f = (f_0, f_1, \ldots, f_n)\mathcal{O}$ be the Jacobi ideal of f.

For an $\mathcal{O}$-module M we define $\ell(M) = \dim_{\mathbb{C}}(M)$.

For basic definitions and results in commutative algebra we refer to Matsumura [Ma].

ACKNOWLEDGEMENT. This work was done at the University of Utrecht and was part of the author's thesis [P,1]. I want to thank Prof. Dr. D. Siersma for his ideas and stimulating support. This work was supported by the Netherlands Foundation for Mathematics SMC with financial aid from the Netherlands Organization for the Advancement of Pure Research ZWO. The Free University of Amsterdam supported the travel to the conference on singularities at Iowa City.

§1. THE TRANSVERSAL TYPE OF f ALONG A CURVE

Let $f\colon (\mathbb{C}^{n+1},0) \to (\mathbb{C},0)$ be a germ of an analytic function with a one dimensional singular locus $(\Sigma,0)$. We analyse the transversal type of f along Σ, with respect to different equivalence relations for isolated singularities, i.e. with respect to R-, K- and μ-type. We shall give a necessary and sufficient condition in the case the transversal R-type of f does not vary along Σ.

DEFINITION 1.1. Let $f,g\colon (\mathbb{C}^{n+1},0) \to (\mathbb{C},0)$ be two germs of analytic functions, then f and g are called R (= *right*) *equivalent* if there exists a germ $h\colon (\mathbb{C}^{n+1},0) \to (\mathbb{C}^{n+1},0)$ of a local analytic isomorphism such that $g = f \circ h$. They are called K (= *contact*) *equivalent* if

$(f^{-1}(0),0)$ and $(g^{-1}(0),0)$ are isomorphic germs of analytic spaces. If moreover f and g have an isolated singularity at 0, then they are called *μ-homotopic*, if there exists a family $\{f_t\}$, $t \in [0,1]$, of germs f_t: $(\mathbb{C}^{n+1},0) \to (\mathbb{C},0)$ of analytic functions with an isolated singular point at 0, such that the Milnor number $\mu(f_t,0)$ is constant, and $f_0 = f$, $f_1 = g$, and F: $\mathbb{C}^{n+1} \times [0,1] \to \mathbb{C}$, defined by $F(z,y) = f_t(z)$ is continuous.

REMARK 1.2. These relations define equivalence relations and the corresponding classes we call *R-class*, *K-class* and *μ-class* resp. or *R*-, *K*- and *μ-type* resp. The following inclusions of classes of a given function exist: R-class $\subset$ K-class $\subset$ μ-class.

DEFINITION 1.3. Let f: $(\mathbb{C}^{n+1},0) \to (\mathbb{C},0)$ be a germ of an analytic function with a singular locus $(\Sigma,0)$, such that $(\Sigma,0)$ is a germ of a non-singular curve. We say that the transversal *R-type of f along Σ at 0 is constant*, if for every choice of local coordinates x, y_1, ..., y_n of $(\mathbb{C}^{n+1},0)$ such that $\Sigma = V(y_1, ..., y_n)$, the R-type of f_t: $(\mathbb{C}^n,0) \to (\mathbb{C},0)$, with $f_t(y) = f(t,y)$ is the same for all t sufficiently small.

In the same way one defines what it means that the K-type (or μ-type) of f along Σ at 0 is constant.

PROPOSITION 1.4. *Let* f: $(\mathbb{C}^{n+1},0) \to (\mathbb{C},0)$ *be a germ of an analytic function with a one dimensional singular locus* $(\Sigma,0)$. *Let* $(\Sigma_1,0)$, ..., $(\Sigma_r,0)$ *be the branches of* $(\Sigma,0)$. *The transversal μ-type of f along* Σ_i *is constant at all points of a punctured neighbourhood of* 0 *in* Σ_i.

PROOF. See [Y] or [Lê] (1.3.1) and (1.3.2).

REMARK 1.5. As a consequence we can associate to every branch $(\Sigma_i,0)$ a well-defined μ-class of an isolated singularity, which is the transversal μ-type of f along $\Sigma_i\backslash\{0\}$. We shall give some examples to show that the situation with respect to transversal R- and K-type is quite different.

EXAMPLE 1.6. Let $f(x,y,z) = y^4 + xy^2z^2 + z^4$, then the singular locus Σ of f is $V(y,z)$. The transversal μ-type is $\tilde{E}_7$ with Milnor number 9. The transversal K-type and R-type vary, since the cross ratio of the four lines of $f_x^{-1}(0)$ varies along Σ.

EXAMPLE 1.7. Let $f(x,y,z) = y^5 + (1 + x)y^2z^2 + z^5$, then the singular locus Σ of f is the x-axis. The transversal μ-type and K-type are constant along Σ $(x \neq -1)$ and of type $T_{5,5,2}$. But the transversal R-type of f along Σ varies.

EXAMPLE 1.8. Let $f(x,y,z) = y^6 + xy^3z^3 + z^6$, then the singular locus of f is the x-axis. Take $\bar{x} = x - y$, then $\bar{f}(\bar{x},y,z) = y^6 + (\bar{x} + y)y^3z^3 + z^6$. The transversal K-type of f along Σ with respect to the coordinates x, y, z is quasi-homogeneous, so $\mu(f_x,0) = \tau(f_x,0)$ the Tjurina number. But for the transversal K-type of $\bar{f}$ along Σ with respect to the coordinates $\bar{x}$, y, z one has $\mu(\bar{f}_{\bar{x}},0) > \tau(\bar{f}_{\bar{x}},0)$. Hence the transversal K- (and R-) type of f depends on the chosen local coordinates.

DEFINITION 1.9. The *analytic stratum* of a germ of an analytic space $(X,0)$ in $(\mathbb{C}^m,0)$ is the set-germ A of points in $(\mathbb{C}^m,0)$ along which a representative X of $(X,0)$ is trivial:

$$A = \{a \in (\mathbb{C}^m,0) \mid (X,a) \cong (X,0)\} \subseteq (\mathbb{C}^m,0)$$

where $\cong$ means isomorphism of germs of analytic spaces. See also [G-H].

The following theorem is due to Ephraim [E].

THEOREM 1.10. *Let $(X,0)$ be a germ of an analytic space in $(\mathbb{C}^m,0)$.*

1) *The analytic stratum $(A,0)$ of $(X,0)$ is the germ of an analytic manifold in $(\mathbb{C}^m,0)$.*
2) *The germ $(X,0)$ is a product along A:*

$$(X,0) \cong (X_0,0) \times A,$$

for some germ of an analytic space $(X_0,0)$ *in* $(\mathbb{C}^m,0)$.

An analogous proposition holds for R-equivalence of functions.

PROPOSITION 1.11. *Let* $f\colon (\mathbb{C}^{n+1},0) \to (\mathbb{C},0)$ *be a germ of an analytic function with a one dimensional singular locus* Σ, *such that* $(\Sigma,0)$ *is a non-singular curve.*

a) *Then the following statements are equivalent:*

   1) *There are local coordinates* $x, y_1, \ldots, y_n$ *such that* $f(x,y) = f(0,y)$ *and* $\Sigma = V(y_1,\ldots,y_n)$.

   2) *There are local coordinates* $\bar{x}, \bar{y}_1, \ldots, \bar{y}_n$ *such that* $f_{\bar{x}}(\bar{y}) = f(\bar{x},\bar{y})$ *and* $\Sigma = V(\bar{y}_1,\ldots,\bar{y}_n)$ *and the R-type of* $f_{\bar{x}}$ *is constant for all* $\bar{x}$ *sufficiently small.*

   3) *The derivative* $(\partial f/\partial z_0)$ *is an element of* $(\partial f/\partial z_1, \ldots, \partial f/\partial z_n)$ *for any choice* $z_0,z_1,\ldots,z_n$ *of local coordinates of* $(\mathbb{C}^{n+1},0)$ *such that* $V(z_0)$ *intersects* Σ *transversally at* 0.

b) *Moreover, if either* a) 1), 2) *or* 3) *holds then the transversal R-type of* f *along* Σ *at* 0 *is constant.*

See for the proof [P,1] Proposition 9.12. From now on we denote $\partial f/\partial z_i$ by f_i.

PROPOSITION 1.12. *Let* $f\colon (\mathbb{C}^{n+1},0) \to (\mathbb{C},0)$ *be a germ of an analytic function with a one dimensional singular locus* $(\Sigma,0)$. *Let* $z_0,z_1, \ldots, z_n$ *be local coordinates of* $(\mathbb{C}^{n+1},0)$, *such that* $\Sigma \cap V(z_0) = \{0\}$. *Then the following two statements are equivalent:*

1) $\dim_{\mathbb{C}}(\mathcal{O}/J_f + ((f_1, \ldots, f_n)\colon f_0)) < \infty$.

2) *The transversal R-type of* f *along* Σ_i *is constant at all points of a punctured neighbourhood of* 0 *in* Σ_i, *for every branch* Σ_i *of* Σ.

PROOF. Let U be an open neighborhood of 0 in $\mathbb{C}^{n+1}$, such that Σ and f are defined on U and $\Sigma \setminus \{0\}$ is non-singular. This is possible, since Σ is one dimensional. Let $\mathcal{O}$ be the sheaf of analytic functions on U. Then $\dim_{\mathbb{C}}(\mathcal{O}/J_f + (f_1, \ldots, f_n)\colon f_0)) < \infty$ is equivalent with $V(J_f + ((f_1, \ldots, f_n)\colon f_0)) = \{0\}$, after possibly shrinking U. Further $V(J_f + ((f_1, \ldots, f_n)\colon f_0)) = \Sigma \cap V((f_1, \ldots, f_n)\colon f_0)$.

The hyperplane $H_t = V(z_0 - t)$ intersects Σ transversally in finitely many points, for all $t \neq 0$ sufficiently small, since $\Sigma \cap V(z_0) = \{0\}$. Let $p \in \Sigma_i \cap H_t$, with $t \neq 0$. Then $p \notin V((f_1, ..., f_n): f_0)$ if and only if $f_0 \in (f_1,...,f_n)\mathcal{O}_p$ if and only if the transversal R-type of f along Σ_i at p is constant, by Proposition 1.11. thus $\Sigma \cap V((f_1, ..., f_n): f_0) = \{0\}$ is equivalent with 2). Hence 1) and 2) are equivalent. This proves the proposition.

§2. CERTAIN NON-ISOLATED HYPERSURFACE SINGULARITIES EXIST

THEOREM 2.1. *Let* $(\Sigma,0)$ *be a germ of a reduced, equidimensional analytic space in* $(\mathbb{C}^{n+1},0)$, *with irreducible components* $\Sigma_1, ..., \Sigma_r$ *of dimension k. Let* $K_1, ..., K_r$ *be K-classes of isolated hypersurface singularities in* $(\mathbb{C}^n,0)$. *Then there exists a germ of an analytic function* f: $(\mathbb{C}^{n+k},0) \to (\mathbb{C},0)$ *with singular locus* Σ *and there exist open dense subsets* U_i *of* Σ_i *such that the transversal K-type of f is constant and equal to* K_i *at every point of* U_i.

REMARK 2.2 1) If $(\Sigma,0)$ is a curve then the open dense subset U_i of Σ_i is a punctured neighbourhood of 0 in Σ_i. 2) The theorem above was posed in [P,1] as a question and proved in the case $K_1 = ... = K_r = A_1$.

In order to prove this theorem we need Bertini's Theorem and a lemma which is an exercise in prime avoidance. Bertini's Theorem is well-known and is a consequence of Sard's Theorem. We omit the proof.

BERTINI'S THEOREM 2.3. *Let* $f, \varphi_1, ..., \varphi_p$ *be germs of analytic functions from* $(\mathbb{C}^m,0)$ to $(\mathbb{C},0)$ *and define* $f_\lambda = f + \Sigma\lambda_i\varphi_i$ *for* $\lambda = (\lambda_1, ..., \lambda_p) \in \mathbb{C}^p$. *Then the singular locus of* $f_\lambda^{-1}(0)$ *is contained in* $V(\varphi_1, ..., \varphi_p)$ *for all* $\lambda \in U$, where *U is a dense subset of* $\mathbb{C}^p$.

The proof of the following lemma is probably known but we could not find a proof.

LEMMA 2.4. *Let R be a Noetherian commutative regular ring and I a radical ideal in R such that the associated primes $p_1, ..., p_r$ of I have the same height n. Then there exist $g_{i1}, ..., g_{in} \in p_i$ such that*

(i) $(g_{i1}, ..., g_{in})\mathcal{R}_{p_i} = I_{p_i}$

(ii) $g_{kj} \notin p_i$ *for all j and $i \neq k$*

PROOF. I is a radical ideal in a Noetherian ring R, hence I has a prime decomposition $I = p_1 \cap ... \cap p_r$, with p_i a prime ideal and $p_i \not\subset p_j$ for all $i \neq j$. All prime ideals p_i have the same height n. Hence by induction we can construct a sequence of elements $g_{i1}, ..., g_{in}$ in p_i such that $g_{i1} \in p_i \setminus (\bigcup_{j\neq i} p_j \cup p_i^2)$ and

$$g_{ik+1} \in p_i \setminus \left[\bigcup_{j\neq i} p_j \cup p_i^2 \cup (R \cap (g_{i1}, ..., g_{ik})R_{p_i})\right]$$

for $i = 1, ..., r$ and $1 \leqslant k < n$, see [Ma] 1.B. Let k_i be the residue field of the local ring R_{p_i}. Then by construction $g_{i1}, ..., g_{in}$ are linear independent over k_i. The localization R_{p_i} of the regular ring R is regular again. Thus R_{p_i} is a regular local ring of Krull dimension = height $p_i = n$.

So $(g_{i1}, ..., g_{in})R_{p_i} = p_i R_{p_i} = I_{p_i}$. And $g_{k_j} \notin p_i$ for all j and $i \neq k$, by construction. This proves the lemma

PROOF OF THEOREM 2.1. 1) Remark that a function f: $(\mathbb{C}^n,0) \to (\mathbb{C},0)$ with an isolated singularity and Milnor number μ is $(\mu+2)$-determined, see [M].

2) Let I be the ideal, defining Σ, in the local ring $\mathcal{O}$ of germs of analytic functions on $(\mathbb{C}^{n+k},0)$. I is radical and all associated primes have the same height n, since by assumption Σ is reduced and equidimensional and of dimension k. From the proof of Lemma 2.4 we conclude that there exist functions $g_{i1}, ..., g_{in}$ in $\mathcal{O}$ for $i = 1, ..., r$ such that

i) $g_{i1}, ..., g_{in}$ are part of local coordinates at every point of an open dense subset V_i of Σ_i.

ii) g_{kj} is not identically zero on Σ_i for all j and all $k \neq i$. Let $U_i = V_i \setminus \cup\{V(g_{kj}) \mid 1 \leq k \neq i \leq r \text{ and } 1 \leq j \leq n\}$ then U_i is an open dense subset of Σ_i and $g_{i1}, \ldots, g_{in}$ are part of local coordinates at every point of U_i.

3) Let f_i: $(\mathbb{C}^n,0) \to (\mathbb{C},0)$ be a function with an isolated singularity at 0 and K-type equal to K_i. Let μ_i be the Milnor number of f_i, it depends only on K_i. Define $N = \max\{\mu_i + 2 \mid 1 \leq i \leq r\}$ and let

$$F_0 = \sum_{i=1}^{r} \prod_{k \neq i} g_{k1}^N f_i(g_{i1}, \ldots, g_{in}).$$

Let $\varphi_1, \ldots, \varphi_p$ be generators of the ideal I^N and let:

$$F_\lambda = F_0 + \Sigma\lambda_i\varphi_i.$$

Then by Bertini's theorem, there exists a λ such that F_λ: $(\mathbb{C}^{n+k},0) \to (\mathbb{C},0)$ has a singular locus contained in $V(I^N) = \Sigma$.

4) At every point a of U_i and transversal slice (S,a) at (Σ_i,a), the functions $g_{i1}, \ldots, g_{in}$ restricted to S are local coordinates of (S,a) by 2). Let τ: $(\mathbb{C}^n,0) \to (\mathbb{C}^{n+k},a)$ be the embedding with image (S,a) such that $\tau^*(g_{ij}) = y_j$ and $y_1, \ldots, y_n$ the standard coordinates of $\mathbb{C}^n$. Consider $F_\lambda \circ \tau$: $(\mathbb{C}^n,0) \to (\mathbb{C},0)$ and let

$$G = \sum_{j \neq i} \prod_{k \neq j} g_{k1}^N f_j(g_{j1}, \ldots, g_{jn}) + \Sigma\lambda_i\varphi_i \quad \text{and}$$

$$u = \prod_{k \neq i} g_{k1}^N.$$

Then $F_\lambda \circ \tau = (u \circ \tau) \cdot f_i + G \circ \tau$. Of course, u and G depend on i and τ depends on i and a. Now $G \circ \tau \in (y_1, \ldots, y_n)^N$ and $N \geq \mu_i + 2$ hence $F_\lambda \circ \tau$ is R-equivalent with $(u \circ \tau) f_i$ by 1). Further, $u \circ \tau$ is a unit, since g_{k1} is not zero at a for all $k \neq i$, by definition of u_i. Hence $F_\lambda \circ \tau$ is K-equivalent with f_i.

5) F_λ is singular at all points of Σ, hence the singular locus of F_λ is equal to Σ, by 3).

This proves the theorem.

REMARK 2.10. It is still an open question whether we can replace the K-classes by R-classes.

§3. SERIES OF ISOLATED SINGULARITIES

Let $f\colon (\mathbb{C}^{n+1},0) \to (\mathbb{C},0)$ define a germ of an analytic function with a one dimensional singular locus Σ. Let $z_0, z_1, \ldots, z_n$ be local coordinates of $(\mathbb{C}^{n+1},0)$ such that $\Sigma \cap V(z_0) = \{0\}$. In this section we show that $f + z_0^k$ has an isolated singularity at 0 for $k >> 0$. We shall give a formula for its Milnor number in case the transversal R-type of f at every point of $\Sigma_i \setminus \{0\}$ is constant for every branch Σ_i of Σ.

PROPOSITION 3.1. *Let $f\colon (\mathbb{C}^{n+1},0) \to (\mathbb{C},0)$ be a germ of an analytic function with singular locus Σ of dimension d. Suppose $z_0, \ldots, z_n$ are local coordinates such that $\Sigma \cap V(z_0, \ldots, z_{d-1}) = \{0\}$ as a germ at 0. Then $V(f_d, \ldots, f_n)$ is a complete intersection of dimension d.*

PROOF. Define $f_\lambda = f + \Sigma_{i=0}^{d-1}\lambda_i z_i^2$. Then by Bertini's theorem there exists a λ such that $f_\lambda^{-1}(0)$ has a singular locus Σ_λ contained in $V(z_0^2, \ldots, z_{d-1}^2)$. So $\Sigma_\lambda = \Sigma_\lambda \cap V(z_0,\ldots,z_{d-1})$. Hence

$$\Sigma_\lambda = V(f_0 + 2\lambda_0 z_0, \ldots, f_{d-1} + 2\lambda_{d-1}z_{d-1}, f_d, \ldots, f_n) \cap V(z_0, \ldots, z_{d-1})$$

$$= V(f_0, \ldots, f_n) \cap V(z_0, \ldots, z_{d-1}) = \Sigma \cap V(z_0, \ldots, z_{d-1})$$

which is equal to $\{0\}$ by assumption. Thus

$$f_0 + 2\lambda_0 z_0, \ldots, f_{d-1} + 2\lambda_{d-1}z_{d-1}, f_d, \ldots, f_n$$

is an $\mathcal{O}$-regular sequence and a fortiori $f_d, \ldots, f_n$ is an $\mathcal{O}$-regular sequence. Hence $V(f_d, \ldots, f_n)$ is a complete intersection of dimension d.

REMARK 3.2. There is a tedious proof of this proposition in [P,1], in the case $d = 1$.

PROPOSITION 3.3. *Let $f\colon (\mathbb{C}^{n+1},0) \to (\mathbb{C},0)$ be a germ of an* *analytic function with a one dimensional singular locus* Σ *and* z_0 *a coordinate such that* $\Sigma \cup V(z_0) = \{0\}$. *Then* $f + z_0^k$ *defines an isolated singularity for* $k >> 0$.

PROOF. The proof is the same as in [Lê] and is due to B. Teissier. A stronger condition is required there: $V(z_0)$ is assumed to be an admissible hyperplane, i.e. $f^{-1}(0) \cap V(z_0)$ has an isolated singularity.

Let $\Gamma = V(f_1, ..., f_n)$, then Γ is one dimensional, by Proposition 3.1, since $\Sigma \cap V(z_0) = \{0\}$ and Σ is one dimensional. Let $p_1, ..., p_s$ be the minimal prime ideals in $\mathcal{O}$ lying over $(f_1, ..., f_n)\mathcal{O}$. Let $\Gamma_i = V(p_i)$, then $\Gamma_1, ..., \Gamma_s$ are the branches of Γ. There is for every i at most one k such that $f_0 + kz_0^{k-1} \in p_i$. Otherwise, there are k and l with $k < l$ such that $f_0 + kz_0^{k-1}$, $f_0 + z_0^{l-1} \in p_i$. Hence $z_0^{k-1}(k + lz_0^{l-k}) \in p_i$. So $z_0^{k-1} \in p_i$, since $(k + lz_0^{l-k})$ is a unit in $\mathcal{O}$. But then also $f_0 \in p_i$ and we have that $(z_0^k, f_0, f_1, ..., f_n) \subseteq p_i$. Thus $\Gamma_i \subseteq \Sigma \cap V(z_0) = \{0\}$ and this is a contradiction. Therefore there exists a k_0 such that for all $k \geq k_0$ and all i we have that $f_0 + kz_0^{k-1} \notin p_i$ so $f_0 + kz_0^{k-1} \notin p_1 \cup ... \cup p_s = \cup \mathrm{Ass}(\mathcal{O}_\Gamma)$. Hence $f_0 + kz_0^{k-1}$ is not a zero-divisor of $\mathcal{O}_\Gamma$. Γ is one dimensional hence

$$V(f_0 + kz_0^{k-1}, f_1, ..., f_n) = \{0\}$$

and $f + z_0^k$ has an isolated singularity for all $k \geq k_0$.

NOTATION 3.4. Let $f: (\mathbb{C}^{n+1},0) \to (\mathbb{C},0)$ be a germ of an analytic function with a one dimensional singular locus Σ. Let $J = J_f = (f_0, f_1, ..., f_n)\mathcal{O}$ be the Jacobi ideal of f in $\mathcal{O}$. Let J_k be the Jacobi ideal of $f + (1/k+1)z_0^{k+1}$ and let μ_k be its Milnor number, i.e., $\mu_k = \ell(\mathcal{O}/J_k)$.

Let $J = q_0 \cap q_1 \cap ... \cap q_r$ be a primary decomposition of J, where q_0 is a $\mathcal{M}$-primary component of J, if J has one, and take $q_0 = \mathcal{O}$ otherwise. The primary components $q_1, ..., q_r$ are uniquely defined for the given J. Let $I = q_1 \cap ... \cap q_r$. The singular locus Σ of f is one dimensional. So $\mathcal{O}/I$ has krull dimensional one and has no embedded component and $\Sigma = V(I)$. We give Σ the analytic structure defined by I, i.e., $\mathcal{O}_\Sigma = \mathcal{O}/I$. Let $z_0, z_1, ..., z_n$ be local coordinates of $\mathbb{C}^{n+1}$ at 0. Take $x = z_0$ and suppose $\Sigma \cap V(x) = \{0\}$. This implies that x is not a zero-divisor on $\mathcal{O}_\Sigma$.

In such a case is $e_x(\Sigma)$, the multiplicity of Σ with respect to x at 0, equals $\ell(\mathcal{O}_\Sigma/(x))$ and $ke_x(\Sigma) = \ell(\mathcal{O}_\Sigma/(x^k))$.

In this situation we have the following theorem.

THEOREM 3.5. *Let f: $(\mathbb{C}^{n+1},0) \to (\mathbb{C},0)$ be a germ of an analytic function with a one dimensional singular locus Σ. Let $x,z_1, ..., z_n$ be local coordinates such that $\Sigma \cap V(x) = \{0\}$. Suppose the transversal R-type of f is constant at every point of a punctured neighbourhood of 0 in Σ_i, for every branch Σ_i of Σ. Then, the Milnor number μ_k of $f + (1/k+1)z_0^{k+1}$ for $k >> 0$ is equal to*

$$\mu_k = \ell(I/J) + ke_x(\Sigma) + \ell(\mathcal{O}/I + ((f_1, ..., f_n): f_0)).$$

For the proof of this theorem we need several lemma's.

LEMMA 3.6. $\mu_k = \ell(I/I \cap J_k) + ke_x(\Sigma)$ *for* $k >> 0$.

PROOF. The following sequence is exact:

$$0 \to (I + J_k)/J_k \to \mathcal{O}/J_k \to \mathcal{O}/(I + J_k) \to 0,$$

and consists for $k >> 0$ of $\mathcal{O}$-modulus of finite length, by Proposition 3.3. Now $I/I \cap J_k \simeq (I + J_k)/J_k$ and $\mu_k = \ell(\mathcal{O}/J_k)$. Furthermore $I + J_k = I + x^k)$, thus $\ell(\mathcal{O}/I + J_k) = ke_x(\Sigma)$ by 3.4.

LEMMA 3.7. *Suppose $\ell(\mathcal{O}/I + ((f_1, ..., f_n): f_0)) < \infty$ then $I \cap J_k = f_0.I + (f_1,...,f_n)$.*

PROOF. Let $J' := f_0.I + (f_1, ..., f_n)\mathcal{O}$. Suppose $a \in I \cap J_k$ then $a = \Sigma a_i f_i + a_0 x^k \in I$ for some $a_0,a_1, ..., a_n \in \mathcal{O}$. Now $J \subseteq I$, hence $a_0 x^k \in I$, so $a_0 \in I$, since x is not a zero- divisor of $\mathcal{O}/I$.

Since $J = I \cap q_0$ with q_0 an $\mathcal{M}$-primary ideal or equal to $\mathcal{O}$. We have that $\ell(I/J) < \infty$. So we can find $n_1 \in \mathbf{N}$ such that $x^{n_1} I \subseteq J$. So, for all $k \geqslant n_1$, $a_0 x^k \in (f_0,f_1, ..., f_n)\mathcal{O}$. We assumed

$$\ell(\mathcal{O}/I + ((f_1, \dots, f_n) : f_0)) < \infty,$$

so there exists $n \in \mathbf{N}$ such that

$$x^{n_2} \in I + ((f_1, \dots, f_n) : f_0),$$

hence $x^{n_2} f_0 \in f_0.I + (f_1, \dots, f_n)\mathcal{O}$. Therefore $a_0 x^k \in J'$ for $k \geq n_1 + n_2$. Thus $a \in J'$ and we have proved $I \cap J_k \subseteq J'$ for $k \geq n_1 + n_2$. There exists $n_3 \geq \mathbf{N}$ such that $x^{n_3} I \subseteq J'$, since $(I/I \cap J_k) < \infty$ for $k >> 0$, by Lemma 3.6, and $I \cap J_k \subset J' \subset I$ for $k \geq n_1 + n_2$. But then

$$J' \subseteq (f_0 + x^k)I + (f_1, \dots, f_n)\mathcal{O} + x^k.I \subseteq I \cap J_k + x^k.I \subseteq I \cap J_k + MJ'$$

for $k \geq n_3 + 1$. Hence $J' \subseteq I \cap J_k$ for $k \geq n_3 + 1$., by Nakayama's lemma. Thus $J' = I \cap J_k$ for all $k >> 0$.

PROOF OF 3.5. The transversality assumption implies by Proposition 1.12 $\ell(\mathcal{O}/I + ((f_1, \dots, f_n) : f_0)) < \infty$. So we can apply Lemmas 3.6 and 3.7.

$$I \cap J_k = f_0.I + (f_1, \dots, f_n)\mathcal{O} = J'$$

The module J/J' is generated by f_0 and one easily shows that the annihilator of J/J' is $I + ((f_1, \dots, f_n) : f_0)$. So $J/I \cap J_k = J/J' \simeq \mathcal{O}/I + ((f_1, \dots, f_n) : f_0)$. The following sequence is exact

$$0 \to J/I \cap J_k \to I/I \cap J_k \to I/J \to 0$$

Combining this with Lemma 3.6 we get the desired result:

$$\mu_k = \ell(I/J) + ke_x(\Sigma) + \ell(\mathcal{O}/I + ((f_1, \dots, f_n) : f_0)) \quad \text{for} \quad k >> 0.$$

§4. HYPERSURFACES WITH A COMPLETE INTERSECTION CURVE AS SINGULAR LOCUS

Let $f : (\mathbb{C}^{n+1}, 0) \to (\mathbb{C}, 0)$ be a germ of an analytic function with a one dimensional locus Σ, which is a complete intersection. Suppose

f has A_1 singularities transversally to Σ. In this situation we give a more detailed formula for the Milnor number of $f + x^k$ and a formula for the Euler characteristic of the Milnor fibre of f.

NOTATION 4.1. It follows from [P,1] that the ideal I, defining Σ, and f have the following properties. Since f has transversal A_1-singularities $I = \mathrm{rad}(J_f)$.

The ideal I is generated by an $\mathcal{O}$-sequence $g_1, \ldots, g_n$, since Σ is complete intersection. Moreover, $f \in I^2$, hence $f = \Sigma\, h_{kl}g_kg_l$ for some $h_{kl} \in \mathcal{O}$ with $h_{kl} = h_{lk}$.

Define $\delta_f = \ell(\mathcal{O}_\Sigma/\det(h_{kl}))$, then δ_f is finite.

THEOREM 4.2. *Let* f: $(\mathbb{C}^{n+1},0) \to (\mathbb{C},0)$ *be a germ of an analytic function with a one dimensional singular locus* Σ, *which is a complete intersection. Suppose* f *has* A_1 *singularities transversally to* Σ. *Let* $x, z_1, \ldots, z_n$ *be local coordinates such that* $\Sigma \cap V(x) = \{0\}$. *Then*

$$\ell(\mathcal{O}/I + ((f_1, \ldots, f_n): f_0)) = \delta_f + \mu(\Sigma,0) + e_x(\Sigma) - 1$$

and the Milnor number μ_k *of* $f + (1/k+1)x^{k+1}$ *is equal to*

$$\ell(I/J) + \delta_f + \mu(\Sigma,0) + (k+1)e_x(\Sigma) - 1, \quad k >> 0$$

PROOF. The second formula follows from the first one, by Theorem 3.5, since the transversal R-type is A_1 on every branch of Σ. We can write $f = \Sigma h_{kl}g_kg_l$ with $h_{kl} = h_{lk}$, so $f_i = \Sigma\varphi_{ik}g_k$ for $i = 0,1, \ldots, n$, where

$$\varphi_{ik} \equiv 2\Sigma h_{kl}\,\frac{\partial g_l}{\partial z_i}\ (\mathrm{mod}\ I).$$

Let φ be the $((n+1) \times n)$-matrix with entries φ_{ik} and φ^i the $(n \times n)$-matrix obtained from φ by deleting the $(i+1)$th row, for $i = 0,1, \ldots, n$. Let $\Delta_i = (-i)^i \det \varphi^i$. We need the following lemma.

LEMMA 4.3. $((f_1, \ldots, f_n): f_0) = (\Delta_0, f_1, \ldots, f_n)\mathcal{O}$.

PROOF. Consider the commutative diagram:

$$\begin{array}{ccccc}
 & & \Lambda^2 \mathcal{O}^{n+1} & \xrightarrow{\Lambda^2\varphi} & \Lambda^2 \mathcal{O}^{n} \\
 & & \Big\downarrow df & & \Big\downarrow g \\
\mathcal{O} & \xrightarrow{\Delta} & \mathcal{O}^{n+1} & \xrightarrow{\varphi} & \mathcal{O}^{n} \\
 & & \Big\downarrow df & & \Big\downarrow g \\
 & & \mathcal{O} & = & \mathcal{O} \\
 & & \Big\downarrow & & \Big\downarrow \\
 & & 0 & & 0
\end{array}$$

where the columns are the beginnings of the Koszul complexes of the sequence $df = (f_0, f_1, \ldots, f_n)$ resp. $g = (g_1, \ldots, g_n)$. The map $\Delta\colon \mathcal{O} \to \mathcal{O}^{n+1}$ has components $\Delta_0, \ldots, \Delta_n$, with $\Delta_i = (-1)^i \det(\varphi^i)$. The complex $\mathbf{K}(g,\varphi)$ constructed in [P,2] has components $K_0(g,\varphi) = \mathcal{O}^n$, $K_1(g,\varphi) = \mathcal{O}^{n+1} \oplus \Lambda^2 \mathcal{O}^n$ and $K_3(g,\varphi) = \mathcal{O} \oplus \Lambda^2 \mathcal{O}^{n+1} \oplus \Lambda^3 \mathcal{O}^n$. The support of I/J is $\{0\}$, hence of codimension $n + 1$, so grade $(I/J) = n + 1$ and the complex $\mathbf{K}(g,\varphi)$ is exact, by [P,2]. $\Sigma\Delta_i f_i = g\varphi\Delta = 0$, so $\Delta_0 \in ((f_1, \ldots, f_n)\colon f_0)$. Diagram chasing and using the fact that the right column is exact, since $g_1, \ldots, g_n$ is a $\mathcal{O}$-regular sequence, one obtains $((f_1, \ldots, f_n)\colon f_0) \subseteq (\Delta_0, f_1, \ldots, f_n)\,\mathcal{O}$. So they are equal. This proves Lemma 4.3.

We continue the proof of Theorem 4.2.

$I + ((f_1, \ldots, f_n)\colon f_0) = I + (\Delta_0)\,\mathcal{O}$, by Lemma 4.3, since $(f_1, \ldots, f_n)\mathcal{O} \subseteq I$. The derivative dg is an $((n+1) \times n)$-matrix. Let $(dg)^0$ be the $(n \times n)$-matrix obtained from dg by deleting the first row. Let h be the matrix with entries h_{kl}. Then $\varphi^0 \equiv 2h(dg)^0 \pmod{I}$

$$\Delta_0 = \det(\varphi^0) \equiv 2^n \det(h) \cdot \det(dg)^0 \pmod{I}.$$

Thus $\ell(\mathcal{O}/I + (\Delta_0)) = \ell(\mathcal{O}/I + (\det(h))) + \ell(\mathcal{O}/I + (\det(dg)^0))$, by [K], V, §2, exercise 2c. The first term at the right hand side is by definition δ_f. The map $\tilde{g} = (x, g_1, \ldots, g_n)\colon (\mathbb{C}^{n+1},0) \to (\mathbb{C}^{n+1},0)$ defines a zero-dimensional complete intersection with Milnor number $\mu(\tilde{g})$ and $\mu(\tilde{g}) = e_x(\Sigma) - 1$. Further $\ell(\mathcal{O}/I + (\det(dg)^0)) = \mu(\Sigma,0) + \mu(\tilde{g})$, see [Lo] (5.11). Hence $\ell(\mathcal{O}/I + ((f_1, \ldots, f_n)\colon f_0)) = \delta_f + \mu(\Sigma,0) + e_x(\Sigma) - 1$. This proves the theorem.

PROPOSITION 4.4. *Let $f\colon (\mathbb{C}^{n+1},0) \to (\mathbb{C},0)$ be a germ of an analytic function with a one dimensional singular locus Σ. Suppose f as A_1 singularities transversally to Σ. Let x be a coordinate such that $f^{-1}(0) \cap V(x)$ has an isolated singularity at* 0. *Then*

$$\chi(F) = \chi(F_k) + (-1)^{n+1}(k+1)e_x(\Sigma) \quad \textit{for } k >> 0,$$

where F and F_k are the Milnor fibres of f resp. $f + (1/k+1)x^{k+1}$ and χ denotes the Euler characteristic.

PROOF. See [Y] or [Lê] (2.2.2). A general theorem is proved by Yomdin and Lê with arbitrary transversal singularities.

PROPOSITION 4.5. *Let $f\colon (\mathbb{C}^{n+1},0) \to (\mathbb{C},0)$ be a germ of an analytic function with a one dimensional singular locus Σ, which is a complete intersection. Suppose f has A_1 singularities transversally to Σ. Then*

$$\chi(F) = 1 + (-1)^n[\ell(I/J) + \delta_f + \mu(\Sigma,0) - 1]$$

PROOF. The Milnor fibre of an isolated hypersurface singularity is a bouquet of μ n-spheres, where μ is the Milnor number. So $\chi(F_k) = 1 + (-1)^n\mu_k$. If we choose a general coordinate function x then $f^{-1}(0) \cap V(x)$ has an isolated singularity. Combining the formulas for μ_k in Theorem 4.2 and for $\chi(F)$ in Proposition 4.4 with the above remark, we get the desired result.

REMARK 4.6. a) In the situation of Proposition 4.5 there is (f_t,Σ_t) a deformation of the pair (f,Σ) such that $f_t^{-1}(0)$ has Σ_t as singular locus and Σ_t is a nonsingular curve, i.e. the Milnor fibre of Σ, moreover f_t has only critical points of type A_1 outside Σ_t and only so called A_∞ and D_∞ singularities on Σ_t, for all $t \neq 0$.

Furthermore

$$\delta_f = \#\{D_\infty \text{ points of } f_t\}$$
$$\ell(I/J) = \#\{D_\infty \text{ points of } f_t\} + \#\{A_1 \text{ points of } f_t\},$$

see [P,1]. Siersma [S,2] computed the homotopy type of the Milnor fibre F of f and has the following result:

(i) F has the homotopy type of a bouquet of b_n n-spheres where $b_n = \ell(I/J) + \delta_f + \mu(\Sigma,0) - 1$, in case $\delta_f > 0$.

(ii) F has the homotopy type of a bouquet of one $(n-1)$-sphere and b_n n-spheres, where $b_n = \ell(I/J) + \mu(\Sigma,0)$, in case $\delta_f = 0$.

b) Van Straten [St.1] was able to compute the Betti numbers of F in the situation of Proposition 4.5, by computing the Gauss-Manin system of f.

c) De Jong [J] computed the homotopy type of F in case Σ is a line and transversal singularities of type A_2, A_3, D_4, E_6, E_7 or E_8.

d) Vannier [V] proved a theorem in the spirit of Lê-Ramanujam (μ constant implies topological triviality) for analytic functions with a one dimensional singular locus.

§5. CONCLUDING REMARKS AND QUESTIONS

5.1. Although we give a formula for the Milnor number of $f + x^{k+1}$, what we need is a formula for $f + \varphi$, with φ an element of $\mathcal{O}$ of high enough order and $\Sigma \cap V(\varphi) = \{0\}$, $\ell(\mathcal{O}_\Sigma/(\varphi)) >> 0$ and φ is not "degenerate".

For instance the series $W^\#_{1,k}$ is obtained from the non-isolated singularity $f = (x^2 + y^3)^2$ as follows. Let $\nu(\varphi) = \ell(\mathcal{O}_\Sigma/(\varphi))$. In this case is $\Sigma = V(x^2 + y^3)$.

$$W^\#_{1,2q-1} : (x^2 + y^3)^2 + axy^{4+q}, \quad \mu(W^\#_{1,2q-1}) = 14 + 2q = 3 + \nu(axy^{4+q}).$$

$$W^\#_{1,2q} : (x^2 + y^3)^2 + ax^2y^{3+q}, \quad \mu(W^\#_{1,2q}) = 15 + 2q = 3 + \nu(ax^2y^{3+q}).$$

Where $a = a_0 + a_1y$ and $a_0, a_1 \in \mathbb{C}$, $a_0 \neq 0$. See [A] page 163.

We expect a formula of the form:

$$\mu(f + \varphi) = c(f) + \nu(\varphi),$$

where $c(f)$ is a constant only depending on f and which must be equal to $(-1)^n(\chi(F) - 1)$, by Proposition 4.4.

If φ is a kth-power of a coordinate function and $f = (x^2 + y^3)^2$, we can get only the Milnor numbers $3 + 2k$ or $3 + 3k$ depending whether the coordinate hyperplane intersects $V(x)$ transversally or not.

We show in some examples how φ can be "degenerate". Let $f = y^2 \in \mathbb{C}\{x,y\}$ then $\Sigma = V(y)$. If $\varphi = 2x^ky + x^{2k}$ then φ has arbitrary high order and $\nu(\varphi) = 2k$, but $f + \varphi = (y + x^k)^2$ has a non-isolated singularity. One can show that $f + t\varphi$ has an isolated singularity for all but finitely many t if $\Sigma \cap V(\varphi) = \{0\}$. If $\varphi = x^ky + x^\ell$ then

$$\mu(f + \varphi) = \begin{cases} 2k-1 & \text{if } 2k \leqslant \ell \\ \ell-1 & \text{if } 2k > \ell, \end{cases}$$

whereas $\nu(\varphi) = \ell$ is independent of k.

5.2. Van Straten [St,2] takes another point of view in his thesis. Series of normal surface singularities are obtained from a non-isolated singularity, which is a weakly normal surface. The resolution graphs of the series have the same center but different chains of $\mathbf{P}^1$'s with self intersection -2.

5.3. Examples suggest that in the case f has a one dimensional singular locus Σ with transversal A_1 singularities. One obtains a series of singularities $\chi_{i_1,\ldots,i_r}$ with r indices, where r is the number of branches of Σ. Adjacencies $\chi_{i_1,\ldots,i_r} \leftarrow \chi_{j_1,\ldots,j_r}$ should exist if $i_k \leqslant j_k$ for all $1 \leqslant k \leqslant r$.

5.4. See also Wall's [W] remarks concerning series.

REFERENCES

[A] V. I. Arnol'd: Singularity theory, selected papers, Lecture Note Ser. **53**, Cambridge Univ. Press. Cambridge, 1981.

[E] R. Ephraim: Isosingular loci and the cartesian product-structure of complex-analytic singularities, Trans. Am. Math. Soc. **241** (1978), 357-371.

[G-H] T. Gaffney and H. Hauser: Characterizing singularities of varieties and of mappings, Invent. Math. **81** (1985), 427-447.

[J] T. de Jong: Some classes of line singularities, to appear in Math. Zeitschrift (1988).

[K] E. Kunz: Introduction to commutative algebra and algebraic geometry, Birkhäuser, Boston etc., 1985.

[Lê] Lê D. T.: Ensembles analytiques complexes avec lieu singulier de dimension un (d'après I. N. Iomdin), Publ. Math. de l'Un. Paris VII, ed. Lê D. T., (1980), 87-94.

[Lo] E. Looijenga: Isolated singular points on complete intersections, Lecture Note Ser. **77**, Cambridge Univ. Press. Cambridge, 1984.

[M] J. Mather: Stability of C^∞-mappings III, Finitely determined map germs, IHES Publ. Math. **35** (1968), 127-156.

[Ma] H. Matsumura: Commutative algebra, Math. Lecture Note Ser. **56**, Benjamin, New York, 1970.

[P,1] R. Pellikaan: Hypersurface singularities and resultions of Jacobi modules, Thesis, Rijksuniversiteit Utrecht, 1985.

[P,2] R. Pellikaan: Projective resolutions of the quotient of two ideals, Indagationes Mathematicae **91** (1988), 65-84.

[S,1] D. Siersma: Isolated line singularities, in Proc. Symp. Pure Math. **40**, Singularities at Arcata 1981, ed. P. Orlik, part 2, 485-496.

[S,2] D. Siersma: Singularities with critical locus a 1-dimensional complete intersection and transversal type A_1, Topology and its Applications, 27 (1987), 51-73.

[St,1] D. van Straten: On the Betti numbers of the Milnor fibre of a certain class of hypersurface singularities, in Proceedings of Lambrecht Conf., 1985, Lecture Notes in Math. **1273**, Springer (1988).

[St,2] D. van Straten: Weakly normal surface singularities and their improvements, Thesis, Rijksuniversiteit Leiden, 1986.

[V] J.-P. Vannier: Familles à un paramètre de fonctions analytiques à lieu singulier de dimension un, C.R. Acad. Sc. Paris **303** (1986), 367-370.

[W] C. T. C. Wall: Notes on the classification of singularities, Proc. London Math. Soc. **48** (1984), 461-573.

[Y] I. N. Yomdin: Variétés complexes avec singularités de dimension un, Siberian Math. J. **15** (1974), 1061-1082.

DEPARTMENT OF MATHEMATICS
AND COMPUTER SCIENCE
UNIVERSITY OF TECHNOLOGY
P.O. BOX 513
5600 MB EINDHOVEN
THE NETHERLANDS

Contemporary Mathematics
Volume **90**, 1989

QUASIHOMOGENEOUS SINGULARITIES WITH TRANSVERSAL TYPE A_1

Dirk Siersma

ABSTRACT. We study holomorphic map germs

$$f\colon (\mathbb{C}^{n+1},0) \to (\mathbb{C},0)$$

with the properties:

(i) The critical set Σ is an isolated complete intersection curve.

(ii) The transversal singularity of f at points of $\Sigma - \{0\}$ is of type A_1.

(iii) f is quasihomogeneous.

We compute the Poincaré polynomials of several graded vectorspaces, associated to f and give formulas for the middle Betti number of the Milnor fibre and other discrete invariants in terms of the weights, the degree of f and the degrees of the functions, which define the reduced critical locus.

§1. INTRODUCTION

(1.1) We study holomorphic map germs $f : (\mathbb{C}^{n+1},0) \to \mathbb{C}$ with the following properties:

(i) the critical set Σ is a 1-dimensional icis (isolated complete intersection singularity).

(ii) the transversal singularity of f at points of $\Sigma - \{0\}$ is of type A_1.

(iii) f is quasihomogeneous.

0271-4132/89 $1.00 + $.25 per page

(1.2) Singularities with a 1-dimensional critical locus and transversal type A_1 are a natural class to study. They have properties, which are rather close to the properties of isolated singularities and there seems to be a relation with series of isolated singularities. This relation is up to now not completely understood.

We use this relationship for giving names to the singularities:

A_∞	$y_1^2 + ... + y_n^2$	in $(x,y_1, ..., y_n)$-space
D_∞	$xy_1^2 + y_2^2 + ... + y_n^2$	in $(x,y_1, ..., y_n)$-space
$T_{\infty,q,r}$	$xyz + y^q + z^r$	in (x,y,z)-space
$T_{\infty,\infty,\infty}$	xyz	in (x,y,z)-space

(1.3) I first examined the simplest case: Σ is a smooth curve in the paper: Isolated line singularities, [Si-1]. Since that time there are results in the case that Σ is a 1-dimensional isolated complete intersection in the following areas:

- analytic theory and deformations: Pellikaan [Pe-1,2,3,4,5], de Jong [Jo-1]
- topology of the Milnor fibre: Siersma [Si-2], [Si-3], de Jong [Jo-2].
- relative De Rham cohomology: Van Straten [vS-1]
- improvements (in the surface case): Stevens [Sv], Van Straten [vS-2]

We first describe some of those results.

From now on we consider only the case that Σ is a 1-dimensional (reduced) isolated complete intersection.

(1.4) Let $g_1, ..., g_n$ define the complete intersection Σ as a reduced analytic variety. Let $I = (g_1, ..., g_n)$. Then we have:

$$f \text{ is singular on } \Sigma \Leftrightarrow f \in I^2 \qquad \text{[Pe-1] (I(1.6))}$$

In this case we can write:

$$f = \Sigma\, h_{ij} g_i g_j \quad \text{with} \quad h_{ij} \in \mathcal{O}_{n+1} \text{ and } h_{ij} = h_{ji} .$$

Consider the following ideals:

$$I(f) = \left[\frac{\partial f}{\partial x_0}, \ldots, \frac{\partial f}{\partial x_n} \right] \qquad \text{Jacobian ideal}$$

$$\tau_e(f) = \{\xi(f) \mid \xi \in \mathrm{Der}_{\mathbb{C}} \mathcal{O} \text{ and } \xi(I) \subset I\} \qquad \text{tangent space}$$

and following quotients:

$I/J(f)$: Jacobi module

$I^2/\tau_e(f)$: transversal slice

$\mathcal{O}_\Sigma/\det h_{ij}$: Hessian ring

with the dimensions:

$j(f) = \dim I/J(f)$ Jacobi number

$c_e(f) = \dim I^2/\tau_e(f)$ codimension

$\delta(f) = \dim \mathcal{O}_\Sigma/(\det h_{ij})$ Hesse number

Equivalent are: [Pe-1]:

(a) $c_e(f) < \infty$

(b) $j(f) < \infty$

(c) The transversal type of f along $\Sigma - \{0\}$ is A_1.

(1.5) Moreover Pellikaan showed [Pe-1] I(7.20) that there is an approximation f_s of f such that on an open neighbourhood V of 0 in $\mathbb{C}^{n+1}$:

(i) There is a smooth curve Σ_s and f_s has only A_∞ and D_∞ singularities on $V \cap \Sigma_s$

(ii) f_s has only A_1-singularities in $V \setminus \Sigma_s$

(iii) $j(f) = {}^{\#}A_1 + {}^{\#}D_\infty$ and $\delta(f) = {}^{\#}D_\infty$ (where ${}^{\#}A_1$ denotes the number of A_1-points of f_s and ${}^{\#}D_\infty$ denotes the number of D_∞-points of f_s)

and according to [Si-3] (Lemma 3.6):

(iv) the Milnor fibrations of f and f_s are equivalent.

(1.6) Siersma considered the topology of the Milnor fibre of f and showed [Si-3] (theorem 6.2):

The homotopy type of the Milnor fibre F of f is a bouquet of spheres:

<u>case A</u>: $^{\#}D_\infty > 0$: $F \overset{h}{\simeq} S^n \vee \ldots \vee S^n$

<u>case B</u>: $^{\#}D_\infty = 0$: $F \overset{h}{\simeq} S^{n-1} \vee S^n \vee \ldots \vee S^n$

and the middle Betti numbers are:

Case A: $b_n(F) = \mu(\Sigma) + {}^{\#}A_1 + 2^{\#}D_\infty - 1$

Case B: $b_n(F) = \mu(\Sigma) + {}^{\#}A_1$.

(1.7) Next Van Straten [vS-1] considered the relative De Rham complex sheaf:

$$\Omega_f^{\cdot} = \Omega^{\cdot}/df \wedge \Omega^{\cdot -1}$$

For dim $\Sigma = 1$ and transversal type A_1 one has:

a) $H^n(\Omega_f^{\cdot})$ is concentrated in $\{0\}$
b) $H^{n-1}(\Omega_f^{\cdot})$ is a constructible sheaf concentrated on Σ or $\Sigma - \{0\}$.
c) $H^k(\Omega_f^{\cdot}) = 0 \quad k \neq 0, n-1, n$.

Van Straten [vS-1] proved the following result (in case Σ 1-icis and transversal type A_1)

a) $H^n(\Omega_f^{\cdot})$ is a free $\mathbb{C}\{f\}$-module of dimension $b_n(F)$
b) $H^{n-1}(\Omega_f^{\cdot})$ is a free $\mathbb{C}\{f\}$-module of dimension $b_{n-1}(F)$
c) $b_n(F) = \dim_{\mathbb{C}}(\Omega^{n+1}/df \wedge \Omega^n + d\mathcal{H}^n)$ where

$\mathcal{H}^n = \mathrm{Ker}(df \wedge : \Omega^n \to \Omega^{n+1})/df \wedge \Omega^{n-1}$: n^{th} Koszul cohomology group.

(1.8) The purpose of this paper is to consider the quasihomogeneous singularities and express the occuring invariants in terms of weights and degrees.

Quasihomogeneous isolated singularities have been studied intensively by Arnol'd [Ar-1], see also [AGV], chapter 12. He determined the Poincaré series of $\mathcal{O}/J(f)$ and the Milnor number of f in terms of the weights and degree of f. From the topological point of view the formula for the Milnor number was derived by Milnor-Orlik [Mi-Or].

More precisely a holomorphic function $f : (\mathbb{C}^{n+1},0) \to (\mathbb{C},0)$ is called quasihomogeneous of degree d with weights $w_0, \ldots, w_n$ if for any $\lambda > 0$ we have

$$f(\lambda^{w_0}x_0, \ldots, \lambda^{w_n}x_n) = \lambda^d f(x_0, \ldots, x_n)$$

For quasihomogeneous *isolated singularities* [Ar-1]:

The Poincaré series $P(\mathcal{O}/J(f)) = \prod_{i=0}^{n} \dfrac{1 - t^{d-w_i}}{1 - t^{w_i}}$

The Milnor number $\mu(f) = \dim \mathcal{O}/J(f) = \prod_{i=0}^{n} \dfrac{d - w_i}{w_i}$

The local algebra $\mathcal{O}/J(f)$ has exactly one generator of maximal degree $d^{max} = (n+1)d - 2\Sigma\, w_i$ (see also Saito [Sa-2]).

(1.9) In our case: Σ 1-icis and transversal type A_1, we replace $\mathcal{O}/J(f)$ by several other vectorspaces:

- The Jacobi module $I/J(f)$
- The transversal slice $I^2/\tau_e(f)$
- The Hessian algebra $\mathcal{O}_\Sigma/(\det h_{ij})$
- The De Rham vectorspace $\Omega^{n+1}/df \wedge \Omega^n + d\mathcal{H}^n$

and compute their Poincaré series in terms of the weights $w_0, \ldots, w_n$ and the degrees $d, a_1, \ldots, a_n$ of f and $g_1, \ldots, g_n$ and also determine the maximal degree d^{max}.

We also give formulas for

- the jacobinumber $j(f)$
- the codimension $c_e(f)$
- the Hesse number $\delta(f)$
- the middle Betti number $b_n(F)$
- $\#A_1$ and $\#D_\infty$

We end with some remarks about modality and simpleness.

(1.10) SUMMARY: Let Σ be a 1-icis, given by the ideal $I = (g_1, \ldots, g_n)$ and $f : (\mathbb{C}^{n+1},0) \to (\mathbb{C},0)$ a quasihomogeneous function with weights $w_0, \ldots, w_n$ and degree d, transversal type A_1 and let $g_1, \ldots, g_n$ be homogeneous of degree $a_1, \ldots, a_n$. Let $W = \Sigma_{i=0}^n w_i$ and $A = \Sigma_{j=1}^n a_j$. Remember

$$P(\mathcal{O}_\Sigma)(t) = \frac{\Pi(1 - t^{a_j})}{\Pi(1 - t^{w_i})} \qquad i = 0, \ldots, n; \;\; j = 1, \ldots, n$$

and write

$$P_{reg}(t) = \frac{\Pi(1 - t^{d-w_i})}{\Pi(1 - t^{w_i})} \qquad i = 0, \ldots, n.$$

Then we have the following *Poincaré polynomials*:

$$P_1(t) = P(I/J(f))(t) = P_{reg}(t) + [t^{(n+1)d-W-A} - 1]P(\mathcal{O}_\Sigma)(t)$$

$$P_2(t) = P(I^2/\tau_e(f))(t) = P_{reg}(t) +$$
$$+ [t^{(n+1)d-W-A} + \Sigma\, t^{d-w_i} - \Sigma t^{a_j} - t^d - 1]P(\mathcal{O}_\Sigma)(t) - t^{A-W+d}$$

$$P_3(t) = P(\mathrm{Coker}\ h_f)(t) = \left[\Sigma t^{a_j} - \Sigma t^{d-a_j}\right] P(\mathcal{O}_\Sigma)(t)$$

$$P_4(t) = P(T^1_\Sigma)(t) = t^d\left[\Sigma t^{-a_j} - \Sigma t^{-w_i} + 1\right] P(\mathcal{O}_\Sigma)(t) + t^{A-W+d}$$

$$P_5(t) = P(\Omega^{n+1}/df \wedge \Omega^n + d\mathcal{H}^n)(t) =$$

$$= \begin{cases} P_{reg}(t) + t^{nd-W-A}(t^d-1)P(\mathcal{O}_\Sigma)(t) & \text{if } \#D_\infty > 0 \\ P_{reg}(t) + t^{nd-W-A}(t^d-1)P(\mathcal{O}_\Sigma)(t) + t^{A-W+d} & \text{if } \#D_\infty = 0. \end{cases}$$

For the *dimensions* we have:

$$j(f) = \Pi \frac{d - w_i}{w_i} - [(n+1)d - W - A] \cdot \frac{\Pi a_j}{\Pi w_i}$$

$$c_e(f) = \Pi \frac{d - w_i}{w_i} - [(2n+1)d - 2W - 2A] \frac{\Pi a_j}{\Pi w_i} - 1$$

$$\delta(f) = (\mathrm{nd} - 2A) \frac{\Pi a_j}{\Pi w_i} = \#D_\infty$$

$$\tau(\Sigma) = [A - W] \frac{\Pi a_j}{\Pi w_i} + 1$$

$$b_n(F) = \Pi \frac{d - w_i}{w_i} - d \frac{\Pi a_j}{\Pi w_i} \quad \text{if} \quad \#D_\infty > 0 \quad (\delta(f) > 0)$$

$$b_n(F) = 1 + \Pi \frac{d - w_i}{w_i} - d \frac{\Pi a_j}{\Pi w_i} \quad \text{if} \quad \#D_\infty = 0 \quad (\delta(f) = 0)$$

$$\#A_1 = \Pi \frac{d - w_i}{w_i} - [(2n+1)d - W - 3A] \frac{\Pi a_j}{\Pi w_i}$$

The *maximal degrees* are:

$$d_{\text{reg}}^{\max} = (n+1)d - 2W$$

$$d_1^{\max} = (n+1)d - 2W - \min\{a_j\}$$

$$d_2^{\max} \leqslant d_1^{\max}$$

$$d_3^{\max} = d - \min{}^*\{a_j\} + A - W$$

$$d_4^{\max} = d - W + A - \min_{i \neq j} \{w_i + w_j\}$$

$$d_5^{\max} = d_1^{\max}$$

§2. THE POINCARÉ POLYNOMIAL OF I/J_f

(2.1) Let $\mathcal{O} = \mathbb{C}[x_0, \ldots, x_n]$ or $\mathbb{C}\{x_0, \ldots, x_n\}$ or $\mathbb{C}[[x_0, \ldots, x_n]]$. Let $(w_0, \ldots, w_n)$ be the weights of $x_0, \ldots, x_n$. We suppose $w_i \in \mathbf{N}$ and $d \in \mathbf{N}$ such that $d \geqslant 2w_i$ for all i.

$f \in \mathcal{O}$ is called *quasihomogeneous of degree d* if all monomials $x_0^{m_0} \ldots x_n^{m_n}$ of f have the property

$$m_0 w_0 + \ldots + m_n w_n = d.$$

This defines a grading on $\mathcal{O}$.

For a graded $\mathbb{C}$-vectorspace A we define the *Poincaré series* of A:

$$P(A)(t) = \sum_{k=0}^{\infty} (\dim A_k) t^k$$

Well known facts about Poincaré series are:

a) $\dim A = P(A)(1)$ if A is finite dimensional.

b) If $0 \to A \to B \to C \to 0$ is exact and the morphisms have degree 0 then $P(A) - P(B) + P(C) = 0$.

c) $P(A \otimes B) = P(A)P(B)$.

Examples:

a) $P(\mathbb{C}[[z]])(t) = 1 + t + \ldots + t^k + \ldots = (1/1-t) = Q(t)$ (definition)

b) Let $\mathcal{O} = \mathbb{C}[[x_0, \ldots, x_n]]$ then:

$$P(\mathcal{O})(t) = Q(t^{w_0}) \ldots Q(t^{w_n}) = \prod_0^n \frac{1}{(1 - t^{w_i})} .$$

(2.2) *Grading on* $\Omega^{\cdot}$ (the complex of differential forms with coefficients in $\mathcal{O}$). Let $I = (i_0, \ldots, i_p)$ and $J = \{0, \ldots, n\} - I = \{j_{p+1}, \ldots, j_n\}$.

The basis elements of $\Omega^{\cdot}$ are φdz_I where $\varphi = x_0^{m_0} \ldots x_n^{m_n}$ and $dz_I = dz_{i_0} \wedge \ldots \wedge dz_{i_p}$.

Let $f : \mathbb{C}^{n+1} \to \mathbb{C}$ be quasihomogeneous of degree d. We define the following grading on $\Omega^{\cdot}$ (which depends on d):

codegree: $\operatorname{codeg}(dz_I) = \sum_{k=0}^{p} (d - w_{i_k})$

$$\operatorname{codeg}(\varphi dz_I) = -\deg(\varphi) + \operatorname{codeg}(dz_I)$$

degree: $\deg(dz_I) = \sum_{k=p+1}^{n} (d - w_{j_k})$

$$\deg(\varphi dz_I) = \deg(\varphi) + \deg(dz_I)$$

Remark that: $\deg(\varphi dz_I) + \operatorname{codeg}(\varphi dz_I) = (n+1)d - W$ where $W = \Sigma_{k=0}^{n} w_k$.

LEMMA. $\operatorname{codeg}(\alpha \wedge \beta) = \operatorname{codeg}(\alpha) + \operatorname{codeg}(\beta)$.

PROPOSITION. *The multiplication with df in the Koszul cohomology complex*

$$0 \to \Omega^0 \overset{\wedge df}{\to} \Omega^1 \overset{\wedge df}{\to} \Omega^2 \to \dots \to \Omega^n \overset{\wedge df}{\to} \Omega^{n+1} \to 0$$

is a quasihomogeneous map of degree 0.

REMARK. $P(\Omega^k)(t) = \sum_{|I|=k} t^{\deg(dz_I)} P(\mathcal{O})(t)$ where $d = \deg(f)$.

Define

$$P_{reg}(t) = P(\Omega^{\cdot})(t) = \sum_{k=0}^{n+1} (-1)^{n+1-k} P(\Omega^k)(t)$$

$$= \sum_{k=0}^{n} \sum_{|I|=k} (-1)^{n+1-k} t^{\deg(dz_I)} P(\mathcal{O})(t)$$

$$= (1-t^{d-w_0}) \dots (1-t^{d-w_n}) P(\mathcal{O})(t) = \prod_{i=0}^{n} \frac{1-t^{d-w_i}}{1-t^{w_i}}.$$

(2.3) We consider the Koszul cohomology groups $\mathcal{H}^k$ of the Koszul cohomology complex. The top dimensional group is:

$$\mathcal{H}_{n+1} = \operatorname{Coker}(\wedge df : \Omega^n \to \Omega^{n+1}) = \frac{\Omega^{n+1}}{df \wedge d\Omega^n} =: \Omega_f .$$

N.B. $\Omega_f \cong \dfrac{\mathcal{O}}{J(f)}$ where $J(f) = \left[\dfrac{\partial f}{\partial x_0}, \dots, \dfrac{\partial f}{\partial x_n}\right]$ the Jacobian ideal.

Ω_f has the induced grading from Ω^{n+1} and the $\mathbb{C}$-isomorphism. $\Omega_f \cong \mathcal{O}/J(f)$ is quasihomogeneous of degree 0. $\mathcal{O}/J(f)$ plays an important role in the study of singularities.

(2.4) First we consider the case that f has an isolated singularity at 0. In this case, it is well known (cf.: Greuel [Gr]) that the Koszul cohomology complex is exact in dimensions smaller than $n + 1$.

PROPOSITION (Arnol'd [Ar-1], [AGV]). *Let f be a quasihomogeneous isolated singularity, then*

a. $P(\mathcal{O}|J(f))(t) = P(\Omega_f)(t) = P_{reg}(t) = \prod_{i=0}^{n} \dfrac{1 - t^{d-w_i}}{1 - t^{w_i}}$

b. $P_{reg}(t)$ is a symmetric polynomial:

$$P_{reg}(t) = t^{(n+1)d-2W} \cdot P_{reg}\left[\frac{1}{t}\right] \text{ with } W = \Sigma w_i.$$

c. The degree $d^{\max}$ of $P_{reg}(t)$ is: $d^{\max} = (n+1)d - 2W.$

d. $\mu(f) = \dim_{\mathbb{C}} \mathcal{O}/J(f) = \prod_{i=0}^{n} \dfrac{d - w_i}{w_i}$

PROOF a.

$$P_{reg}(f) = \sum_{k=0}^{n+1} (-1)^{n+1-k} P(\Omega^k) = \sum_{k=0}^{n+1} (-1)^{n+1-k} P(\mathcal{H}^k) = P(\mathcal{H}^{n+1}) = P(\Omega_f).$$

(2.5) Next we consider the case that $f : \mathbb{C}^{n+1} \to \mathbb{C}$ has a one dimensional singular locus Σ, which is a complete intersection and with transversal type A_1 (cf. the introduction).

Let I be the ideal which defines Σ as a reduced variety, $I = (g_1, \ldots, g_n)$.

Let $\deg g_1 = a_1, \ldots, \deg(g_n) = a_n$.

We set $\Sigma_{k=1}^n a_k = A$ and $\Sigma_{k=0}^n w_i = W$ (as before).

In this case it is known (cf. Van Straten [vS-1]§3) that the Koszul cohomology complex is exact, except possibly in dimensions n and $n + 1$.

$$\mathcal{H}^{n+1} = \Omega_f = \Omega^{n+1}/df \wedge \Omega^n \cong \mathcal{O}/J(f)$$
$$\mathcal{H}^n = \mathrm{Ker}(df\wedge : \Omega^n \to \Omega^{n+1})/df \wedge \Omega^{n-1}.$$

Pellikaan [Pe-1] gave an explicit isomorphism

$$\mathcal{H}^n \cong \omega_\Sigma = \mathrm{Ext}^n_{\mathcal{O}}(\mathcal{O}_\Sigma, \mathcal{O})$$

According to Van Straten [vS-1] this isomorphism has the following description:

Define $\omega_i \in \Omega^1$ by $df = \Sigma \omega_i g_i$. Then

$$\mathcal{H}^n = \omega_1 \wedge .... \wedge \omega_n \quad \mathcal{O}_\Sigma$$

Since

i) codeg($\alpha \wedge \beta$) = codeg(α) + codeg(β)

ii) codeg(df) = 0 and codeg($d\theta$) = codeg(θ) + d

iii) codeg(ω_i) = a_i

We have: $\text{codeg}(\omega_1 \wedge ... \wedge \omega_n) = \sum_{i=1}^{n} a_i = A$

So: $\deg(\omega_1 \wedge ... \wedge \omega_n) = (n+1)d - W - A.$

From the Koszul cohomology complex and the nonvanishing $\mathcal{H}^n$ and $\mathcal{H}^{n+1}$ we find:

$$P(\mathcal{O}/J(f))(t) = P_{\text{reg}}(t) + t^{(n+1)d-W-A}P(\mathcal{O}_\Sigma(t))$$

(2.6) *The Poincaré series of* $\mathcal{O}_\Sigma$ can be found in several places (cf. Stanley [St1,2] or Wagreich [Wg]:

$$P(\mathcal{O}_\Sigma)(t) = \frac{\prod_{j=1}^{n} (1 - t^{a_j})}{\prod_{i=0}^{n} (1 - t^{w_i})}$$

which we abbreviate to

$$P(\mathcal{O}_\Sigma)(t) = \prod \frac{(1 - t^{a_j})}{(1 - t^{w_i})}$$

(2.7) THEOREM. *Let* $f : \mathbb{C}^{n+1} \to \mathbb{C}$ *be quasihomogeneous of transversal type* A_1; $\Sigma(f) = \Sigma$ *a* 1-*icis. Then*:

a. $P(I/J(f))(t) = P_{\text{reg}}(t) + (t^{(n+1)d-W-A} - 1)P(\mathcal{O}_\Sigma)(t)$

$$= \prod \frac{1 - t^{d-w_i}}{1 - t^{w_i}} - (1 - t^{(n+1)d-W-A})\prod \frac{(1 - t^{a_j})}{(1 - t^{w_i})}$$

($i = 0, ..., n$; $j = 1, ..., n$).

b. $P(I/J(f))$ is a symmetric polynomial:

$$P(t) = t^{(n+1)d-2W}P\left(\frac{1}{t}\right)$$

c. The degree $d^{\max}$ of $P(I/J(f))$ is: $d^{\max} = (n+1)d - 2W - \min\{a_i\}$.

d. $$j(f) = \Pi \frac{d - w_i}{w_i} - [(n+1)d - W - A] \cdot \Pi \frac{a_j}{w_i} \quad \begin{matrix}(i=0,\dots,n)\\(j=1,\dots,n)\end{matrix}$$

e. $j(f) = 0$ if and only if f is of type A_∞.

PROOF.

a. This follows from the above computations.

b. $\dim_{\mathbb{C}}(I/J_f) < \infty$ since f has transversal type A_1, so $P(I/J(f))$ is a polynomial. Moreover:

$$P\left(\frac{1}{t}\right) = \frac{(-1)^{n+1}t^W}{(-1)^{n+1}t^{(n+1)d-W}} \cdot P(t)$$

So: $t^{(n+1)d-2W}P\left(\frac{1}{t}\right) = P(t)$

and the coefficient of t^k is equal to the coefficient of $t^{(n+1)d-2W-k}$.

c. The numerator is generalized symmetric and equal to:

$$\left(1 - \Sigma t^{d-w_i} + \sum_{i,j} t^{2d-w_i-w_j} \pm \dots\right) - \left(1 - t^{a_0} - \Sigma t^{a_i} + \sum_{i,j} t^{a_i+a_j} \pm \dots\right)$$

where $a_0 = (n+1)d - W - A$. We have to find out if $t^{a_0}, t^{a_1}, \dots, t^{a_n}$ cancel against some other term. Remember that we consider only $(d; w_0, w_1, \dots, w_n)$ such that $w_i \leqslant d/2$ for all i.

Since $\det h_{ij}$ has degree $nd - 2A$ we must have $2A \leqslant nd$. So we have:

$$d - w_k \geqslant d/2 \geqslant \min\{a_i\},$$

where the minimum is taken over $i = 1, \dots, n$.

Suppose now $\min\{a_i\} = d/2$ then it follows that

(i) $a_1 = \dots = a_n = d/2$ and so $\#D_\infty = 0$.

Let now $w_0 \leqslant w_1 \leqslant \ldots \leqslant w_n \leqslant d/2$. For the canceling of $t^{\min\{a_i\}}$ it is also necessary that

ii) $w_1 = \ldots = w_n = d/2$.

But then $a_0 = (n+1)d - w_0 - n(d/2) - n(d/2) = d - w_0$. So also t^{a_0} cancels. In this case we have even that $P(I/J(f))$ is identically zero: The conditions (i) and (ii) imply the existence of constants such that

$$\begin{cases} g_1 = b_1 x_0^k + a_{11}x_1 + \ldots + a_{1n}x_n \\ \vdots \\ g_n = b_n x_0^k + a_{n1}x_1 + \ldots + a_{nn}x_n \end{cases}$$

Since Σ is reduced and $\dim \Sigma = 1$ we can change the coordinates homogeneously such that

$$g_1 = x_1, \ldots, g_n = x_n.$$

So Σ is a smooth line and f must be of type A_∞ since $^{\#}D_\infty = 0$.

What about a_0? In case f is not of type A_∞, we have

$$a_0 = (n+1)d - W - A > (n+1)d - (n+1)\frac{d}{2} - n\frac{d}{2} = \frac{d}{2} \geqslant \min\{a_i\}$$

So the minimal degree of the terms of $P(t)$ is $\min\{a_i\}$. By symmetry $d^{\max} = (n+1)d - 2W - \min\{a_i\}$.

d. $j(f) = P(I/J(f))(1)$

$$= \lim_{t\to 1}\left[\Pi\frac{1-t^{d-w_i}}{1-t}\cdot\frac{1-t}{1-t^{w_i}} - \frac{1-t^{a_0}}{1-t}\cdot\Pi\frac{1-t^{a_j}}{1-t}\cdot\frac{1-t}{1-t^{w_i}}\right]$$

$$= \Pi\frac{d-w_i}{w_i} - a_0\frac{\Pi a_i}{\Pi w_i}$$

e. In part c. of the proof we have already shown that $j(f) = 0$ implies f is of type A_∞.

§3. THE POINCARÉ POLYNOMIAL OF COKER h_f and of $I^2/I^2 \cap J(f)$

(3.1) As before $I = (g_1, \ldots, g_n)$ defines a reduced complete intersection of dimension one. We have

$$f \in I^2 \Leftrightarrow f \quad \text{is singular on} \quad \Sigma.$$

Consider derivations $\xi \in \mathrm{Der}_{\mathbb{C}}(\mathcal{O})$ with the property $\xi(I) \subset I$. For $f \in I^2$, Pellikaan [Pe-1] (p. 31) or [Pe-4] defined the tangent spaces:

$$\tau_e(f) = \tau_{I,e}(f) = \{\xi(f) \mid \xi(I) \subset I\}$$

$$\tau(f) = \tau_I(f) = \{\xi(f) \mid \xi(I) \subset I \text{ and } \xi(0) = 0\}$$

and he showed in our case:

$$\tau_e(f) = I^2 \cap J(f)$$

$$\tau(f) = I^2 \cap mJ(f) \quad (m \text{ is max. ideal of } \mathcal{O})$$

Next he defined:

$$c_e(f) = \dim \frac{I^2}{I^2 \cap J(f)} \quad \text{(extended codimension)}$$

$$c(f) = \dim \frac{I^2}{I^2 \cap mJ(f)} \quad \text{(codimension)}$$

In most cases $c_e(f) = c(f)$, more precisely:

- If Σ is singular $c_e(f) = c(f)$
- If Σ is smooth $c_e(f) = c(f)$ if f is of type A_∞
  $c_e(f) = c(f) + 1$ for other types.

We shall treat the extended case only.

(3.2) In this section we are interested in the Poincaré polynomial of $I^2/I^2 \cap J(f)$ and its relation to the Poincaré polynomial of $I/J(f)$.

We follow the proof of theorem 5.14 of Pellikaan [Pe-1] (p. 59) cf. also [Pe-5] who showed the relation:

$$j(f) = c_e(f) + \tau(\Sigma,0) + \delta(f)$$

where $\tau(\Sigma,0)$ is the Tjurina number of Σ and $\delta(f)$ the Hesse number (to be defined later).

Remark first: $\dfrac{I^2}{I^2 \cap J(f)} \cong \dfrac{I^2 + J(f)}{J(f)}$

In Pellikaan's work we find the following diagrams and constructions (p. 60-61):

$$0 \to \frac{I^2 + J(f)}{J(f)} \to \frac{I}{J(f)} \to \frac{I}{I^2 + J(f)} \to 0$$

is exact (all maps induced by inclusions). For $f \in I^2$

$$\begin{array}{ccccccccc}
 & & & & & & 0 & & \\
 & & & & & & \downarrow & & \\
 & & & & 0 & \to & T^1_\Sigma & & \\
 & & & & \downarrow & & \downarrow & & \\
 & & (\mathcal{O}_\Sigma)^m & \overset{\phi}{\to} & \dfrac{I}{I^2} & \to & \dfrac{I}{I^2 + J(f)} & \to & 0 \\
 & & \downarrow dg & & \downarrow & & \downarrow & & \\
0 & \to & \mathrm{Hom}_{\mathcal{O}_\Sigma}\left[\dfrac{I}{I^2}, \mathcal{O}_\Sigma\right] & \overset{h_f}{\to} & \dfrac{I}{I^2} & \to & \mathrm{Coker}\, h_f & \to & 0 \\
 & & \downarrow & & \downarrow & & \downarrow & & \\
 & & T^1_\Sigma & & 0 & & 0 & & \\
 & & \downarrow & & & & & & \\
 & & 0 & & & & & &
\end{array}$$

is exact and commutative

(3.3) In the case of Σ a reduced complete intersection we can be more explicit:

$$
\begin{array}{ccccccccc}
 & & & & & & 0 & & \\
 & & & & & & \downarrow & & \\
 & & & & 0 & \rightarrow & T^1_\Sigma & & \\
 & & & & \downarrow & & \downarrow & & \\
 & & (\mathcal{O}_\Sigma)^{n+1} & \overset{\phi}{\rightarrow} & (\mathcal{O}_\Sigma)^n & \overset{\alpha}{\rightarrow} & \dfrac{I}{I^2 + J(f)} & \rightarrow & 0 \\
 & & \downarrow dg & & \downarrow & & \downarrow & & \\
0 & \rightarrow & (\mathcal{O}_\Sigma)^n & \overset{h_f}{\rightarrow} & (\mathcal{O}_\Sigma)^n & \rightarrow & \text{Coker } h_f & \rightarrow & 0 \\
 & & \downarrow & & \downarrow & & \downarrow & & \\
 & & T^1_\Sigma & & 0 & & 0 & & \\
 & & \downarrow & & & & & & \\
 & & 0 & & & & & &
\end{array}
$$

where the maps are defined as follows:

ϕ) Let $(\partial f/\partial x_j) = \Sigma\phi_{ij}g_i$. The map $\mathcal{O}^{n+1} \rightarrow \mathcal{O}^n$ with matrix (ϕ_{ij}) induces $\phi : (\mathcal{O}_\Sigma)^{n+1} \rightarrow (\mathcal{O}_\Sigma)^n$.

α) $\alpha(\xi_1, \ldots, \xi_n) = \Sigma\xi_i g_i$ defines a map $(\mathcal{O}_S)^n \rightarrow I/I^2$.

h_f) Let $f = \Sigma h_{ij}g_i g_j$ then $h_f(\xi_1, \ldots, \xi_n) = \Sigma h_{ij}\xi_i g_j$. $\delta_f = \dim \text{Coker } h_f$

dg) $dg(\xi_0, \ldots, \xi_n) = \sum_{k=0}^{n} \left(\dfrac{\partial g_i}{\partial x_k}\right)\xi_k$

and Coker $dg = T^1_\Sigma$ is the so-called module of first order deformations and $\dim_{\mathbb{C}} T^1_\Sigma = \tau(\Sigma,0)$ is the Tjurina number.

(3.4) From these exact sequences we find:

$$P(I/J(f)) = P(I^2/I^2 \cap J(f)) + P(\text{Coker } h_f) + P(T^1_\Sigma).$$

For shorthand:

$$P_1 = P_2 + P_3 + P_4.$$

We have to define a good graded structure for Coker h_f and T^1_Σ.

(3.5) First we treat Coker h_f. Remember: deg $g_i = a_i$; deg $f = d$. Since $f = \Sigma h_{ij} g_i g_j$ we have deg $h_{ij} = d - a_i - a_j$. We put on $(\mathcal{O}_\Sigma)^n$ new gradings such that all maps in the following diagram have degree zero.

$$\begin{array}{ccccccc} & & \dfrac{I}{I^2} & \to & \dfrac{I}{I^2 + J(f)} & \to & 0 \\ & & \cong\downarrow\alpha & & \downarrow & & \\ 0 \to & \mathcal{O}^n_\Sigma[d-a_1, \ldots, d-a_n] & \xrightarrow{h_f} & \mathcal{O}^n_\Sigma[a_1, \ldots, a_n] & \to & \text{Coker } h_f \to 0 & (*) \\ & & & \downarrow & & & \\ & & & 0 & & & \end{array}$$

Nb. $(\varphi, 0, \ldots, 0) \in \mathcal{O}^n_\Sigma[a_1, \ldots, a_n]$ has degree $-a_1 + \deg(\varphi)$, etc. We give Coker h_f the grading induced from $\mathcal{O}^n_\Sigma[a_1, \ldots, a_n]$.

Let $a_1 \leqslant a_2 \leqslant \cdots \leqslant a_n < d$. We define $\min^*\{a_i\}$ as follows: Delete all a_j's with $a_j = d/2$ and all pairs $\{a_p, a_q\}$, with $a_p + a_q = d$. Take the minimum of the remaining a_i's and call that $\min^*\{a_i\}$.

(3.6) PROPOSITION. *Let $f \in I^2$ be quasihomogeneous and of transversal type A_1. Let $P_3 = P$ (Coker h_f). Then:*

a) $$P_3(t) = \left[\left(t^{a_1} - t^{d-a_1}\right) + \cdots + \left(t^{a_n} - t^{d-a_n}\right)\right] \cdot \Pi \frac{(1 - t^{a_j})}{(1 - t^{w_i})}$$

b) *P_3 is a polynomial, which is symmetric:*

$$P_3(t) = t^{A+d-W} P_3\left(\frac{1}{t}\right)$$

c) *The degree $d_3^{\max}$ of P_3 is*

$$d_3^{\max} = d - \min^*\{a_i\} + A - W$$

d) $\delta_f = \dim \text{Coker } h_f = (nd - 2A)\Pi \dfrac{a_j}{w_i}$.

PROOF. a)

$$P(\text{Coker } h_f) = P(\mathcal{O}^n_\Sigma[a_1, ..., a_n]) - P(\mathcal{O}^n_\Sigma[d - a_1, ..., d - a_n])$$
$$= \left[t^{a_1} + \cdots + t^{a_n}\right]P(\mathcal{O}_\Sigma) - \left[t^{d-a_1} + \cdots + t^{d-a_n}\right]P(\mathcal{O}_\Sigma)$$

b) $\dim(\text{Coker } h_f) < \infty$ so P_3 is a polynomial; the symmetry is clear.

c) We have cancelling in the formula for P_3 when $a_j = d/2$ and when $a_p + a_q = d$. Use also the fact that all coefficients of P_3 are non-negative.

(3.7) REMARK. Pellikaan [Pe-1] (p. 82) mentions

$$\dim_{\mathbb{C}} \text{Coker } h_f \cong \dim_{\mathbb{C}} \mathcal{O}_\Sigma/\det(h_{ij})$$

For the obvious graded structure of $\mathcal{O}_\Sigma/\det(h_{ij})$ with

$$\deg(\det(h_{ij})) = nd - 2A$$

we have:

$$P(\mathcal{O}_\Sigma/\det(h_{ij}))(t) = P(\mathcal{O}/(g_1, ..., g_n, \det(h_{ij}))(t) = (1-t^{nd-2A})\prod \frac{(1 - t^{a_j})}{(1 - t^{w_i})}$$

which shows that the grading on Coker h_f is different from the obvious grading on $\mathcal{O}_\Sigma/\det(h_{ij})$.

Of course $\delta_f = (nd - 2A)\cdot\prod(a_j/w_i)$ in both cases.

(3.8) Next we consider T^1_Σ. A formula for the Poincaré polynomial for quasihomogeneous curves occurs in the work of Scherbak and Lyashko, cf. Arnol'd [Ar-2]:

$$P(T^1_\Sigma)(t) = \left[\Sigma t^{-a_j} - \Sigma t^{-w_j} + 1\right] \prod \frac{1 - t^{a_j}}{1 - t^{w_i}} + t^{A-W}$$

With respect to the grading, T^1_Σ is there identified with

$$\mathcal{O}_\Sigma \frac{\partial}{\partial y_1} \oplus \cdots \oplus \mathcal{O}_\Sigma \frac{\partial}{\partial y_n} \text{ and } \deg\left[\frac{\partial}{\partial y_q}\right] = -a_q .$$

In our notation: $T^1_\Sigma = \mathcal{O}^n_\Sigma[-a_1, ..., -a_n]$.

Arnol'd's paper [Ar-2] contains also a Lyashko formula for $P(T^1_\Sigma)$ in case of a higher dimensional complete intersection. A proof of the formula can be found in a paper of Alexandrov [Al-2] as a consequence of the Corollary on page 49. See also Alexandrov [Al-2].

In the case we consider, we use on T^1_Σ the grading of

$$\mathcal{O}^n_\Sigma[d-w_0, \ldots, d-a_n]$$

from the exact sequence:

$$\mathcal{O}^{n+1}_\Sigma[d - w_0, \ldots, d - w_n] \xrightarrow{dg} \mathcal{O}^n_\Sigma[d - a_1, \ldots, d - a_n] \to T^1_\Sigma \to 0$$

(3.9) COROLLARY. *Let Σ be a 1-icis and T^1_Σ supplied with the above grading. Write $P_4 = P(T^1_\Sigma)$. Then:*

a) $P_4(t) = t^d\left[\Sigma t^{-a_j} - \Sigma t^{-w_i} + 1\right] \Pi \dfrac{1 - t^{a_j}}{1 - t^{w_i}} + t^{A-W+d}$

b) *P_4 is a polynomial.*

c) *The degree d_4^{max} of P_4 is* $d_4^{max} = d - W + A - \min_{i\neq j}\{w_i+w_j\}$

d) $\tau(\Sigma) = \dim T^1_\Sigma = [A - W]\dfrac{\Pi a_j}{\Pi w_i} + 1.$

(3.10) Giusti showed [Gi-1], [Gi-3] the formula:

$$\mu(\Sigma) = 1 + \frac{\Pi a_j}{\Pi w_i}$$

in the case that Σ was a quasihomogeneous complete intersection of dimension one. Together with assertion (3.9d) this shows again $\tau(\Sigma) = \mu(\Sigma)$, which is due to Greuel [Gr]. Greuel and Hamm [Gr-Ha] gave a general formula for $\mu(\Sigma)$ for isolated quasihomogeneous complete intersections, without the condition $\dim \Sigma = 1$.

(3.11) PROPOSITION. *Let $f \in I^2$ with transversal type A_1 and $P_2 = P(I^2/\tau_e(f))$. Then:*

a) $$P_2(t) = \Pi \frac{1 - t^{d-w_i}}{1 - t^{w_i}} + \left[t^{(n+1)d-W-A} + \Sigma t^{d-w_i} + \Sigma t^{a_j} - t^d - 1\right] \cdot \frac{\Pi(1 - t^{a_j})}{\Pi(1 - t^{w_i})} - t^{A-W+d}$$

b) P_2 *is a polynomial*

c) *The degree* $d_2^{\max}$ *of* P_2 *has the property:* $d_2^{\max} \leq d_1^{\max}$

d) $c_e(f) = \Pi \dfrac{d - w_i}{w_i} - [(2n+1)d - 2A - 2W] \dfrac{\Pi a_j}{\Pi w_i} - 1.$

PROOF. a) follows from $P_2 = P_1 - P_3 - P_4$. b) follows from $\dim (I^2/\tau_e(f)) < \infty$. c) $d_2^{\max} \leq d_1^{\max}$ follows from a) and the fact that all coefficients must be nonnegative. d) $c_e(f) = j(f) - \delta(f) - \tau(\Sigma)$.

NB. I expect that in most of the cases $d_2^{\max} = d_1^{\max}$.
With respect to $d_4^{\max} \leq d_1^{\max}$ we have:

$$A + \min\{a_i\} \leq (n+1) \frac{d}{2}$$

$$W \leq (n-1) \frac{d}{2} + \min_{i \neq j} \{w_i + w_j\}$$

So: $A + W + \min\{a_i\} \leq nd + \min_{i \neq j}(w_i + w_j)$ and

$$d - W - \min_{i \neq j}\{w_i + w_j\} \leq (n+1)d - 2W - \min\{a_i\}).$$

So: $d_4^{\max} \leq d_1^{\max}$. In case $n \geq 2$, we have even $d_4^{\max} < d_1^{\max}$.

Let $n = 1$. If $d_4^{\max} = d_1^{\max}$ we have $d - 2W = 2d - 2W - A$ and so must $d = 2$ and it follows $f = c \cdot g^2$ in which case $c_e(f) = 0$ and $P_2 = 0$.

§4. THE COHOMOLOGY OF THE MILNOR FIBRE

(4.1) PROPOSITION. *Let* $f \in I^2$ *with transversal type* A_1 *and quasihomogeneous of degree* d. *Let* F *be the Milnor fibre of* f. *The Euler characteristic of* F *is*:

$$\chi(F) = 1 + (-1)^n \left[\frac{\Pi(d - w_i)}{\Pi w_i} - d \frac{\Pi a_j}{\Pi w_i} \right].$$

PROOF. According to Pellikaan [Pe-1] Proposition (10.11); p. 113:

$$\chi(F) = 1 + (-1)^n[j(f) + \delta(f) + \mu(\Sigma) - 1].$$

By (2.7d), (3.6d) and (3.9d) we have

$$j(f) + \delta(f) + \mu(\Sigma) - 1 =$$

$$= \frac{\Pi(d-w_i)}{\Pi w_i} - \frac{\Pi a_j}{\Pi w_i}[(n+1)d-W-A] + [nd-2A]\frac{\Pi a_j}{\Pi w_i} + \frac{\Pi a_j}{\Pi w_i}[A-W] =$$

$$= \frac{\Pi(d-w_i)}{\Pi w_i} + \frac{\Pi a_j}{\Pi w_i}\{-(n+1)d+W+A+nd-2A+A-W\} = \frac{\Pi(d-w_i)}{\Pi w_i} - d\frac{\Pi a_j}{\Pi w_j},$$

(4.2) Let again $f \in I^2$ with transversal type A_1.

From the topological description [Si-3] we know that the Milnor fibre F of f is homotopy equivalent to a bouquet of spheres:

case A: ${}^{\#}D_\infty > 0 : F \overset{h}{\simeq} S^n \vee \cdots \vee S^n$.

case B: ${}^{\#}D_\infty = 0 : F \overset{h}{\simeq} S^{n-1} \vee \cdots \vee S^n$.

The number of n-spheres in case A is

$$\mu_n(f) = 2\,{}^{\#}D_\infty + {}^{\#}A_1 + \mu(\Sigma) - 1 = \frac{\Pi(d-w_i)}{\Pi w_i} - d\,\frac{\Pi a_j}{\Pi w_i}$$

In Case B:

$$\mu_n(f) = {}^{\#}A_1 + \mu(\Sigma) = \frac{\Pi(d-w_i)}{\Pi w_i} - d\,\frac{\Pi a_j}{\Pi w_i} + 1.$$

These formulas follow from the above proposition (4.1). Instead of Proposition (10.11) of Pellikaan, we can also use his proposition (7.20) [Pe-1] p. 87, which states:

$$j(f) = {}^{\#}A_1 + {}^{\#}D_\infty \quad \text{and} \quad \delta(f) = {}^{\#}D_\infty$$

So: $2\,{}^{\#}D_\infty + {}^{\#}A_1 + \mu(\Sigma) - 1 = j(f) + \delta(f) + \mu(\Sigma)$. The condition $D_\infty = 0$ is for quasihomogeneous singularities equivalent with $nd = 2A$.

(4.3) PROPOSITION. ${}^{\#}D_\infty = (nd - 2A)\dfrac{\Pi a_j}{\Pi w_i}$

$${}^{\#}A_1 = \Pi\,\frac{d-w_i}{w_i} - [d - W + A]\,\frac{\Pi a_j}{\Pi w_i}$$

PROOF.

$${}^{\#}D_\infty = \delta(f) = (nd - 2A)\,\frac{\Pi a_j}{\Pi w_i}$$

$$\#A_1 + \#D_\infty = j(f) = \Pi \frac{d - w_i}{w_i} - [(n+1)d - W - A] \frac{\Pi a_i}{\Pi w_i}.$$

(4.4) We can also use the approach of van Straten [vS-1], see also (2.5).

On the cohomology groups $\mathcal{H}^k$ of the Koszul complex

$$\to \Omega^{k-1} \overset{\wedge df}{\to} \Omega^k \overset{\wedge df}{\to} \Omega^{k+1} \to \dots \to \Omega^{n+1} \to 0$$

we consider also the action of the differential operator d of the De Rham complex:

$$\dots \to \Omega^{k-1} \overset{d}{\to} \Omega^k \overset{d}{\to} \Omega^{k+1} \to \dots \to \Omega^{n+1} \to 0$$

So we get:

$$\dots \to \mathcal{H}^{k-1} \overset{d}{\to} \mathcal{H}^k \overset{d}{\to} \mathcal{H}^{k+1} \to \dots \to \mathcal{H}^{n+1} \to 0$$

Van Straten showed [vS-1] (Theorem 3 and its corollary): If $\#D_\infty > 0$ then $d : \mathcal{H}^n \to \mathcal{H}^{n+1}$ is injective. If $\#D_\infty = 0$ then $d : \mathcal{H}^n \to \mathcal{H}^{n+1}$ has a 1-dimensional kernel represented by the closed form ω.

Moreover: $H^{n+1}(\mathcal{H}^\cdot) = \mathcal{H}^{n+1}/d\mathcal{H}^n = \Omega^{n+1}/df \wedge \Omega^n + d\mathcal{H}^n$ and:

$$b_n(F) = \dim_{\mathbb{C}} H^{n+1}(\mathcal{H}^\cdot) = \dim_{\mathbb{C}} \Omega^{n+1}/df \wedge \Omega^n + d\mathcal{H}^n.$$

(4.5) Next we compute the Poincaré polynomial of $H^{n+1}(\mathcal{H}^\cdot)$. Since $\mathcal{H}^{n+1} = \Omega^{n+1}/df \wedge \Omega^n$ we have from (2.7a)

$$P(\mathcal{H}^{n+1})(t) = P_{reg}(t) + t^{(n+1)d-W-A} P(\mathcal{O}_\Sigma)(t)$$

Also $\mathcal{H}^n = \omega_1 \wedge \dots \wedge \omega_n \mathcal{O}_\Sigma$ (see (2.5)). Consider now:

$$0 \to \operatorname{Ker} d \to \mathcal{H}^n \overset{d}{\to} \mathcal{H}^{n+1} \to \mathcal{H}^{n+1}/d\mathcal{H}^n \to 0$$

Remember $\omega = \omega_1 \wedge \dots \wedge \omega_n$

$$\operatorname{codeg}(\omega) = A;\ \deg(\omega) = (n+1)d - W - A$$
$$\operatorname{codeg}(d\omega) = d + A;\ \deg(d\omega) = nd - W - A.$$

So the degree of the De Rham differential operator d is $-d$.

a) In case ${}^{\#}D_\infty > 0$: Ker $d = 0$ and

$$\begin{aligned} P(\mathcal{H}^{n+1}/d\mathcal{H}^n)(t) &= P(\mathcal{H}^{n+1})(t) - P(d\mathcal{H}^n)(t) \\ &= P_{reg}(t) + t^{(n+1)d-W-A}P(\mathcal{O}_\Sigma)(t) - t^{nd-W-A}P(\mathcal{O}_\Sigma)(t) \\ &= P_{reg}(t) + t^{nd-W-A}(t^d - 1)P(\mathcal{O}_\Sigma)(t) \end{aligned}$$

b) In case ${}^{\#}D_\infty = 0$ ($\Leftrightarrow nd = 2A$): Ker $d \cong \mathbb{C}\omega_1 \wedge \dots \wedge \omega_n$. So:

$$P(\mathcal{H}^{n+1}/d\mathcal{H}^n)(t) = P_{reg}(t) + t^{A-W}[t^d - 1]P(\mathcal{O}_\Sigma)(t) + t^{d-A-W}.$$

(4.6) PROPOSITION. *Let $f \in I^2$ with transversal type A_1 and*

$$P_5 = P(\mathcal{H}^{n+1}/d\mathcal{H}^n) = P(\Omega^{n+1}/df \wedge \Omega^n + d\mathcal{H}^n).$$

Then:

a) $$P_5(t) = \Pi \frac{1-t^{d-w_i}}{1-t^{w_i}} + t^{nd-W-A}(t^d - 1) \frac{\Pi(1-t^{a_j})}{\Pi(1-t^{w_i})} \quad \text{if } {}^{\#}D_\infty > 0$$

$$P_5(t) = \Pi \frac{1-t^{d-w_i}}{1-t^{w_i}} + t^{nd-W-A}(t^d-1) \frac{\Pi(1-t^{a_j})}{\Pi 1-t^{w_i})} + t^{d-W+A} \text{ if } {}^{\#}D_\infty = 0$$

b) *P_5 is a polynomial.*

c) *The degree d_5^{max} of P_5 is*

$$d_5^{max} = d_1^{max} = (n+1)d - 2W - \min\{a_i\}.$$

d) *The nth Betti number $b_n(F)$ of the Milnor fibre of f is*

$$b_n(F) = \frac{\Pi(d - w_i)}{\Pi w_i} - d\frac{\Pi a_j}{\Pi w_i} \quad \textit{if } {}^{\#}D_\infty > 0$$

$$b_n(F) = 1 + \frac{\Pi(d - w_i)}{\Pi w_i} - d\frac{\Pi a_j}{\Pi w_i} \quad \textit{if } {}^{\#}D_\infty = 0.$$

PROOF. a) is already shown. b) is true because

$$\dim_{\mathbb{C}}(\mathcal{H}^{n+1}/ d\mathcal{H}^n) = b_n(F) < \infty.$$

c) $$P_5(t) - P_1(t) = (1 - t^{nd-W-A}) \frac{\Pi(1 - t^{a_j})}{\Pi(1 - t^{w_i})}$$

with max. degree: $nd - W - A + A - W = nd - 2W$.

$d_1^{\max} = (n+1)d - 2W - \min\{a_i\} > nd - 2W$ since $f \in I^2$ and so $d > \min\{a_i\}$.
d) $b_n(F) = P_5(1)$.

§5. REMARKS ABOUT MODALITY AND ABOUT SIMPLE QUASIHOMOGENEOUS SINGULARITIES

(5.1) For *isolated singularities* $f : (\mathbb{C}^n,0) \to \mathbb{C}$ Arnol'd defined the *modality* of f to be the least number such that a sufficient small neighbourhood of $j^k f$ (in a sufficient k-jet space) is covered by a finite number of orbits (of the group $\mathfrak{D}$ of local diffeomorphisms).

The germ f is called *simple* if its modality is zero, that is if its neighbourhood intersects only a finite number of orbits. It is well known, that up to stably equivalence the simple hypersurface singularities are given by the following list:

$A_k : x^{k+1}$ $(k \geqslant 1)$
$D_k : x^2y + y^{k+1}$ $(k \geqslant 4)$
$E_6 : x^3 + y^4$
$E_7 : x^3 + xy^3$
$E_8 : x^3 + y^5$

(5.2) For quasihomogeneous singularities there is also the concept of *inner modality* (i-modality); cf. Arnol'd [Ar-1], [AGV], §13. The i-modality m_0 of a quasihomogeneous function germ f is the number of diagonal and superdiagonal basis monomials of the local algebra $\mathcal{O}/J(f)$.

$$m_0(f) = \dim_{\mathbb{C}} \frac{\mathcal{O}_{\geqslant d}}{J(f) \cap \mathcal{O}_{\geqslant d}}$$

where $\mathcal{O}_{\geqslant d} = \{f \in \mathcal{O} \mid \deg(f) \geqslant d\}$. One should think about the i-modality as the codimension of the orbit of f intersected with $\mathcal{O}_{\geqslant d}$ in $\mathcal{O}_{\geqslant d}$. f is called *i-simple* if $m_0(f) = 0$ or equivalently

$$d_{\max} < d = \deg(f).$$

Since for isolated singularities $d_{\max} = (n+1)d - 2W$ this inequality reduces to:

$$nd < 2W$$

which plays an important role in the classification of all i-simple isolated singularities.

For quasihomogeneous hypersurface singularities modality and i-modality coincide and i-simple is the same as simple. See Kushnirenko-Gabrielov [Ga] and Varchenko (cf. [AVG] p. 222).

(5.3) Also for *quasihomogeneous isolated complete intersection singularities* there are notions of modality and i-modality. The i-modality of a complete intersection Σ can be calculated as the number of monomials of positive degree in a monomial basis of T^1_Σ. (Arnol'd [Ar-2]).

Arnol'd mentions the conjecture that the i-modality is equal to the modality (of the contact group action). He also remarks that this conjecture is partially confirmed by a theorem (due to I. G. Scherbak) that i-modality 0 complete intersections $\mathbb{C}^3 \to \mathbb{C}^2$ coincide with the simple curves of Giusti [Gi-2,4].

It is implicit in the work of Giusti [Gi-2,4] and Wall [Wa-1,2] that i-simple singularities of $g : \mathbb{C}^{n+1} \to \mathbb{C}^n$ (the case we consider) occur only in case $n = 2$ and are equal to the simple singularities of Giusti's list.

(5.4) Next we discuss modality for the singularities of this paper: $f : \mathbb{C}^{n+1} \to \mathbb{C}$ quasihomogeneous, $\deg(f) = d$ with singular locus a curve and transversal type A_1.

The first idea is to work with $I^2/\tau_e(f)$. This "transversal slice" is nicely related to the action of $\mathfrak{D}_\Sigma$ on I^2, where $\mathfrak{D}_\Sigma$ consists of the local diffeomorphisms which preserve Σ. A disadvantage is that $I^2/\tau_e(f)$ can be trivial in several cases, e.g.

$$f = g^2 \qquad \text{in the curve case}$$

and its "suspension"

$$f = g^2 + z_2^2 + \dots + z_n^2 \qquad \text{in general.}$$

We have the feeling that it is nicer to work with the Jacobi-module $I/J(f)$. We then also have information about T^1_Σ which is following (3.2-4) together with $I^2/\tau_e(f)$ and $\mathrm{Coker}(h_f)$ a direct summand of $I/J(f)$. Moreover $j(f) = \dim_{\mathbb{C}} I/J(f) = 0$ if and only if f is of type A_∞.

We define the inner-modality of f by:

$$m_0(f) = \dim_{\mathbb{C}} \frac{I \cap \mathcal{O}_{\geq d}}{J(f) \cap \mathcal{O}_{\geq d}}$$

and call f *i-simple* if $m_0(f) = 0$ or equivalently

$$d_1^{\max} < d$$

Since $d_1^{\max} = (n+1)d - 2W - \min\{a_i\}$ this inequality reduces to

$$nd < 2W + \min\{a_i\}$$

(5.5) As a first result we have:

PROPOSITION. *f is i-simple $\Rightarrow$ Σ is i-simple.*

PROOF. Let $d_\Sigma^{\max}$ be the maximal degree of T^1_Σ with the grading:

$$T^1_\Sigma = \mathcal{O}^n_\Sigma[-a_1, \ldots, -a_n]$$

(this is the grading used by Arnol'd and Scherbak). So Σ *i*-simple $\Leftrightarrow$ $d_\Sigma^{\max} < 0$. Remark that $d_4^{\max} = d_\Sigma^{\max} + d$. So $d_1^{\max} < d$ implies $d_\Sigma^{\max} < 0$ since $d_4^{\max} \leq d_1^{\max}$.

(5.6) REMARK. Perhaps it is possible to use proposition (5.5) to list all *i*-simple singularities. We have not done that up to now. In the next sections we study some special cases.

(5.7) *The curve case* ($n = 1$). Let $f : \mathbb{C}^2 \to \mathbb{C}$ be quasihomogenous with transversal type A_1. We can write

$$f = h \cdot g^2$$

with h and g square free and without common factors. Define: $f_{\text{red}} = h \cdot g : \mathbb{C}^2 \to \mathbb{C}$. f_{red} has an isolated singularity.

PROPOSITION. *f i-simple $\Leftrightarrow$ f_{red} simple.*

PROOF. Let $\deg(g) = a$ and $\deg(h) = b$. So $d = \deg(f) = 2a + b$

$$d_1^{\max} = 2d - 2(w_0 + w_1) - a = 3a + 2b - 2(w_0 + w_1)$$

Now: f i-simple $\Leftrightarrow d_1^{\max} < d \Leftrightarrow a + b < 2(w_0 + w_1)$.
Since $\deg(f_{\text{red}}) = a + b$ and for isolated singularities

$$d^{\max} = 2 \deg(f_{\text{red}}) - 2(w_0 + w_1)$$

We have:

$$f_{\text{red}} \text{ simple} \Leftrightarrow d^{\max} < a + b \Leftrightarrow a + b < 2(w_0 + w_1).$$

REMARK. Since the only simple isolated curve singularities are

$$A_k (k \geq 1), \quad D_k (k \geq 4), \quad E_6, \quad E_7, \quad E_8$$

we have the following possibilities for i-simple f:

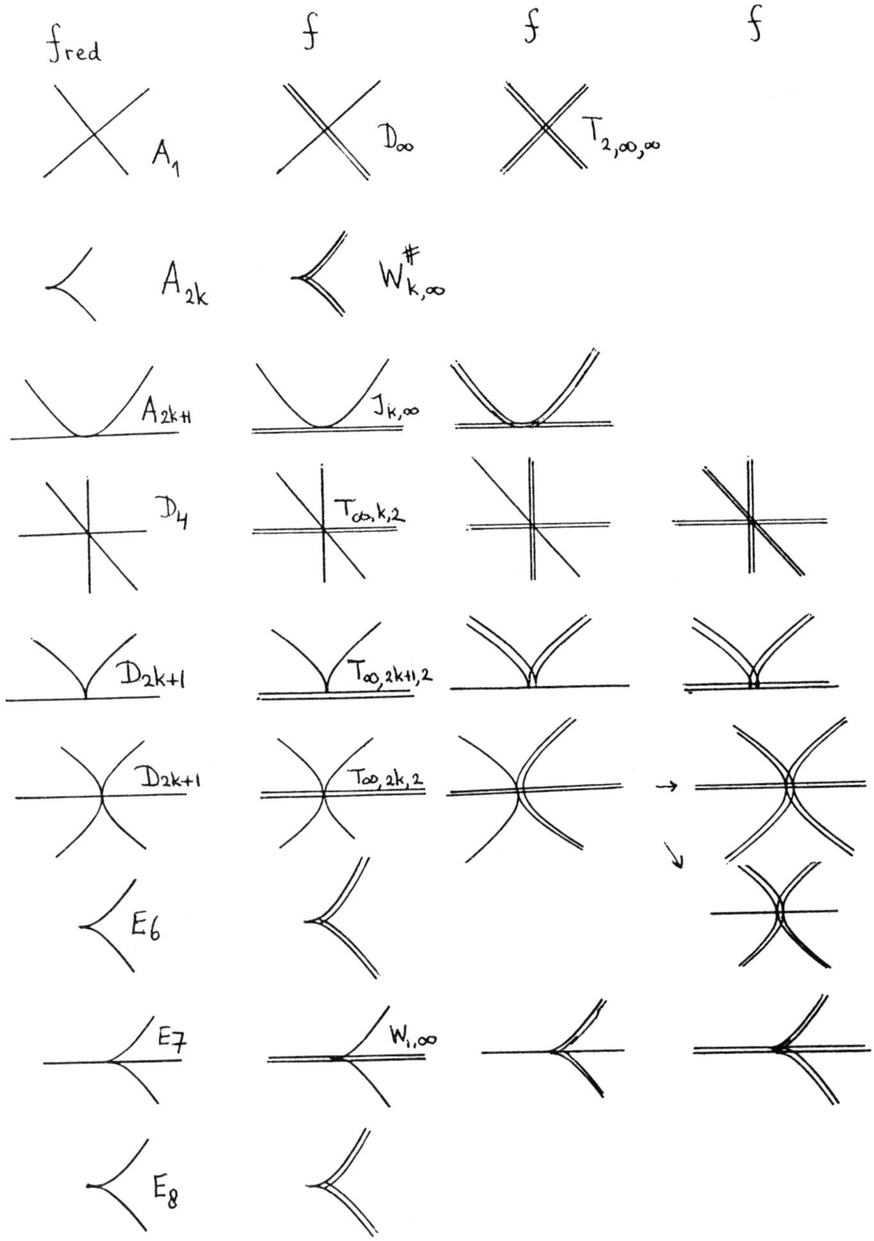

(5.8) *Case: isolated line singularities.* The singular locus is now a smooth line and transversal type A_1. In [Si-1] we gave the following list of simple isolated line singularities. They are all quasihomogeneous

Name	formula	w_0	w_1	w_2	d	d_1^{max}
A_∞	y^2	1	1	–	2	–
D_∞	xy^2	1	1	–	3	1
$J_{k,\infty}(k\geq 2)$	$y^3+x^ky^2$	1	k	–	$3k$	$3k$-2
$T_{\infty,k,2}(k\geq 4)$	$x^2y^2+y^k$	k-2	2	–	$2k$	$2k$-2
$Z_{k,\infty}(k\geq 1)$	$xy^3+x^{k+2}y^2$	1	k+1	–	$3k$+4	$3k$+3
$W_{1,\infty}$	$y^4+y^2x^3$	2	3	–	12	11
$T_{\infty,q,r}\begin{cases} r\geq 3 \\ q\geq 3\end{cases}$	$xyz+y^q+z^r$	qr-q-r	r	q	qr	qr-min$\{q,r\}$
$Q_{k,\infty}$	$xz^2+y^3+x^ky^2$	1	$2k$	$3k$-1	$6k$	$6k$-2
$S_{1,\infty}$	$y^2z+xz^2+x^2y^2$	2	3	4	10	9

Nonsimple boundary cases are:

	formula	w_0	w_1	w_2	d	d_1^{max}
	$y^4+Ay^3x^2+y^2x^4$	1	2	–	8	8
	$x^3y^2+Axy^4+\epsilon y^5$	r	1	–	5	5
$S_\infty^{\#}$	$yz^2+xz^2+Axy^3+x^3y^2$	1	2	3	7	7

This list shows f simple [Si-1] $\Rightarrow$ f is i-simple.

For $n = 1$ it is not difficult to show:

$$f \text{ simple [Si-1]} \Leftrightarrow f \text{ is } i\text{-simple.}$$

(5.9) *Case: Singular locus: plane curve* A_1. In [Si-2] we have the following list of simple singularities.

Name	formula	w_0	w_1	w_2	d	d_1^{max}
$T_{\infty,\infty,2}$	x^2y^2	1	1	–	4	2
	$x^2y^2(y+x^n)$	1	n	–	$3n$+2	$3n$+1
$T_{\infty,\infty,r}$	$xyz+z^r$	1	r-2	1	r	r-1
	$xz^2+yz^2+x^2y^2$	2	2	3	8	7

Also in this case: f is simple [Si-2] $\not\Rightarrow$ f i-simple.

(5.10) *Case:* $f : \mathbb{C}^3 \to \mathbb{C}$. According to (5.5): If f is i-simple then Σ is simple. All simple 1 dim. complete intersections are contained in Giusti's lists [Gi-2,4]. We consider now functions f of degree d, where the critical locus is a simple complete intersection, which is not a plane curve.

Σ	formula of Σ	w_0	w_1	w_2	W	a_1	a_2	A	d^1_{max}
S_μ	$(x^2+y^2+z^{\mu-3},yz)$	μ-3	μ-3	2	2μ-4	$2(\mu$-3)	μ-1	3μ-7	$3d-5\mu+9$
T_7	$(x^2+y^3+z^3,yz)$	3	2	2	7	6	4	10	$3d$-18
T_8	$(x^2+y^3+z^4,yz)$	6	4	3	13	12	7	10	$3d$-33
T_9	$(x^2+y^3+z^5,yz)$	15	16	6	31	30	16	46	$3d$-78
U_7	$(x^2+yz,xy+z^3)$	4	5	3	12	8	9	17	$3d$-32
U_8	(x^2+yz+z^3,xy)	3	4	2	9	6	7	13	$3d$-24
U_9	$(x^2+yz,xy+z^4)$	5	7	3	15	10	12	22	$3d$-40
W_8	(x^2+z^3,y^2+xz)	6	5	4	15	12	10	22	$3d$-40
W_9	(x^2+yz^2,y^2+xz)	5	4	3	12	10	8	18	$3d$-32
Z_9	(x^2+z^3,y^2+z^3)	3	3	2	8	6	6	12	$3d$-22
Z_{10}	(x^2+yz^2,y^2+z^3)	7	6	4	17	14	12	26	$3d$-46

	Condition		Condition
S_μ	$d<\frac{1}{2}(5\mu-9)$	U_9	$d<20$
T_7	$d<9$	W_8	$d<20$
T_8	$d<16\frac{1}{2}$	W_9	$d<16$
T_9	$d<39$	Z_9	$d<11$
U_7	$d<16$	Z_{10}	$d<23$
U_8	$d<12$		

A necessary condition for being a transversed A_1 singularity with the given critical locus is: $\delta(f) \geqslant 0$. So

$$2d - 2A \geqslant 0$$
$$d \geqslant A$$

From the above list it follows that $d^1_{max} < d$ and $d \geqslant A$ give always a contradiction.

CONCLUSION. There are no *i*-simple maps $\mathbb{C}^3 \to \mathbb{C}$ with a singular critical locus Σ, (as above). The only *i*-simple maps $\mathbb{C}^3 \to \mathbb{C}$ must have singular locus a plane curve, embedded in $\mathbb{C}^3$.

REFERENCES

[Al-1] A.G. Alexandrov: The De Rham complex of a quasi-homogeneous complete intersection. Funktsional'nyi Analiz i ego Phrilozheniya 17:1 (1983), 63-64; Functional Analysis and its Applications 17:1 (1983), 42-43.

[Al-2] A.G. Alexandrov: Normal forms of one-dimensional quasi homogeneous complete intersections. Math. USSR Sbornik 45:1 (1983), 1-30.

[Ar-1] V.I. Arnol'd: Normal forms of functions in the neighbourhood of degenerate critical points. Uspekhi Mat. Nauk 29:2 (1974), 11-49; Russian Math. Surveys 29:2 (1974), 18-50.

[Ar-2] V.I. Arnol'd: On some problems in singularity theory. Geometry and Analysis, Papers dedicated to the memory of V.K. Patodi. Bombay, 1980. Also in Proceedings of the Indian Acad. of Sc. (Math. Sc.), 9:1.

[AGV] V.I. Arnol'd, S.M. Gusein-Zade, A.N. Varchenko: Singularities of Differentiable Maps, Volume I. Monographs in Mathematics; 82; Birkhauser Verlag 1985. original Russian version published by Science, Moscow, 1982.

[Ga] A.M. Gabrielov: Intersection matrices for certain singularities. Funkt. An. 7:3 (1973), 18-32 (Russian), Funct. An. 7, 182-193 (English).

[Gi-1] M. Giusti: Sur les singularités isolées d'intersections complète quasi-homogènes. Ann. Insti. Fourier, Grenoble 27:3 (1977), 163-192.

[Gi-2] M. Giusti: Classification des singularités isolées simples d'intersections complètes. These: Université ParisVII (23 juin 1981), 1-68 [=Prépublication du Centre Mathématiques de l'Ecole Polytechnique (1977)].

[Gi-3] M. Giusti: Intersections complète quasi-homogènes: calcul d'invariants. These: Université Paris VII (23 juin 1981), 69-99. [Prépublication du Centre Mathématiques de l'Ecole Polytechnique (1979)].

[Gi-4] M. Giusti: Classification des singularitiés isolées d'intersections complète simples, Note aux C.R. Acad. Sc. Paris t. 284 (17.1.77).

[Gr] G.-M. Greuel: Der Gauss-Manin Zusammenhang isolierter Singularitäten von vollständigen Durchschnitten. Math. Ann. 214 (1975), 235-266.

[Gr-Ha] G.-M. Greuel, H. Hamm: Invarianten quasihomogener vollständiger Durchschnitten. Inventiones Math. 49 (1978), 67-86.

[Jo-1] T. de Jong: Nonisolated Hypersurface Singularities. Thesis Kath. Universiteit Nijmegen, 1988.

[Jo-2] T. de Jong: Some Classes of Line Singularities. Math. Zeitschrift, 1988.

[Lo] E.J.N. Looijenga: Isolated Singular Points on Complete Intersections, LMS-Lecture Note Series 77, Cambridge University Press 1984.

[Mi-Or] J. Milnor, P. Orlik: Isolated singularities defined by weighted homogeneous polynomials. Topology 9 (1970), 385-393.

[Pe-1] R. Pellikaan: Hypersurface singularities and resolutions of Jacobi modules. Thesis Rijksuniversiteit Utrecht, 1985.

[Pe-2] R. Pellikaan: Projective resolutions of the quotient of two ideals. Proc. Kon. Nederlandse Ac. Wet. A 91(1), 65-84.

[Pe-3] R. Pellikaan: Series of isolated singularities. This volume.

[Pe-4] R. Pellikaan: Finite determinacy of functions with non-isolated singularities. Proc. London Math. Soc. (3) 57 (1988), 357-382.

[Pe-5] R. Pellikaan: Deformations of Hyperfaces with a 1-dimensional singular locus.

[Sa-1] K. Saito: Quasihomogene isolierte Singularitäten von Hyperflächen. Invent. Math. 14 (1971), 123-142.

[Sa-2] K. Saito: Einfach elliptische Singularitäten. Invent. Math. 23 (1974), 289-325.

[Si-1] D. Siersma: Isolated line singularities, Proceedings of Symposia in Pure Mathematics Volume 40 (1983), Part 2, 485-496.

[Si-2] D. Siersma: Hypersurfaces with singular locus a plane curve and transversal type A_1. Preprint 406, Rijksuniversiteit Utrecht (1986). To appear in: Proceedings Singularity Semester Warsaw 1985.

[Si-3] D. Siersma: Singularities with critical locus a 1-dimensional complete intersection and transversal type A_1. Topology and its Applications 27 (1987), 51-73.

[St-1] R.P. Stanley: Hilbert functions of graded algebras. Advances in Mathematics 28 (1978), 57-83.

[St-2] R.P. Stanley: Invariants of finite groups and their application to combinatorics. Bull. AMS, 1:3 (1979), 475-511.

[Sv] J. Stevens: Improvements of non-isolated surface singularities. Preprint 455, Rijksuniversiteit Utrecht (1987). To appear in Journal of the London Math. Society.

[vS-1] D. van Straten: On the Betti numbers of the Milnor fibre of a certain class of hypersurface singularities. In: Singularities, Representation of Algebras and Vector Bundles. Proceedings of a Symposium held in Lambrecht (Germany) 1985, 203-220 (1987).

[vS-2] D. van Straten: Weakly Normal Surface Singularities and their Improvements. Thesis Rijksuniversiteit Leiden (1987).

[Wg] P. Wagreich: The structure of quasihomogeneous singularities. Proceedings of Symposia in Pure Mathematics, Volume 40, Part 2 (1983), 593-611.

[Wh] J.M. Wahl: Derivations, automorphisms and deformations of quasihomogeneous singularities. Proceedings of Symposia in Pure Mathematics, Volume 40, Part 2 (1983), 613-624.

[Wa-1] C.T.C. Wall: Notes on the classification of singularities. Proc. London Math. Soc. (3), 48 (1984), 461-513.

[Wa-2] C.T.C. Wall: Classification of unimodal isolated singularities of complete intersections. Proceedings of Symposia in Pure Mathematics, Volume 40, Part 2 (1983), 625-640.

MATHEMATISCH INSTITUT
RIJKSUNIVERSITEIT UTRECHT
P.O. BOX 80.010
3508 TA UTRECHT
THE NETHERLANDS

Contemporary Mathematics
Volume **90**, 1989

DYNKIN GRAPHS AND COMBINATIONS OF SINGULARITIES ON PLANE SEXTIC CURVES

Tohsuke Urabe

ABSTRACT

Possible combinations of singularities on sextic curves in the two-dimensional projective space $\mathbf{P}^2$ can be described systematically by a law from the view point of Dynkin graphs.

1. INTRODUCTION

Possible combinations of singularities on sextic curves in the two-dimensional projective space $\mathbf{P}^2$ can be described systematically by a law from the view point of Dynkin graphs. In this article we show this law and give it a proof. This law is similar to the one for quartic surfaces in [12]. Therefore this article is a continuation of Urabe [12], [13]. We assume that every variety is algebraic and is defined over the complex number field **C**.

DEFINITION 1.1. A disjoint finite union of connected Dynkin graphs of type A, D, or E is called a Dynkin graph. The following procedure is called an *elementary transformation* of such a graph:

(1) Replace each component by the extended Dynkin graph of the corresponding type.

0271-4132/89 $1.00 + $.25 per page

(2) Choose in an arbitrary manner at least one vertex from each component (of the extended Dynkin graph) and then remove these vertices together with edges issuing from them. (Bourbaki [4]).

Note that a connected Dynkin graph of type A, D or E corresponds to the singularity on a curve, whose local defining equality is the following (Durfee [7]):

$$A_k;\ x^2 + y^{k+1} = 0\ (k \geqslant 1),\quad D_\ell;\ x^2y + y^{\ell-1} = 0\ (\ell \geqslant 4),$$

$$E_6;\ x^3 + y^4 = 0,\quad E_7;\ x^3 + xy^3 = 0,\quad E_8;\ x^3 + y^5 = 0.$$

If a Dynkin graph G contains a_k of connected components of type A_k, b_ℓ of components of type D_ℓ, and c_m of components of type E_m ($k \geqslant 1$, $\ell \geqslant 4$, $m = 6,7,8$), we denote $G = \Sigma\, a_kA_k + \Sigma\, b_\ell D_\ell + \Sigma\, c_mE_m$.

THEOREM 1.2. *Let* $G = \Sigma_{k\geqslant 1}a_kA_k + \Sigma_{\ell\geqslant 4}b_\ell D_\ell + \Sigma_{m=6}^{8}c_mE_m$ *(a finite sum) be a Dynkin graph. Set* $r = \Sigma\, a_kk + \Sigma\, b_\ell\ell + \Sigma\, c_mm$. *Then the following conditions* (A) *and* (B) *are equivalent.*

(A) *There exists a sextic curve in the projective space of dimension* 2 *whose combination of singularities just agrees with G and moreover one of the following conditions* ⟨1⟩, ⟨2⟩, ⟨3⟩, ⟨4⟩ *holds for the root lattice* $Q = Q(G)$ *of type G. By* $d(Q)$ *we denote the discriminant of Q.*

⟨1⟩ $r = 17$, $2d(Q)$ *is a square number, and for every prime number p,* $\epsilon_p(Q) = 1$.

⟨2⟩ $r = 16$, *and for every prime number p,* $\epsilon_p(Q) = (-2, d(Q))_p$.

⟨3⟩ $r = 15$, *and for every prime number p,* $-2d(Q) \notin \mathbf{Q}_p^{*2}$ *or* $\epsilon_p(Q) = (-1, -1)_p$.

⟨4⟩ $r \leqslant 14$.

(B) *G coincides with a Dynkin graph which is obtained from one of the following 4 basic Dynkin graphs by elementary transformations repeated twice.*

$$2E_8 + A_1, \quad D_{16} + A_1, \quad A_{17}, \quad D_{10} + E_7$$

(Here the vertex in the component A_1 of $2E_8 + A_1$ and the one in the component A_1 of $D_{16} + A_1$ are marked ones. All the other vertices are non-marked ones. (See Definition 1.4.))

REMARKS.

1. r = rank Q = the number of vertices in G.

2. The symbol $\epsilon_p(Q) \in \{+1, -1\}$ denotes the Hasse symbol of the inner product space $Q \otimes \mathbf{Q}$ over $\mathbf{Q}$. The symbol $(\ ,\ )_p$ is the Hilbert symbol. $\mathbf{Q}_p$ is the field of p-adic numbers and $\mathbf{Q}_p^{*2} = \{a^2 \mid a \in \mathbf{Q}_p, a \neq 0\}$. (Serre [11]).

3. If $G = G' + G''$ for some Dynkin graphs G' and G'', then $d(Q(G)) = d(Q(G'))d(Q(G''))$ and $\epsilon_p(Q(G)) = \epsilon_p(Q(G'))\epsilon_p(Q(G''))(d(Q(G')), d(Q(G'')))_p$.

4. $d(Q(A_k)) = k + 1$, $d(Q(D_\ell)) = 4$, $d(Q(E_6)) = 3$, $d(Q(E_7)) = 2$, and $d(Q(E_8)) = 1$.

5. $\epsilon_p(Q(A_k)) = (-1, k + 1)_p$ and $\epsilon_p(Q(D_\ell)) = \epsilon_p(Q(E_m)) = 1$.

6. Assume that $a,b \in \mathbf{Q}$ can be written in the form $a = p^\alpha u$, $b = p^\beta v$ with p-adic units $u,v \in \mathbf{Q}_p$.

In case $p \neq 2$, $(a,b)_p = (-1)^{\alpha\beta\chi(p)}(u/p)^\beta(v/p)^\alpha$. Here $(\ /p)$ is Legendre's quadratic residue symbol.

In case $p = 2$, $(a,b)_p = (-1)^{\chi(u)\chi(v)+\alpha\omega(v)+\beta\omega(u)}$. Here $\chi(n) \equiv (n-1)/1$, $\omega(n) \equiv (n^2-1)/8 \pmod 2$.

7. There are 4 kinds of positive finite even lattices with discriminant 2 and of rank 17. The elements η in these lattices with $\eta^2 = 2$ form root systems of type $2E_8 + A_1$, $D_{16} + A_1$, A_{17}, $D_{10} + E_7$ respectively. This characterizes the basic 4 graphs.

8. If there is a sextic curve in $\mathbf{P}^2$ whose combination of singularities is G, then $r \leq 19$.

In order to deal with cases to which we cannot apply Theorem 1.2, we would like to propose another notion.

DEFINITION 1.3. Assume that applying the following procedure to the Dynkin graph G, we have obtained the Dynkin graph G'.

Then we call the following procedure a *connection* of Dynkin graphs.

(1) Attach an integer to each vertex of G by the following rule: Now let $\alpha_1, \alpha_2, \ldots, \alpha_k$ be the fundamental system of roots associated with a connected component G_0 of G. Let $\Sigma_{i=1}^{k} n_i \alpha_i$ be the associated maximal root. Then the attached integer to the vertex corresponding to α_i is n_i.

(2) Add one vertex and some edges to each component of G and make it the extended Dynkin graph of the corresponding type. Attach moreover the integer 1 to each new vertex.

(3) Choose in an arbitrary manner subsets A, B of the set of vertices of the extended graph satisfying the following conditions:

   $\langle a \rangle$ $A \cap B = \emptyset$

   $\langle b \rangle$ Choose arbitrarily a component $\tilde{G}_1$ of the extended graph and let V be the set of vertices in $\tilde{G}_1$. Let N be the sum of numbers attached to elements in $B \cap V$. (If $B \cap V = \emptyset$, $N = 0$.) Then, the greatest common divisor of N and the numbers attached to elements in $A \cap V$ is necessarily 1.

(4) Erase out all attached integers.

(5) Remove vertices belonging to A together with the edges issuing from them.

(6) Draw another new vertex called θ. Connect θ and each element in B with an edge.

DEFINITION 1.4. When some of vertices of the Dynkin graph G is distinguished from the others, we call the graph a *marked* Dynkin graph and call the distinguished vertices *marked* vertices. They are subject to the following rules:

(1) If we apply one elementary transformation to G, then thereafter all vertices become non-marked ones.

(2) When one adds one vertex and a few edges to one component of G to make it the extended Dynkin graph in the procedure (2) of a connection, if the added vertex is connected to a marked vertex, then the added vertex is also regarded as a marked one. Otherwise the added vertex is not a marked one.

(3) The added new vertex θ in the procedure (6) of a connection is a non-marked one.

(4) If a marked vertex is connected with θ in the procedure (6) of a connection, then it becomes a non-marked one.

THEOREM 1.5. *Assume that G' is the marked Dynkin graph obtained by an elementary transformation or a connection from one of the basic 4 marked Dynkin graphs in Theorem 1.2. Assume moreover that applying an elementary transformation or a connection to G' once more, we have obtained a Dynkin graph G without marked vertices. Then there exists a sextic curve in the projective space of dimension 2 whose combination of singularities agrees with G.*

REMARK. By Nikulin's lattice embedding theorem (Nikulin [10]) and by surjectivity of the period mapping for K3 surfaces we had showed the following proposition in [14]. However, when we wrote [14], we did not notice the law in this article.

PROPOSITION 1.6. *Let $G = \Sigma a_k A_k + \Sigma b_\ell D_\ell + \Sigma c_m E_m$ be a Dynkin graph. Assume that the number of vertices $r = \Sigma a_k k + \Sigma b_\ell \ell + \Sigma c_m m$ is less than or equal to 15. Then there is a sextic curve in P^2 with singularity G if and only if $G \neq A_4 + 11A_1$, $2A_2 + 11A_1$, $A_2 + 13A_1$.*

Now, the above results are very similar to the ones in Urabe [12], [13]. Indeed the formulations and the proofs are parallel with those in Urabe [12], [13]. This is because the fundamental ideas are the same. Thus in this article we write down important points only and omit most of the proofs. However, most of the contents are the repetitions of those in Urabe [13]. We refer the reader to Urabe [13] for the precise proof.

Below, we would like to give the proof of Theorem 1.2 only. The proof of Theorem 1.5 will be published in the near future together with the proof of the corresponding theorem for quartic surfaces.

In Section 2 we explain that describing possible combinations of rational singularities on plane sextic curves is reduced to the theory of integral symmetric bilinear forms. We use the theory of K3 surfaces and their periods for this reduction. Section 3 is devoted to explanations of bilinear forms. In particular it is explained why the notion of elementary transformations of Dynkin graphs appears in our theory. In the first part of Section 4 conditions related isotropic elements written with the Hasse symbol and the Hilbert symbol are treated. In the latter half of Section 4 classification of certain lattices is discussed. In Section 5 the implication (B) ⇒ (A) is shown.

I would like to thank Professor Richard Randell for organizing an excellent conference for us.

2. PRELIMINARIES

In this section we explain two steps of translation of our problem. First of all, we translate the problem of sextic curves into the one of K3 surfaces. Secondly we reduce the problem to the theory of symmetric bilinar forms by the theory of period for K3 surfaces.

Now let $B \subset \mathbf{P}^2$ be a sextic curve without multiple components in the projective space $\mathbf{P}^2$ of dimension 2. Let $F(z_0,z_1,z_2)$ be the defining homogeneous polynomial of B. We give weight one to each variable z_0, z_1, z_2. Let z_3 be another variable with weight 3. By X we denote the irreducible surface defined by $z_3^2 - F(z_0,z_1,z_2) = 0$ in the weighted projective space $\mathbf{P}(1,1,1,3)$ (Dolgachev [6]). The projection

$$\pi\colon\ \mathbf{P}(1,1,1,3) - \{(0:\ 0:\ 0:\ 1)\} \to \mathbf{P}^2;\ (z_0:\ z_1:\ z_2:\ z_3) \mapsto (z_0:\ z_1:\ z_2)$$

defines a surjective morphism $\pi : X \to \mathbf{P}^2$ of degree 2. The branch locus of π is B. The following is well-known (Arnold [1]).

PROPOSITION 2.1. *A point $x \in X$ is singular if and only if $\pi(x) \in B$ and $\pi(x)$ is a singular point of B. Moreover the isomorphism class of the surface singularity (x,X) and that of the curve singularity $(\pi(x),B)$ determine each other uniquely.*

Thus the study of singularities on the curve B is reduced to that on the surface X.

Now, throughout this article we assume that B has only rational singularities as singularities.

Let $\rho : Z \to X$ be the minimal resolution of singularities of X. Note that $R^1\rho_* \mathcal{O}_Z = 0$ under this assumption. We can easily show the following by standard technics in the surface theory (Barth et al. [3]).

PROPOSITION 2.2. *Z is a* K3 *surface, that is, $H^1(\mathcal{O}_Z) = 0$ and the canonical line bundle K_Z is trivial. The line bundle $L = \rho^*\pi^*\mathcal{O}_{\mathbf{P}^2}(1)$ is numerically effective and of degree* 2.

Here obviously the morphism $\pi\rho$ coincides with the one ϕ_L associated with the line bundle L. But conversely, when does a given numerically effective line bundle L of degree 2 on a K3 surface Z define a surjective morphism $\phi_L: Z \to \mathbf{P}^2$ of degree 2?

We can show the following (Urabe [14]).

PROPOSITION 2.3. *Let L be a numerically effective line bundle of degree* 2 *on a* K3 *surface Z. The following conditions are equivalent.*

(1) *The line bundle L does not define a surjective morphism $\phi_L: Z \to \mathbf{P}^2$ of degree* 2.

(2) *The complete linear system $|L|$ has a fixed component.*

(3) *There is a smooth elliptic curve E and a smooth rational curve Γ with $E \cdot \Gamma = 1$ such that any member in $|L|$ is linearly equivalent to $2E + \Gamma$.*

(4) *There exists an element* $U \in \operatorname{Pic}(Z)$ *with* $U^2 = 0$ *and* $U \cdot L = 1$.

If one of the above conditions holds, then Γ *in* (3) *is the fixed component of* $|L|$.

Because of the recent development of the theory of periods for K3 surfaces, we can reduce the problem to the theory of lattices (i.e., free **Z**-modules of finite rank equipped with symmetric integral bilinear forms). (See Barth et al. [3], Urabe [14], [13].)

THEOREM 2.4. *Let* $G = \Sigma\, a_k A_k + \Sigma\, b_\ell D_\ell + \Sigma\, c_m E_m$ *be a Dynkin graph. The following two conditions are equivalent:*

(1) *There is a reduced sextic curve in* $\mathbf{P}^2$ *with only rational singularities as singularities, the combination of singularities corresponding to* G.

(2) *Let* $Q = Q(G)$ *be the root lattice of type* G. *Let* Λ *denote the even unimodular lattice with signature* (19, 3). *The lattice* $S = \mathbf{Z}\lambda \oplus Q$ ($\lambda^2 = -2$, *orthogonal direct sum) has an embedding* $S \subset \Lambda$ *satisfying the following conditions* ⟨a⟩ *and* ⟨b⟩. *Let* $\tilde{S} = \{x \in \Lambda \mid$ *For some non-zero integer* m, $mx \in S\}$ *denote the primitive hull of* S *in* Λ.

⟨a⟩ *If* $\eta \in \tilde{S}$, $(\eta, \lambda) = 0$, *and* $\eta^2 = 2$, *then* $\eta \in Q$.

⟨b⟩ $\tilde{S}$ *does not contain any element* u *with* $u^2 = 0$ *and* $(u, \lambda) = -1$.

Note that in the above condition (2) the signature of the bilinear form has been reversed compared with the one on the second cohomology group of K3 surfaces.

3. THEORY OF BILINEAR FORMS I

Our problem has been reduced to the theory of lattices. We deal with this theory in this section. We use standard terminologies in it (Milnor-Husemoller [9], Nikulin [10], Serre [11], Urabe [13]). In particular, if a free **Z**-module L of finite rank has a symmetric bilinear form (,) with values in $\mathbf{Q}$, then we say that L is a *quasi-lattice.* If the values of the bilinear form are integers, the pair $(L, (\ ,\ ))$ is called a *lattice.* By $O(L)$ we denote the automorphism group of L. When L is a non-degenerate lattice, we regard L as a submodule of its dual module $L^* = \mathrm{Hom}(L,\mathbf{Z}) \cong \{x \in L \otimes \mathbf{Q} \mid \text{For every } y \in L\ (x,y) \in \mathbf{Z}\}$.

Consider a submodule M of a quasi-lattice L. The *primitive hull* $\{x \in L \mid \text{For some non-zero integer } m,\ mx \in M\}$ of M in L is denoted by $P(M,L)$ or $\check{M}$ when we need not mention L. The orthogonal complement $\{x \in L \mid \text{For every } y \in M\ (x,y) = 0\}$ of M in L is denoted $C(M,L)$ or $M^{\perp}$ when we need not mention L.

LEMMA 3.1.

(1) *Let L be a non-degenerate lattice and M be a primitive non-degenerate sublattice. Then the orthogonal complement $M^{\perp}$ of M in L is also a primitive non-degenerate lattice.*

(2) *Let π: $L \to L/M$ denote the canonical surjection. If we denote the composition*

$$M^{\perp} \xrightarrow{\text{inclusion}} L \xrightarrow{\pi} L/M$$

by f, f is injective and we can define a non-degenerate bilinear form (,) on L/M with values in $\mathbf{Q}$ such that $(x,y) = (f(x),f(y))$ for every $x,y \in M^{\perp}$.

(3) *For every $x,y \in L$, writing $x = x_1 + x_2$ and $y = y_1 + y_2$ where $x_1,y_1 \in M^{\perp} \otimes \mathbf{Q}$ and $x_2,y_2 \in M \otimes \mathbf{Q}$, we have $(\pi(x),\pi(y)) = (x_1,y_1) = (x,y_1) = (x_1,y)$.*

(4) *When L is a unimodular lattice, L/M and the dual $M^{\perp *}$ are isomorphic as quasi-lattices.*

Though we have defined the notion of "root modules" in Urabe [13], here we use a slightly different definition. The definition here is useful for the case $\lambda^2 = -2$.

Let L be a quasi-lattice. Set $R(L) = \{x \in L \mid x^2 = 2\}$ and $\bar{R}(L) = \{x \in L \mid x^2 = 2 \text{ or } 1/2\}$. If L satisfies the following condition (R), we say that L is a *root module*, $\bar{R}(L)$ is the *root system* of L, and

$R(L)$ is the *reduced root system* of L. (Indeed if $\overline{R}(L)$ is finite, it satisfies the axioms of the root system in Bourbaki [4] except that $\overline{R}(L)$ need not span $L \otimes \mathbf{R}$ over $\mathbf{R}$, and so does $R(L)$).

(R) For every $\alpha \in \overline{R}(L)$ and $x \in L$, $2(\alpha,x)/\alpha^2$ is an integer.

Any lattice is a root module. Any submodule of a root module is a root module.

In the sequel L denotes a root module. Let $R = R(L)$ be a *reduced* root system of L. The submodule $Q(R) = \Sigma_{\alpha\in R}\mathbf{Z}\alpha \subset L$ spanned by elements in R is called the *root lattice* of the root module L. Sometimes we denote $Q(R(L)) = Q(L)$.

The following lemma plays an important role in the proof of Proposition 3.7.

LEMMA 3.2. *Let $R = R(L)$ be the reduced root system of a root module L. For every $\alpha \in R$, $\mathbf{R}\alpha \cap Q(R) = \mathbf{Z}\alpha$.*

Note that if $\mu \in \overline{R}(L)$ and $\mu \notin R(L)$, then $2\mu \in R(L)$. Thus $\overline{R}(L)$ is finite if and only if $R(L)$ is finite. We say that L is of *finite type*, if $R(L)$ is a finite set.

It is easy to check the following (Bourbaki [4]).

PROPOSITION 3.3. *Let L be a root module of finite type. Any irreducible component of the root system $\overline{R}(L)$ or of the reduced root system $R(L)$ has one of the following types:*

$$A_k\ (k \geqslant 1),\quad D_\ell\ \ (\ell \geqslant 4),\quad E_6,\quad E_7,\quad E_8,\quad BC_1.$$

Here BC_1 is the non-reduced root system of rank 1 consisting of four elements $\{2\mu,\mu,-\mu,-2\mu\}$ with $\mu^2 = 1/2$.

The reduced root system $R(L)$ never has a component of type BC_1.

DEFINITION 3.4. The following procedure by which we can make a reduced root subsystem R' from a given finite reduced root system R is called an *elementary transformation* of the root system R.

(1) Decompose $R = \oplus_{i=1}^{m} R_i$ into irreducible root systems.

(2) Choose a fundamental system of roots $\Delta_i \subset R_i$ for $1 \leqslant i \leqslant m$. Set $\tilde{\Delta}_i = \Delta_i \cup \{-\eta_i\}$ for $1 \leqslant i \leqslant m$, where η_i is the maximal root associated with Δ_i.

(3) Choose a proper subset $\Delta_i' \subset \tilde{\Delta}_i$ for $1 \leqslant i \leqslant m$.

(4) Set $R' = \oplus_{i=1}^{m} R_i'$ where R_i' is the root system generated by Δ_i'.

Assume that the above R is of type G and R' is of type G'. If one recalls the rule used to make the Dynkin graph from the fundamental system of roots (Bourbaki [4]), it is clear that G' is obtained by an elementary transformation of Dynkin graphs (see Definition 1.1) from the Dynkin graph G.

We would like to apply the above notions to our problem.

LEMMA 3.5. *Let M be an even non-degenerate lattice and λ be an element of M with $\lambda^2 = -2$.*

(1) *$M/\mathbf{Z}\lambda$ is a free module.*

(2) *Let $\pi : M \to M/\mathbf{Z}\lambda$ denote the canonical surjection. Let $\bar{\mu} \in M/\mathbf{Z}\lambda$. The following* ⟨a⟩ *and* ⟨b⟩ *are equivalent.*

⟨a⟩ $\bar{\mu}^2 = 1/2$.

⟨b⟩ *There is a unique element $\mu \in M$ such that $\pi(\mu) = \bar{\mu}$, $\mu^2 = 0$ and $(\mu,\lambda) = -1$.*

(3) *Let $\bar{\eta} \in M/\mathbf{Z}\lambda$. The following* ⟨c⟩ *and* ⟨d⟩ *are equivalent.*

⟨c⟩ $\bar{\eta}^2 = 2$.

⟨d⟩ *There is a unique element $\eta \in M$ such that $\pi(\eta) = \bar{\eta}$, $\eta^2 = 2$ and $(\eta,\lambda) = 0$.*

(4) *Set $\bar{x} = \pi(x)$ for $x \in M$. If (x,λ) is an even integer, then $\bar{x}^2$ is an even integer. If (x,λ) is an odd integer, then $\bar{x}^2 - (1/2)$ is an even integer.*

(5) *$M/\mathbf{Z}\lambda$ is a root module.*

(6) *If moreover M has signature of type $(k,1)$, then $M/\mathbf{Z}\lambda$ is a root module of finite type.*

COROLLARY 3.6. *Let M and λ be the same as in the above lemma.*

(1) *Let* $\overline{Q}(M,\lambda)$ *denote the sublattice generated by the set*

$$\{\lambda\} \cup \{\eta \in M \mid \eta^2 = 2, \quad (\eta,\lambda) = 0\}.$$

Then $\overline{Q}(M,\lambda)/\mathbf{Z}\lambda = Q(M/\mathbf{Z}\lambda)$.

(2) *There is no element* $\overline{x} \in M/\mathbf{Z}\lambda$ *with* $\overline{x}^2 = 1$.

Now we would like to treat the situation in Theorem 2.4 (2). The following proposition is one of the key parts in this article. It explains why the elementary transformation appears in our problem (Urabe [13]).

PROPOSITION 3.7. *Let* Λ *be an indefinite even unimodular lattice and* S *be a non-degenerate sublattice of* Λ. *We assume that the following condition* $I(\Lambda,S)$ *holds.*

$I(\Lambda,S)$: *The orthogonal complement* $S^{\perp} = C(S,\Lambda)$ *of* S *in* Λ *contains an isotropic element.*

Then the following holds.

(1) *There are elements* $u \in S^{\perp}$, $v \in \Lambda$ *with* $u^2 = v^2 = 0$ *and* $(u,v) = 1$. *Set* $H = \mathbf{Z}u + \mathbf{Z}v$ *and* $\Lambda_1 = C(H,\Lambda)$. H *and* Λ_1 *are even unimodular lattices and* $\Lambda = \Lambda_1 \oplus H$ *(orthogonal direct sum). The following inclusion relation holds;* $S \subset \tilde{S} = P(S,\Lambda) \subset \Lambda_1 + \mathbf{Z}u$.

(2) *Let* $\pi : \Lambda \to \Lambda_1$ *denote the orthogonal projection to the* Λ_1*-factor. The restriction* $\pi|\tilde{S}$ *and* $\pi|S$ *are isomorphisms of lattices onto their images.*

(3) *Let* $S_1 = \pi(S) \subset \Lambda_1$ *and* $\tilde{S}_1 = P(S_1,\Lambda_1)$. *Then* $\tilde{S}_1$ *is a non-degenerate sublattice of* Λ_1 *and* $\tilde{S}_1/\pi(\tilde{S})$ *is a finite cyclic group.*

Here assume moreover that S *has signature* $(k,1)$ *for some non-negative integer* k. *Let* $\lambda \in S$ *be an element with* $\lambda^2 = -2$.

(4) *Let* S' *be a sublattice satisfying the following* ⟨a⟩, ⟨b⟩ *and* ⟨c⟩.

⟨a⟩ $\tilde{S}_1 \subset S' \subset \Lambda_1$.

⟨b⟩ S' *is primitive in* Λ_1.

⟨c⟩ S' *is a non-degenerate and it has signature* $(\ell,1)$ *for some integer* $\ell \geq k$.

Then the finite root system $R(\breve{S}/\mathbf{Z}\lambda)$ can be obtained from the finite root system $R(S'/\mathbf{Z}\pi(\lambda))$ by one elementary transformation.

REMARK. $S' = \breve{S}_1$ satisfies ⟨a⟩, ⟨b⟩ and ⟨c⟩.

Now here we consider the situation in Theorem 2.4 (2). By condition ⟨a⟩, the root system $R(\breve{S}/\mathbf{Z}\lambda)$ is of type G. Assume that the condition $I(\Lambda,S)$ is satisfied. Then by the above proposition the root system $R(\breve{S}/\mathbf{Z}\lambda)$ is obtained by one elementary transformation from $R(\breve{S}_1/\mathbf{Z}\lambda_1)$. ($\lambda_1 = \pi(\lambda)$.) Here Λ_1 is an even unimodular lattice with signature (18, 2). Assume moreover that $I(\Lambda_1,S_1)$ is satisfied. We can apply the above proposition once more and we have an even unimodular lattice Λ_2 with signature (17, 1). In this case Λ_2 itself satisfies condition ⟨a⟩, ⟨b⟩ and ⟨c⟩ in (4). Thus we can conclude that the original Dynkin graph G can be obtained from the Dynkin graph of $R(\Lambda_2/\mathbf{Z}\lambda_2)$ by elementary transformations repeated twice, where $\lambda_2 \in \Lambda_2$ is an element with $\lambda_2^2 = -2$.

The remaining parts to show (A) ⇒ (B) in our Theorem 1.2 are the following two.

(1) To write condition $I(\Lambda,S)$ and $I(\Lambda_1,S_1)$ with the Hasse symbol and the Hilbert symbol.
(2) To classify all the possibilities of $R(\Lambda_2/\mathbf{Z}\lambda_2)$ where Λ_2 is an even unimodular lattice with signature (17, 1) and $\lambda_2 \in \Lambda_2$ is an element with $\lambda_2^2 = -2$.

Note in particular that $S \cong S_1$ as lattices. Since any even indefinite unimodular lattice is uniquely determined by its signature (Milnor-Husemoller [9]), we can assume that

$$\Lambda = Q(2E_8) \oplus H \oplus H \oplus H, \quad \Lambda_1 = Q(2E_8) \oplus H \oplus H, \quad \Lambda_2 = Q(2E_8) \oplus H$$

in the above. (H is the hyperbolic plane.)

4. THEORY OF BILINEAR FORMS II

In this article by $d(L) \in \mathbf{Z}$ we denote the *discriminant* of a lattice L. Let V be a finite dimensional vector space over $\mathbf{Q}$ equipped with a symmetric bilinear form with values in $\mathbf{Q}$ (an *inner product space*, in short). We can define the *discriminant* $\bar{d}(V)$ and the *Hasse symbol* $\epsilon_p(V)$. The discriminant $\bar{d}(V)$ is an element in $\mathbf{Q}/\mathbf{Q}^{*2} = \{0\} \cup \mathbf{Q}^*/\mathbf{Q}^{*2}$ ($\mathbf{Q}^* = \mathbf{Q} - \{0\}$). $\bar{d}(V) = 0$ if and only if V is degenerate. If V is non-degenerate, the Hasse symbol $\epsilon_p(V) \in \{+1,-1\}$ is defined for every prime number p and $p = \infty$ (Serre [11]). Obviously $\bar{d}(L \otimes \mathbf{Q}) \equiv d(L) \pmod{\mathbf{Q}^{*2}}$ for any lattice L. We write $\epsilon_p(L)$ instead of $\epsilon_p(L \otimes \mathbf{Q})$ for any lattice L for simplicity. We use the Hilbert symbol $(\ ,\ )_p$ in the sequel (Serre [11]). Note that $(a,b)_p \in \{+1,-1\}$, $(a,b)_p = (b,a)_p$, $(a,bc)_p = (a,b)_p \cdot (a,c)_p$ and $(a,c^2b)_p = (a,b)_p$ for $a,b,c \in \mathbf{Q}^*$.

We can show the following (Serre [11], Urabe [13]).

PROPOSITION 4.1. *Let Q be a positive definite lattice of rank r and λ be an element with $\lambda^2 = -2$. Set $S = \mathbf{Z}\lambda \oplus Q$ (orthogonal direct sum). Assume that there is an embedding $S \subset Q(2E_8) \oplus H \oplus H \oplus H$. If $I(Q(2E_8) \oplus H \oplus H \oplus H, S)$ is satisfied, then there is an embedding $S \subset Q(2E_8) \oplus H \oplus H$. Moreover, the following two conditions are equivalent.*

(1) *$I(Q(2E_8) \oplus H \oplus H \oplus H, S)$ and $I(Q(2E_8) \oplus H \oplus H, S)$ are satisfied.*

(2) *One of the following ⟨1⟩, ⟨2⟩, ⟨3⟩, ⟨4⟩ holds.*

 ⟨1⟩ $r = 17$, *$2d(Q)$ is a square number, and for every prime number p, $\epsilon_p(Q) = 1$.*

 ⟨2⟩ $r = 16$, *and for every prime number p, $\epsilon_p(Q) = (-2,d(Q))_p$.*

 ⟨3⟩ $r = 15$, *and for every prime number p, $-2d(Q) \notin \mathbf{Q}_p^{*2}$ or $\epsilon_p(Q) = (-1,-1)_p$.*

 ⟨4⟩ $r \leqslant 14$.

Next we consider the possibilities for the quasi-lattice $Q(2E_8) \oplus H/\mathbf{Z}\lambda$, where $\lambda \in Q(2E_8) \oplus H$ is an element with $\lambda^2 = -2$. Note that an even unimodular lattice with signature (r,s) exists if and only if $r - s \equiv 0 \pmod 8$.

LEMMA 4.2. *Let r be a positive integer with $r \equiv 1 \pmod 8$. Let M be a positive definite even lattice with discriminant 2 and of rank r. Let $x_0 \in M^*$ be an element with $x_0 \notin M$. Then $x_0^2 - (1/2)$ is an even integer.*

PROOF. It is easy to see that $x_0^2 \equiv \pm(1/2) \pmod{2\mathbf{Z}}$. If $x_0^2 + (1/2)$ is an even integer, then we can conclude that there exists a positive definite even unimodular lattice of rank $r + 1$, which is a contradiction. Q.E.D.

PROPOSITION 4.3. *Let r be a positive integer with $r \equiv 1$ (mod 8). Let Λ be an even unimodular lattice with signature $(r,1)$. For every $\lambda \in \Lambda$ with $\lambda^2 = -2$ set $F_\lambda = \Lambda/\mathbf{Z}\lambda$ (a quasi-lattice) and $M_\lambda = C(\mathbf{Z}\lambda,\Lambda)$ = the orthogonal complement of $\mathbf{Z}\lambda$ in Λ. Then M_λ is a positive definite even lattice with discriminant 2 and of rank r, and this sets up a bijective correspondence among*

(1) *$O(\Lambda)$-orbits of $\lambda \in \Lambda$ with $\lambda^2 = -2$.*
(2) *isomorphism classes of quasi-lattices $F_\lambda = \Lambda/\mathbf{Z}\lambda$, where λ runs over all elements $\lambda \in \Lambda$ with $\lambda^2 = -2$.*
(3) *isomorphism classes of positive definite even lattices with discriminant 2 and of rank r.*

For their root systems $R(F_\lambda) = R(M_\lambda)$.

REMARK. $F_\lambda \cong M_\lambda^*$ by Lemma 3.1 (4).

PROOF. Obviously an item in (1) determines an item in (2), and since $M_\lambda \cong \{x \in F_\lambda | x^2 \text{ is an even integer}\}$, an item in (2) determines an item in (3). Next we show that we can define a correspondence from (3) to (1).

Let M be a representative of an isomorphism class of (3). Let m be a generator of a quasi-lattice $\mathbf{Z}m$ of rank 1 such that $m^2 = -1/2$. Set $\Lambda' = \{(x,tm) \in M^* \oplus \mathbf{Z}m \mid x^2 - (t^2/2) \text{ is an even integer}\}$.

By Lemma 4.2 one sees that Λ' is an even unimodular lattice with signature $(r,1)$. Thus $\lambda = -2m \in \Lambda'$ defines an item in (1).

It is not difficult to check that these correspondences are inverse ones of each other (Urabe [13]). Q.E.D.

Let U, U' be lattices and $h \in U$, $h' \in U'$ be elements. Here we define that the pair (U,h) is *isomorphic* to (U',h'), if there is an isomorphism of lattices $f : U \to U'$ with $f(h) = h'$.

PROPOSITION 4.4 [Kneser [8]]. *Let r be the same as above. Let U be a positive definite* (*odd*) *unimodular lattice of rank $r + 1$ and $h \in U$ be an element with $h^2 = 2$. Set $N_h = C(\mathbf{Z}h,U)$.* N_h is a positive definite (*not necessarily even*) *lattice with discriminant* 2 *and of rank r, and this sets up a bijective correspondence between*

(1) *isomorphism classes of pairs (U,h) where U is a positive definite unimodular lattice of rank $r + 1$ and $h \in U$ is an element with $h^2 = 2$.*

(2) *isomorphism classes of positive definite lattice with discriminant* 2 *and of rank r.*

Let U be a positive definite unimodular lattice of rank 18. An element $h \in U$ with $h^2 = 2$ is called a *special* element, if the orthogonal complement of $\mathbf{Z}h$ in U is an even lattice. Our problem has been reduced to the classification of U's and their special elements by Proposition 4.3 and Proposition 4.4. Set $R'(U) = \{x \in U \mid x^2 = 2$ or $1\}$. By abuse of terminologies we call $R'(U)$ *the root system* of the lattice U, since it satisfies axioms of root systems (Bourbaki [4]). Note that in general $R'(U)$ is different from $R(U)$ and $\bar{R}(U)$ introduced in the previous section.

THEOREM 4.5 [Conway-Sloane [5]]. *There are* 13 *isomorphism classes for positive definite unimodular lattices U of rank* 18. *Their root systems $R'(U)$ are of type*

$$A_{17} + A_1,\ 2A_9,\ A_1 + D_{10} + E_7,\ 3D_6,\ A_{11} + E_6 + B_1,\ D_{16} + B_2$$

$$2D_8 + B_2,\ 2E_8 + B_2,\ A_{15} + B_3,\ 2E_7 + B_4,\ D_{12} + B_6,\ E_8 + B_{10},\ B_{18}$$

respectively.

REMARKS. A root system of type B_1 is the one consisting of two elements $\{\mu, -\mu\}$ with $\mu^2 = 1$, while one of type A_1 consists of two elements $\{\alpha, -\alpha\}$ with $\alpha^2 = 2$. An irreducible component of $R'(U)$ contains an element μ with $\mu^2 = 1$ if and only if it is of type B.

THEOREM 4.6. *There are* 4 *isomorphism classes of positive definite even lattice M of rank* 17 *and with discriminant* 2. *Their (reduced) root systems $R(M) = \overline{R}(M) = R'(M)$ are of type*:

$$2E_8 + A_1,\quad D_{16} + A_1,\quad A_{17},\quad D_{10} + E_7.$$

respectively.

For the first two cases the root system $\overline{R}(M^*)$ of the dual M^* is of type $2E_8 + BC_1$ and $D_{16} + BC_1$ respectively. For the latter two cases M^* contains no element μ with $\mu^2 = 1/2$ and thus $\overline{R}(M^*) = R(M)$.

REMARKS. 1. It is well known that there are 2 kinds of positive definite even unimodular lattice of rank 16. One is $Q(2E_8)$ and the other is called Γ_{16}. The root system of Γ_{16} is of type D_{16}. The lattices $Q(2E_8 + A_1)$ and $\Gamma_{16} \oplus Q(A_1)$ represent 2 of the above 4 isomorphism classes.

2. In the sequel by $Q(B_k)$ we denote the root lattice of type B_k. This is the free **Z**-module of rank k with the quadratic form $\sum_{i=1}^{k} x_i^2$.

PROOF. First recall the following fact: Assume that two elements in $R'(U)$ belong to the same irreducible component and that they have the same length. Then there is a composition w of reflections associated with elements in their irreducible component of $R'(U)$ such that one is sent to the other by w. Thus they are in the same $O(U)$-orbit.

Assume that U contains a special element h. It is easy to see that if $R'(U)$ has a component of type B_k, then $k = 2$. Thus there remains only 7 candidates as U. Among them one of type $2D_8 + B_2$ has the form $U = L \oplus Q(B_2)$ and L is an odd lattice with no element x with $x^2 = 1$. Thus this case is excluded. For the ones corresponding to $2E_8 + B_2$ and $D_{16} + B_2$ it is easy to deduce the conclusion above. As for the remaining four cases $A_{17} + A_1$, $2A_9$, $A_1 + D_{10} + E_7$, $3D_6$, with the help of the information in Conway-Sloane [5], we obtain the conclusion after some calculation by using the above fact.

If $R(M) \neq \bar{R}(M^*)$, then $\bar{R}(M^*)$ has a component of type BC_1 and $R(M) = R(M^*)$ has a component of type A_1. Thus for the latter two cases in the theorem they are equal. Q.E.D.

The implication (A) $\Rightarrow$ (B) in our main theorem Theorem 1.2 is now clear by Proposition 3.7, Proposition 4.1 and Theorem 4.6.

5. PROOF OF THE CONVERSE

In this section we show the implication (B) $\Rightarrow$ (A).

Let L be a root module and S be a submodule. We say that the embedding $S \subset L$ is *full*, if $R(S) = R(\check{S})$ for their reduced root systems. Here $\check{S}$ is the primitive hull of S in L.

PROPOSITION 5.1. *Let R' be a root system obtained by one elementary transformation from a finite root system R consisting of elements α with $\alpha^2 = 2$. Let L be a root module. Assume that a full embedding $Q(R) \subset L$ is given. Then there is a full embeddding $Q(R')$ $\subset L \oplus H$ such that the primitive hull $P(Q(R'), L \oplus H)$ does not contain any element μ with $\mu^2 = 1/2$ and such that u is orthogonal to $Q(R')$, where $H = \mathbf{Z}u + \mathbf{Z}v$ is the hyperbolic plane and $u^2 = v^2 = 0$, $(u,v) = 1$. Moreover, if $Q(R)$ is orthogonal to an element $w \in L$, then $Q(R')$ is orthogonal to $w = w \oplus 0 \in L \oplus H$.*

PROOF. Let $R = \oplus_{i=1}^{n} R_i$ be the irreducible decomposition of the root system R. By the definition of the elementary transformation, for each i with $1 \leqslant i \leqslant n$, there is a fundamental system of root $\Delta_i \subset R_i$ and a proper subset $\Delta_i' \subset \Delta_i \cup \{-\eta_i\}$ (η_i is the maximal root associated with Δ_i) such that if R_i' is the root system generated by Δ_i', then $R' = \oplus_{i=1}^{n} R_i'$. We assume that for $m < i \leqslant n$, R_i is of type A_1 and for $1 \leqslant i \leqslant m$, R_i is not of type A_1. Here we consider a fixed suffix i with $m < i \leqslant n$ for the time being. If $\Delta_i' \neq \emptyset$, then $\Delta_i' = \Delta_i$ or $\Delta_i' = -\Delta_i$. Note that Δ_i and $-\Delta_i$ generate the same root system R_i of type A_1. Thus we can assume moreover that either $\Delta_i' = \emptyset$ or $\Delta_i' = -\Delta_i = \{-\eta_i\}$ for $m < i \leqslant n$. In the sequel we assume this condition.

The set $\cup_{i=1}^{n} \Delta_i'$ is a free basis of $Q(R')$. We define the embedding $\phi : Q(R') \to L \oplus H$ by setting for $\alpha \in \Delta_i'$ with $1 \leqslant i \leqslant n$

$$\phi(\alpha) = \alpha \oplus 0 \quad (\text{if } \alpha \neq -\eta_i)$$

$$\phi(\alpha) = \alpha \oplus u \quad (\text{if } \alpha = -\eta_i)$$

Obviously ϕ preserves bilinear forms and u is orthogonal to $\phi(Q(R'))$. Moreover, we can check that ϕ is a full embedding (Urabe [13]). Thus we have only to check the condition on μ.

Now assume that there is an element μ with $\mu^2 = 1/2$ in the primitive hull $\widetilde{Q(R')} = P(Q(R'), L \oplus H)$. By the fullness $2\mu \in R' \subset Q(R') \subset L \oplus H$. Let $\pi : L \oplus H \to L$ denote the projection. The element $\pi(\mu)$ belongs to $\widetilde{Q(R)}$. Since $\pi(\mu)^2 = \mu^2 = 1/2$, the non-reduced root system $\overline{R}(\widetilde{Q(R)})$ has a component of type BC_1, whose generator is $\pi(\mu)$. By the fullness $2\pi(\mu) \in R = R(Q(R))$ and R has a component of type A_1 whose generator is $2\pi(\mu)$. Thus there is a suffix i with $m < i \leqslant n$ such that $2\pi(\mu) = \pi(2\mu) = \pm\eta_i = \pm\pi(\eta_i - u)$. Note that 2μ, $\eta_i - u \in \widetilde{Q(R')}$ and $\pi|\widetilde{Q(R')}$ is injective. Thus $2\mu = \pm(\eta_i - u)$. We have

$$(\mu,\nu) = \pm(\eta_i - u,\nu)/2 = \mp 1/2 \notin \mathbf{Z}.$$

However, for every $x \in L \oplus H$, $(x,\nu) \in \mathbf{Z}$, which is a contradiction.

This completes the proof. Q.E.D.

LEMMA 5.2. *Let* Λ *be an even lattice,* S *be a sublattice, and* $\lambda \in S$ *be an element. Under these notations,* $P(S,\Lambda)/\mathbf{Z}\lambda \cong P(S/\mathbf{Z}\lambda, \Lambda/\mathbf{Z}\lambda)$ *holds.*

Now let R be a root system whose type is one of the four in Theorem 4.6. There is $\lambda \in Q(2E_8) \oplus H$ with $\lambda^2 = -2$ such that $R = R(L)$ for $L = Q(2E_8) \oplus H/\mathbf{Z}\lambda$. Thus there is a full embedding $Q(R) \subset L$. Let R' be a root system obtained from R by one elementary transformation. By the above proposition there is a full embedding $Q(R') \subset L \oplus H_1$ where $H_1 = \mathbf{Z}u_1 + \mathbf{Z}v_1$ ($u_1^2 = v_1^2 = 0$, $(u_1,v_1) = 1$) is a hyperbolic plane. We can assume that u_1 is orthogonal to $Q(R')$. Let R'' be a root system obtained from R' by one elementary transformation. There is a full embedding $Q(R'') \subset L \oplus H_1 \oplus H_2$ such that there is no element $\bar{\mu} \in \widetilde{Q(R'')}$ with $\bar{\mu}^2 = 1/2$, where $H_2 = \mathbf{Z}u_2 + \mathbf{Z}v_2$ ($u_2^2 = v_2^2 = 0$, $(u_2,v_2) = 1$) is another hyperbolic plane. By the above proposition $Q(R'')$ is orthogonal to u_1 and u_2. Let $\pi : Q(2E_8) \oplus H \oplus H_1 \oplus H_2 \to L \oplus H_1 \oplus H_2$ be the induced surjection by the canonical surjection $Q(2E_8) \oplus H \to L$. Set $S = \pi^{-1}(Q(R''))$. By Corollary 3.6 (1), $S \cong \mathbf{Z}\lambda \oplus Q(R'')$. The lattice S satisfies condition (2) in Proposition 4.1, since u_1 and u_2 are orthogonal to it. Let $\widetilde{S}$ denote the primitive hull of S in $Q(2E_8) \oplus H \oplus H_1 \oplus H_2$. By Lemma 5.2, $\widetilde{S}/\mathbf{Z}\lambda$ is the primitive hull $\widetilde{Q(R'')}$ of $Q(R'')$ in $L \oplus H_1 \oplus H_2$. By the fullness $Q(R'') = Q(\widetilde{S}/\mathbf{Z}\lambda)$. By using Corollary 3.6 (1) again, we have $S = \bar{Q}(S,\lambda) = \bar{Q}(\widetilde{S},\lambda)$, which implies that the condition ⟨a⟩ in Theorem 2.4 is satisfied. On the other hand one can conclude that $\widetilde{S}$ has no element μ with $\mu^2 = 0$ and $(\mu,\lambda) = -1$ by the condition on $\bar{\mu}$ for $\widetilde{Q(R'')}$ and Lemma 3.5 (2). Thus condition ⟨b⟩ in Theorem 2.4 (2) is also satisfied. By Theorem 2.4, one knows that condition (A) in Theorem 1.2 holds.

We have completed all the proof of Theorem 1.2.

REFERENCES

[1] V. I. Arnold, Local Normal Forms of Functions, Invent. Math. 35 (1967), 87-109.

[2] M. Artin, On Isolated Rational Singularities of Surfaces, Amer. J. Math. 88 (1966), 129-136.

[3] W. Barth, C. Peters and A. Van de Ven, Compact Complex Surfaces, Berlin-Heidelberg-New York, Springer, 1984.

[4] N. Bourbaki, Groupes et algèbre de Lie, Chaps. 4-6, Paris, Hermann, 1968.

[5] J. H. Conway and N. J. A. Sloane, On the Enumeration of Lattices of Determinant One, J. Number Theory 15 (1982), 83-94.

[6] I. Dolgachev, Weighted Projective Varieties, in Group Actions and Vector Fields, L.N.M. 956, Berlin-Heidelberg-New York, Springer, (1982), 34-71.

[7] A. Durfee, Fifteen Characterizations of Rational Double Points and Simple Critical Points, Engein. Math. II, 25 (1979), 131-163.

[8] M. Kneser, Klassenzahlen definiter quadratisher Formen, Arch. Math. 8 (1957), 241-250.

[9] J. Milnor and D. Husemoller, Symmetric Bilinear Forms, Berlin-Heidelberg-New York, Springer, 1973.

[10] V. V. Nikulin, Integral Symmetric Bilinear Forms and Some of Their Applications, Math. USSR Izvestija 14 (1980), 103-167.

[11] J. P. Serre, Cours d'arithmètique, Paris, Presses Univ. France, 1970.

[12] T. Urabe, Dynkin Graphs and Combination of Singularities on Quartic Surfaces, Proc. Japan Acad. 61, Ser. A (1985), 266-269.

[13] T. Urabe, Elementary Transformation of Dynkin Graphs and Singularities on Quartic Surfaces, to appear in Inventiones Mathematicae.

[14] T. Urabe, Combination of Rational Singularities on Plane Sextic Curves with the Sum of Milnor Numbers Less than Sixteen, submitted to the Proceedings of the Symposium held at Warsaw in 1985.

DEPARTMENT OF MATHEMATICS
TOKYO METROPOLITAN UNIVERSITY
FUKAZAWA, SETAGAYA, TOKYO 158
JAPAN

Contemporary Mathematics
Volume **90**, 1989

THE INEQUALITY $\mu \geqslant 12p_g - 4$ FOR HYPERSURFACE WEAKLY ELLIPTIC SINGULARITIES

Yi-Jung Xu and Stephen S.-T. Yau

§1. INTRODUCTION

Two of the most important numerical measures of the complexity of a germ of isolated hypersurface singularity are Milnor number μ and the local geometric genus p_g. In 1977 [2], Laufer established a celebrated formula which relates μ and p_g. However this formula does not give inequality between p_g and μ immediately. In 1978 [1], Durfee conjectured that $6p_g \leqslant \mu$. Recently Tomari [7] has proved that $8p_g + 1 \leqslant \mu$ for double points. In private communication, Randell [5]* has announced $6p_g \leqslant \mu$ for nondegenerate hypersurface singularity. In this paper, we shall prove the following theorem.

THEOREM. *Let (V,p) be a weakly elliptic hypersurface singularity. Then $\mu \geqslant 12p_g - 4$.*

We recall that weakly elliptic hypersurface singularities are a general class of hypersurface singularities which have arithmetic

---

Research partially supported by N.S.F. Grant No DMS8411477 and Institut Mittag-Leffler

*The second author was kindly informed by Professor Randell that the proof given in [5] is wrong.

genus p_a equal to one. Notice that as shown in [7], p_g can be arbitrary large. Our theorem cannot be done if we go beyond weakly elliptic singularity as we can see from the following simple example. Let $V = \{(x,y,z)\colon x^4 + y^4 + z^4 = 0\}$. Then $\mu = 27$ and $p_g = 4$.

It follows easily that we have the following Corollary:

COROLLARY. *Let (V,p) be a weakly elliptic hypersurface singularity. Then the signature σ of the Milnor fiber V' is non-positive, more precisely*

$$\sigma \leqslant 2 - 8p_g$$

§2. PRELIMINARIES

Let $f : (\mathbf{C}^3,0) \to (\mathbf{C},0)$ be a holomorphic function such that it has an isolated singularity at origin. Let B_ϵ be the open ball of radius ϵ in $\mathbf{C}^3$. Let $V_t = f^{-1}(t)$, let $S_\epsilon = \partial(\bar{B}_\epsilon)$, and let $K_\epsilon = V_0 \cap S_\epsilon$. Milnor [4] shows that for ϵ sufficiently small, K_ϵ is an connected 3-manifold and $S_\epsilon - K_\epsilon$ fibers over S^1 via the map $f/|f|$. The fibre F is diffeomorphic to $V_\delta \cap B_\epsilon$ for $0 < \delta << \epsilon$, and F has the homotopy type of a wedge of μ 2-spheres. This number μ is called the Milnor number and is the most important topological invariant of the singularity. It is well-known that

$$\mu = \dim_{\mathbf{C}} \mathbf{C}[[x,y,z]] / \left(\frac{\partial f}{\partial x}, \frac{\partial f}{\partial y}, \frac{\partial f}{\partial z} \right).$$

The signature $\sigma(F)$ of an arbitrary real oriented four-manifold F is defined as follows: There is a symmetric bilinear intersection pair $(\ ,\ )$ on $H_n(F;\mathbf{R})$ defined by setting

$$(x,y) = (x' \cup y')[F]$$

where x' and y' in $H^2(F,\partial F;\mathbf{R})$ are Lefschetz dual to x and y in $H^2(F,\mathbf{R})$ and $[F] \in H_4(F,\partial F;\mathbf{R})$ is the orientation class. This bilinear form may be diagonalized, with diagonal entries $+1$, 0, and -1.

The signature of F is the signature of this bilinear form, namely, the number of positive minus the number ofnegative diagonal entries.

Now let us describe the invariants of a resolution. Let V be a Stein space with p as its only singularity. It follows from Hironaka's work that a resolution π: $M \to V$ always exists. It is well-known that $\dim H^i(M, \mathcal{O}) = \dim(R^i\pi_* \mathcal{O})_p$ for $i \geqslant 0$. The geometric genus p_g of the singularity of V is defined as $p_g = \dim H^1(M,\mathcal{O})$. This number is independent of the resolution. (See Theorem A of [9]).

Let π: $M \to V$ be a resolution of the singularity at $p \in V$. Let K^2 denote the self-intersection of the associated canonical divisor K [2], and let $\chi_T(A)$ denote the topological Euler characteristic of the exceptional set $A = \pi^{-1}(p)$. It is well-known that $K^2 + \chi_T(A)$ is independent of the resolution, since every resolution may be obtained from the minimal resolution by a sequence of monoidal transforms, and a monoidal transform increases $\chi_T(A)$ by one and decreases K^2 by one.

The arithmetic genus of a normal two-dimensional singularity (V,p) is the integer $p_a = p_a(V,p)$ defined by

$$p_a(V,p) = \sup_{\substack{D>0 \\ |D| \subset A}} p_a(D).$$

Here, the integer $p_a(D)$ is the virtual genus of the divisor D on M given by

$$p_a(D) = 1 - \chi(D) = 1 - h^0(M, \mathcal{O}_D) + h^1(M, \mathcal{O}_D).$$

§3. WEAKLY ELLIPTIC SINGULARITY

Let π: $M \to V$ be a resolution of normal two dimensional Stein space V. We assume that p is the only singularity of V. Let $\pi^{-1}(p) = A = \cup A_i$, $1 \leqslant i \leqslant n$, be the decomposition of the exceptional set A into irreducible components. Suppose π is the minimal good resolution. Let Γ be the associated weighted dual graph. The vertices of Γ correspond to the A_i. The edges of Γ connecting the vertices corresponding to A_i and A_j, $i \neq j$, corresponds to the points of $A_i \cap A_j$. Finally, associated to each A_i is its genus, g_i, as a Riemann surface, and its weight $A_i \cdot A_i$, the topological self-intersection number. Γ will denote the graph along with the genera and the weights.

A cycle D on A is an integral combination of the $A_i \cdot D = \Sigma d_i A_i$, $1 \leq i \leq n$ with d_i an integer. In this paper "cycle" will always be a cycle on A. There is a natural partial ordering, denoted by $<$, between cycles defined by comparing the coefficients. We shall only be considering cycles $D \geq 0$. We let supp $D = |D| = \cup A_i$, $d_i \neq 0$, denote support of D.

Associated to π is a unique fundamental cycle Z such that $Z > 0$, $A_i \cdot Z \leq 0$ for all A_i, and such that Z is minimal with respect to those two properties.

DEFINITION 3.1. *p is said to be weakly elliptic if $p_a(Z) = 1$. It is well-known that $p_a(Z) = 1$ if and only if $p_a(V,p) = 1$.*

DEFINITION 3.2. *A cycle E is minimally elliptic if $\chi(E) = 0$ and $\chi(D) > 0$ for all cycles D such that $0 < D < E$.*

DEFINITION 3.3. *Let π: $M \to V$ be the minimal resolution of V. p is minimally elliptic if p is weakly elliptic and every connected proper subvariety of A is the exceptional set for a rational singularity.*

DEFINITION 3.4. *Let K be the canonical divisor on M. We define the negative cycle $K' = \Sigma\, k_i A_i$ on A, where $k_i \in Z$, the set of integers, to be a cycle such that $A_i \cdot K' = A_i \cdot K$ for all $A_i \subseteq A$. K' is called the canonical cycle. Notice that such a cycle K' may not exist in general.*

DEFINITION 3.5. [7] *Let A be the exceptional set of the minimal good resolution π: $M \to V$, where V is a normal two dimensional Stein space with p as its only weakly elliptic singularity. If $E \cdot Z < 0$, we say that the elliptic sequence is $\{Z\}$ and the length of* *elliptic sequence is equal to one. Suppose* $E \cdot Z = 0$. *Let* B_1 *be the maximal connected subvariety of* A *such that* $B_1 \supseteq \operatorname{supp} E$ *and* $A_i \cdot Z = 0$ *for all* $A_i \subseteq B_1$. *Since* A *is an exceptional set,* $Z \cdot Z < 0$, *so* B_1 *is properly contained in* A. *Let* Z_{B_1} *be the fundamental cycle on* B_1. *Suppose* $Z_{B_1} \cdot E = 0$. *Let* B_2 *be the maximal connected subvariety of* B_1 *such that* $B_2 \supseteq |E|$ *and* $A_i \cdot Z_{B_1} = 0$ *for all* $A_i \subseteq B_2$. *By the same argument as above,* B_2 *is properly contained in* B_1. *Continuing this process, we finally obtain* B_m *with* $Z_{B_m} \cdot E < 0$. *We call* $\{Z_{B_0} = Z,$ $Z_{B_1}, \ldots, Z_{B_m}$ *the elliptic sequence, and the length of the elliptic sequence is* $m + 1$.

THEOREM 3.1. [10] *Let* $\pi\colon M \to V$ *be the minimal good resolution of normal two dimensional Stein space with* p *as its only weakly elliptic singularity. Suppose* p *is not a minimally elliptic singularity. If* $E \cdot Z < 0$ *and* $|E| \neq A$, *then* K' *does not exist. If* K' *exists, then the elliptic sequences is of the following form*

$$Z_{B_0} = Z, Z_{B_1}, \ldots, Z_{\ell}, Z_{B_{\ell+1}} = Z_E, \quad \ell \geq 0.$$

Moreover

$$-K' = \sum_{i=1}^{\ell} Z_{B_i} + E.$$

PROPOSITION 3.2. [11] *Let* Γ *be a weighted dual graph including genera for the vertices, associated to the minimal good resolution of weakly elliptic singularity. Suppose* K' *exists. Let* $Z = Z_{B_0}, \ldots, Z_{B_\ell}, Z_E$ *be the elliptic sequence. Then*

$$Z \cdot Z \leq Z_{B_1} \cdot Z_{B_1} \leq \cdots \leq Z_{B_\ell} \cdot Z_{B_\ell} \leq Z_E \cdot Z_E.$$

If $Z_{B_i} \cdot Z_{B_i} = Z_{B_{i+1}} \cdot Z_{B_{i+1}}$, *then* $A_j \cdot A_j = -2$ *for all* $A_j \subseteq B_i$, $A_j \subsetneq B_{i+1}$, $0 \leq i \leq \ell$.

PROPOSITION 3.3. [11] *Let* Γ *be a weighted dual graph* *including genera for the vertices associated to the minimal good resolution of weakly elliptic singularity. Suppose K' exists. Let $Z = Z_{B_0}, \ldots, Z_{B_\ell}, Z_E$ be the elliptic sequence. If $Z \cdot Z = -1$, then Γ is of the following form*

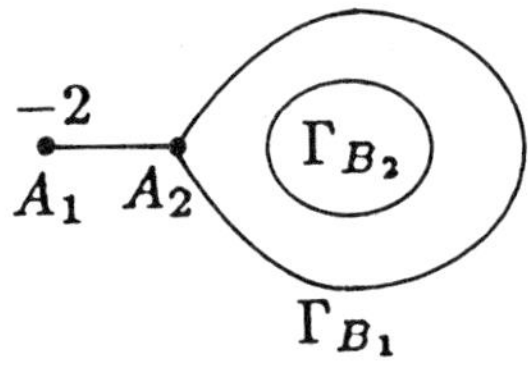

$Z = A_1 + Z_{B_1}$ $A_2 \cdot Z_{B_1} = -1$ $z_2 = 1$

PROPOSITION 3.4. *Let Γ be a weighted dual graph including genera for the vertices associated to the minimal good resolution of weakly elliptic singularity. Suppose K' exists. Let $Z = Z_{B_0}, \ldots, Z_{B_\ell} Z_E$ be the elliptic sequence. If $Z \cdot Z = -2$, $Z_{B_1} \cdot Z_{B_1} = -1$, then Γ must be one of the following forms:*

(1) $A_2 \subseteq B_1$, $A_2 \subsetneq B_2$, $Z = A_1 + Z_{B_1}$, $A_2 \cdot Z_{B_1} = -1$, $z_2 = 1$

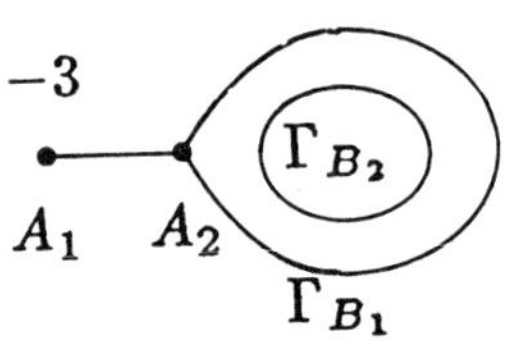

(2) $Z = 2A_1 + D$, D is a positive cycle, $|D| = B_1$, $z_2 = 3$

$$A_2 \cdot Z_{B_1} = 0, \quad A_2 \subseteq B_1 \quad A_2 \subsetneq B_2$$

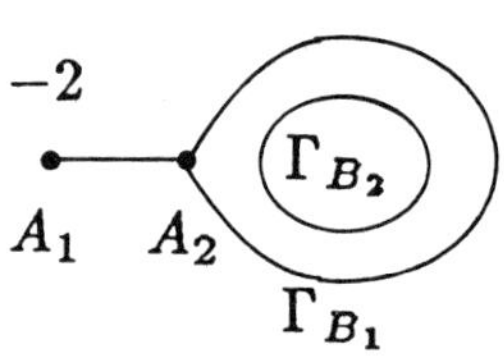

PROPOSITION 3.5. [11] *Let Γ be a weighted dual graph including genera for the vertices associated to the minimal good resolution of weakly elliptic singularity. Suppose K' exists. Let $Z = Z_{B_0}, \ldots, Z_{B_\ell}, Z_E$ be the elliptic sequence. If $Z \cdot Z = -2 = Z_{B_1} \cdot Z_{B_1}$,*

then Γ *must be one of the following forms:*

(1) $Z = A_1 + Z_{B_1} + A_2$, $A_3, A_4 \subseteq B_1$, $A_3, A_4 \subsetneq B_2$; $z_3 = z_4 = 1$

$$A_3 \cdot Z_{B_1} = A_4 \cdot Z_{B_1} = -1$$

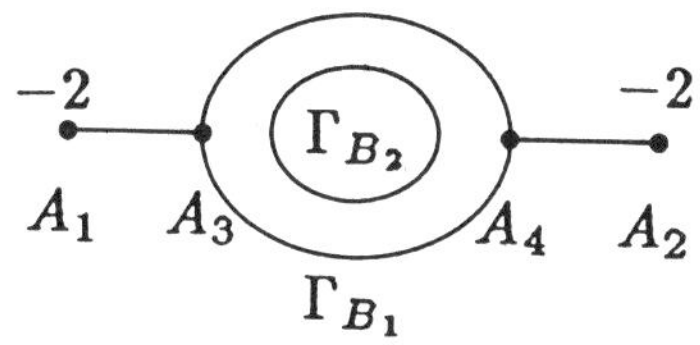

(2) $Z = {}^{1}_{1}Z_{B_1}$, $A_3 \subseteq B_1$, $A_3 \subsetneq B_2$; $z_3 = 1 \cdot A_3 \cdot Z_{B_1} = -2$

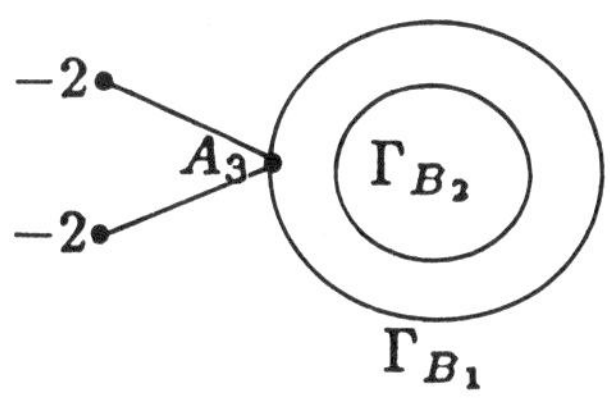

(3) $Z = 1 \;\; 2 \;\; 2 \ldots Z_{B_1}$ $A_2 \subseteq B_1$, $A_2 \subsetneq B_2$; $z_2 = 1$, $A_2 \cdot Z_{B_1} = -2$.

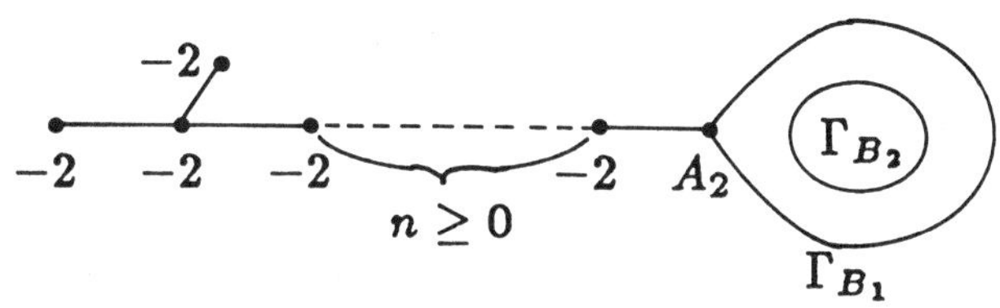

(4) $Z = 1 \; 2 \; D$, $|D| = B_1$ D *is a positive cycle*, $z_2 = 2$.

$$A_2 \cdot Z_{B_1} = 0. \quad A_2 \subseteq B_1, A_2 \subsetneq B_2.$$

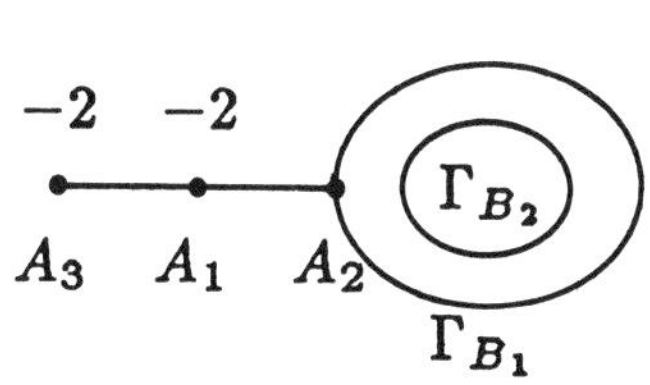

(5) $Z = 2D$, $|D| = B_1$, *D is a positive cycle,* $z_2 = 3$.

$$A_2 \cdot Z_{B_1} = 0, \quad A_2 \subseteq B_1, \quad A_2 \subsetneq B_2$$

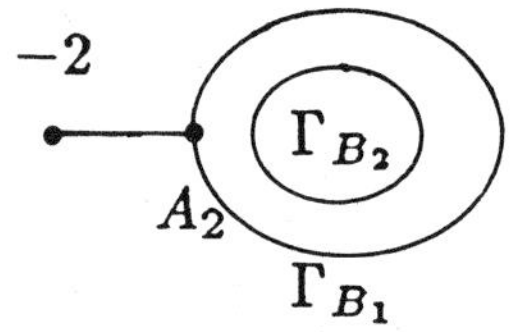

where Γ_{B_1}, *is the graph of* B_i.

PROPOSITION 3.6. [11] *Let* Γ *be a weighted dual graph including genera for the vertices associated to the minimal good resolution of weakly elliptic singularity. Suppose* K' *exists. Let* $Z = Z_{B_0}, \ldots, Z_{B_\ell}, Z_E$ *be the elliptic sequence. If* $Z \cdot Z = -3$ *and* $Z_{B_1} \cdot Z_{B_1} = -1$, *then* Γ *must be one of the following forms*:

(1) $Z = 1\ Z_{B_1}$, $A_2 \subseteq B_1$, $A_2 \subsetneq B_2$; $z_2 = 1$, $A_2 \cdot Z_{B_1} = -1$

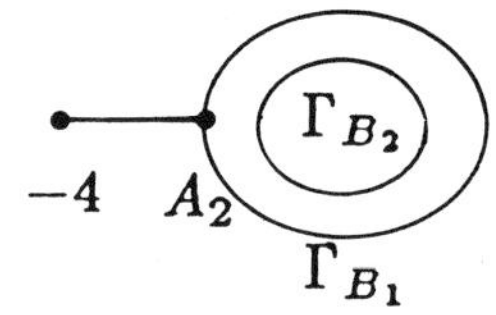

(2) $Z = 1\ 2\ D$, $|D| = B_1$, $A_2 \subseteq B_1$, $A_2 \subsetneq B_2$, $z_2 = 2$.

$$A_2 \cdot Z_{B_1} = 0$$

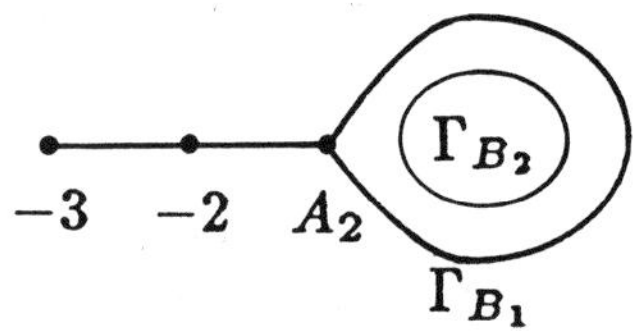

(3) $Z = 1\ 2\ 3\ D$, $|D| = B_1$, $A_2 \subseteq B_1$, $A_2 \subsetneq B_2$; $z_2 = 3$

$$A_2 \cdot Z_{B_1} = 0$$

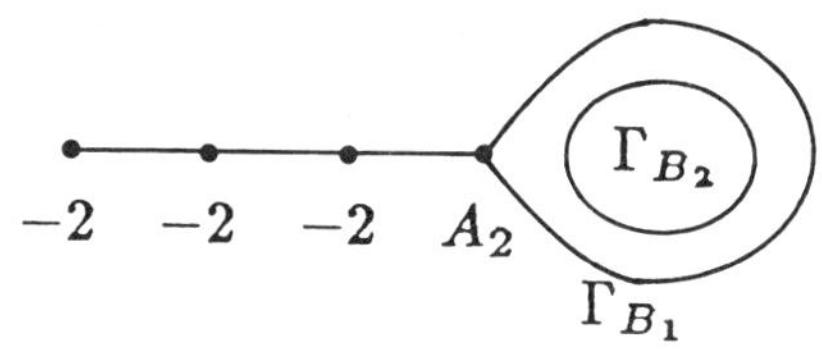

(4) $Z = 3\ D$, $|D| = B_1$, $A_2 \subseteq B_1$, $A_2 \subsetneq B_2$, $z_2 = 5$

$$A_2 \cdot Z_{B_1} = 0$$

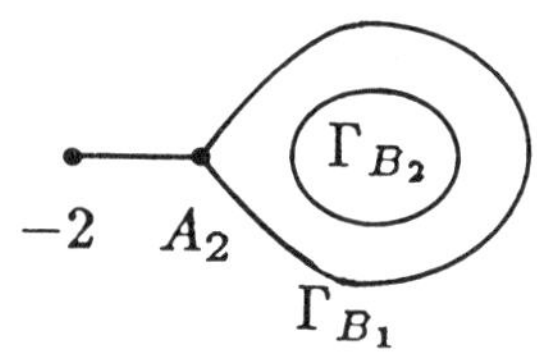

where Γ_{B_i} *is the graph of* B_i.

PROPOSITION 3.7. [11] *Let* Γ *be a weighted dual graph including genera for the vertices associated to the minimal good resolution of weakly elliptic singularity. Suppose* K' *exists and* $Z = Z_{B_0}, \dots, Z_{B_\ell}, Z_E$ *be the elliptic sequence. If* $Z \cdot Z = -3$ *and* $Z_{B_1} \cdot Z_{B_1} = -2$, *then* Γ *must be one of the following forms*:

(1) $Z = 1D21$, $z_3 = 1$, $z_4 = 2$, $D > 0$. $|D| = B_1$,

$$A_3 \cdot Z_{B_1} < 0,\ A_4 \cdot Z_{B_1} = 0,\ A_3, A_4 \subsetneq B_2,\ A_3 \neq A_4.$$

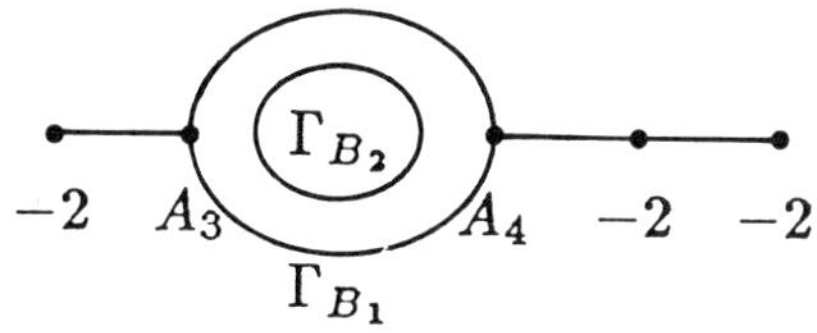

(2) $$Z = 1D2,\ z_3 = 1,\ z_4 = 3,\ D > 0,\ |D| = B_1,$$

$$A_3 \cdot Z_{B_1} < 0,\ A_4 \cdot Z_{B_1} = 0,\ A_3, A_4 \subseteq B_1,\ A_3, A_4 \subsetneq B_2,\ \ A_3 \neq A_4.$$

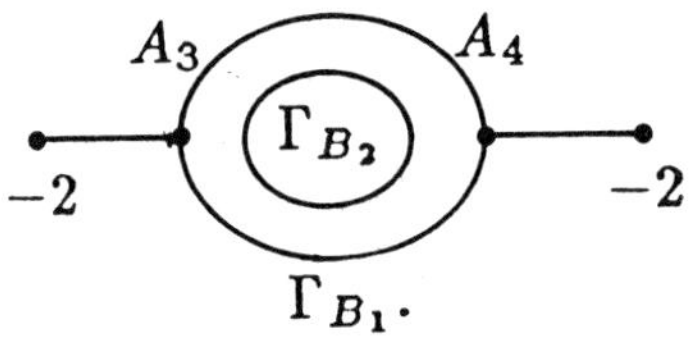

(3) $Z = 1Z_{B_1}1,\ z_3 = z_4 = 1,\ A_3 \cdot Z_{B_1} = A_4 \cdot Z_{B_1} = -1,$

$A_3, A_4 \subseteq B_1,\ A_3, A_4 \subsetneq B_2,\ \ A_3 \neq A_4.$

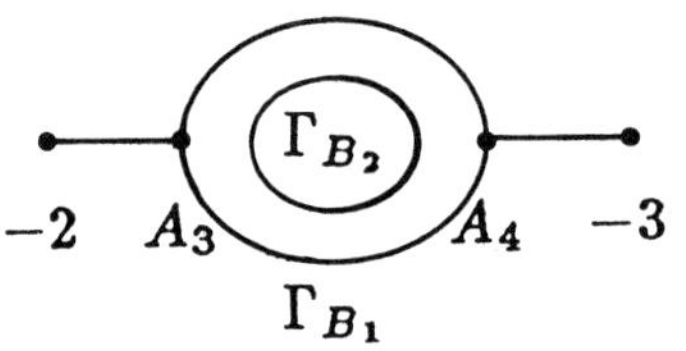

(4) $Z = {}_{1}^{1}Z_{B_1},\ z_3 = 1,\ A_3 \cdot Z_{B_1} = -2,\ A_3 \subseteq B_1,\ A_3 \subsetneq B_2.$

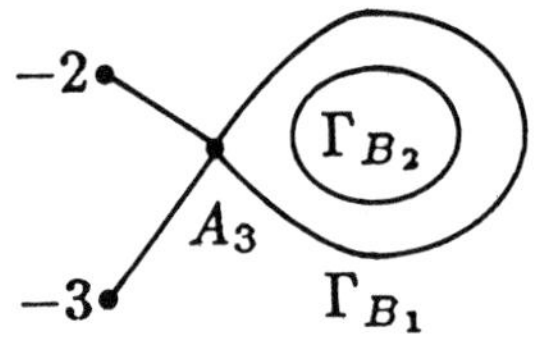

(5) $Z = \begin{smallmatrix}1\\1\end{smallmatrix}\ 2\ 2\ ...2\ Z_{B_1},\ z_2 = 1,\ A_2 \cdot Z_{B_1} = -2,\ A_2 \subseteq B_1,\ A_2 \subsetneq B_2,$

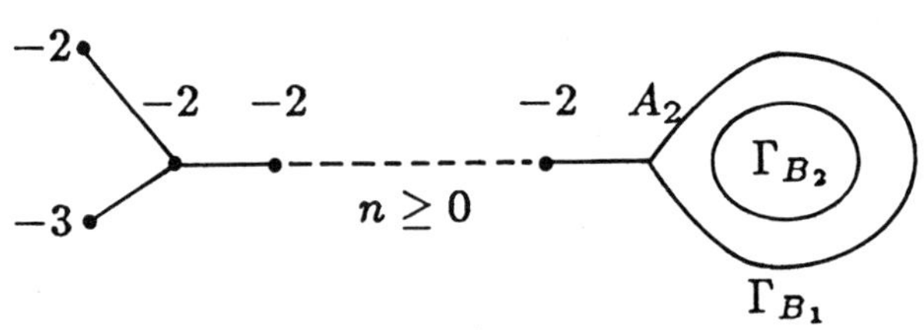

(6) $Z = 1\ 2\ D$, *D is a positive cycle*, $|D| = B_1$, $z_2 = 2$

$$A_2 \subseteq B_1,\ A_2 \subsetneq B_2,\ A_2 \cdot Z_{B_1} = 0$$

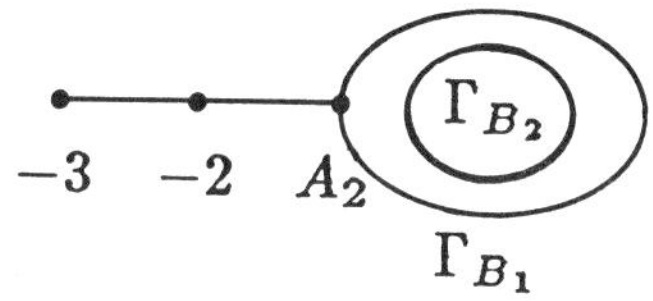

(7) $Z = 1\ 2\ 3\ D$, *D is a positive cycle*, $|D| = B_1$, $z_2 = 3$,

$$A_2 \cdot Z_{B_1} = 0,\ \ A_2 \subseteq B_1,\ \ A_2 \subsetneq B_2$$

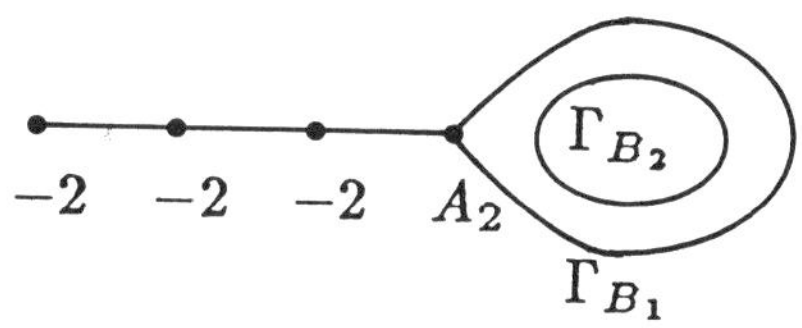

(8) $Z = 3D,\ D > 0,\ |D| = B_1,\ \ z_2 = 5,\ A_2 \cdot Z_{B_1} = 0,$

$$A_2 \subseteq B_1,\ \ A_2 \subsetneq B_2$$

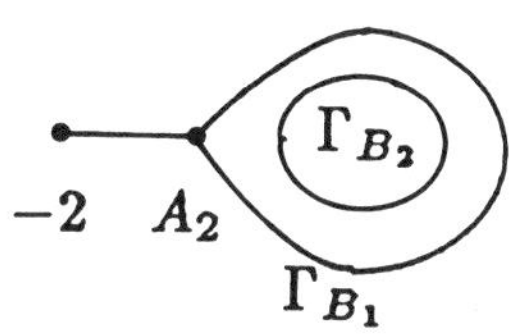

where Γ_{B_i} *is the graph of* B_i.

PROPOSITION 3.8. [11] *Let* Γ *be a weighted dual graph including genera for the vertices associated to the minimal good resolution of weakly elliptic singularity. Suppose* K' *exists. Let* $Z = Z_{B_0}, \dots, Z_{B_\ell}, Z_E$ *be the elliptic sequence. If* $Z \cdot Z = -3$ *and* $Z_{B_1} \cdot Z_{B_1} = -3$, *then* Γ *must be one of the following forms:*

(1) $Z = 1\ Z_{B_1}\ 1,\ z_4 = z_5 = z_6 = 1,\ A_4 \cdot Z_{B_1} = A_5 \cdot Z_{B_1} = A_6 \cdot Z_{B_1} = -1$

$$A_4, A_5, A_6 \subseteq B_1,\ A_4, A_5, A_6 \subsetneq B_2,\ A_4 \neq A_5 \neq A_6 \neq A_4\,.$$

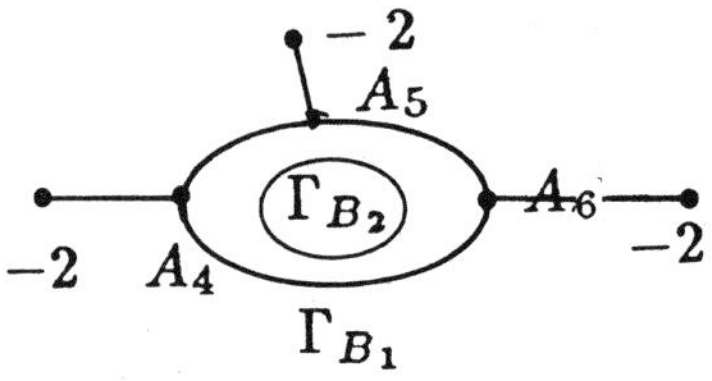

(2) $Z = {}^{1}_{1}\, Z_{B_1}\ 1,\ z_4 = z_5 = 1,\ A_4 \cdot Z_{B_1} = -2,\ A_5 \cdot Z_{B_1} = -1.$

$$A_4, A_5 \subseteq B_1,\quad A_4, A_5 \subsetneq B_2,\quad A_4 \neq A_5.$$

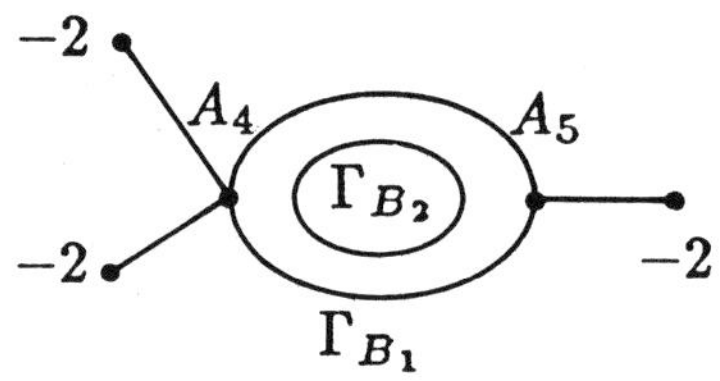

(3) $Z = 1{}^{1}_{1}\, Z_{B_1},\ z_4 = 1,\ A_4 \cdot Z_{B_1} = -3,\ A_4 \subseteq B_1,\ \ A_4 \subsetneq B_2.$

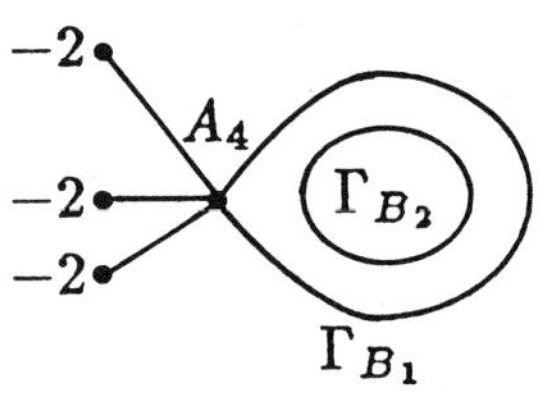

(4) $Z = 1\ 2^1\ 2 \ldots 2\ {}^1Z_{B_1},\ z_3 = 1,\ A_3 \cdot Z_{B_1} = -3.$

$$A_3 \subseteq B_1,\quad A_3 \subsetneq B_2.$$

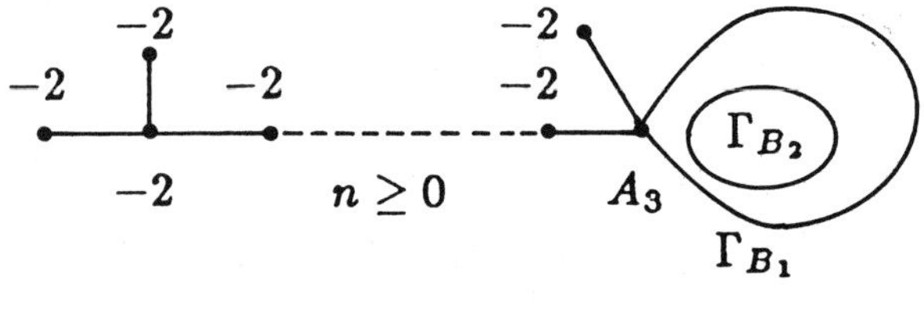

(5) $Z = 1\ Z_{B_1}\ 2\ldots2\ 2\ {}^{1}_{1},\ z_3 = z_4 = 1.\ \ A_3 \cdot Z_{B_1} = -1.\ \ A_4 \cdot Z_{B_1} = -2.$

$$A_3, A_4 \subseteq B_1,\ A_3, A_4 \subsetneq B_2.\ A_3 \neq A_4$$

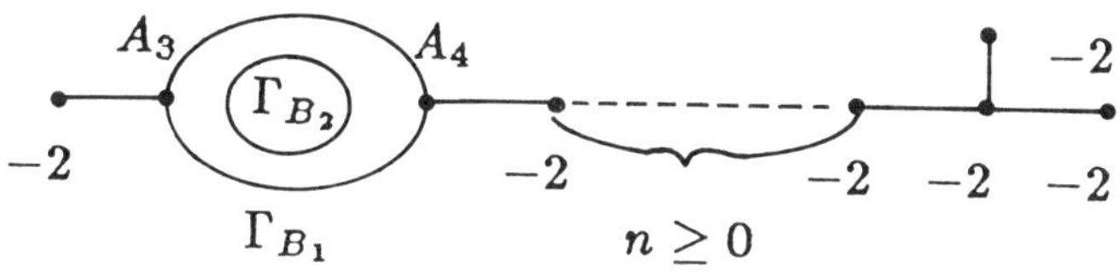

(6) $Z = 1\ D\ 2\ 1,\ D > 0.\ |D| = B_1,\ z_3 = 1.\ z_4 = 2.$

$A_3 \cdot Z_{B_1} < 0.\ A_4 \cdot Z_{B_1} = 0.\ A_3, A_4 \subseteq B_1.\ A_3, A_4 \subsetneq B_2.\ A_3 \neq A_4.$

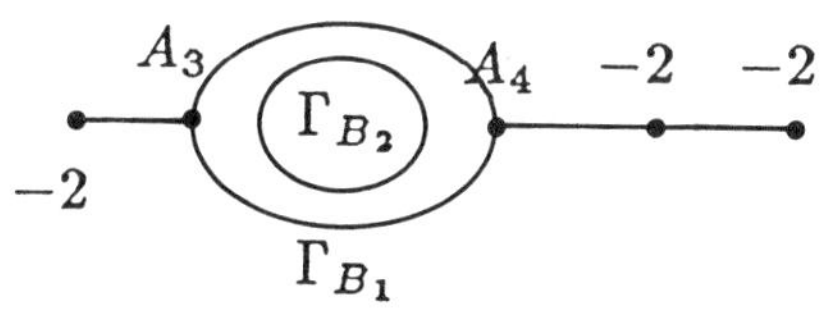

(7) $Z = 1\ D\ 2,$ *D is a positive cycle*, $|D| = B_1.\ z_3 = 1.\ z_4 = 3.$

$A_3 \cdot Z_{B_1} < 0.\ A_4 \cdot Z_{B_1} = 0.\ A_3, A_4 \subseteq B_1,\ A_3, A_4 \subsetneq B_2.\ A_3 \neq A_4.$

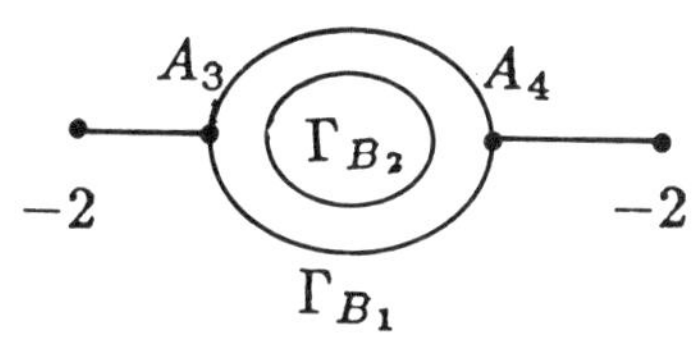

(8) $Z = \begin{smallmatrix}1&2\\1&2\end{smallmatrix}\ 3\ Z_{B_1},\ z_3 = 1.\ A_3 \cdot Z_{B_1} = -3.\ A_2 \subseteq B_1,\ A_2 \subsetneq B_2.$

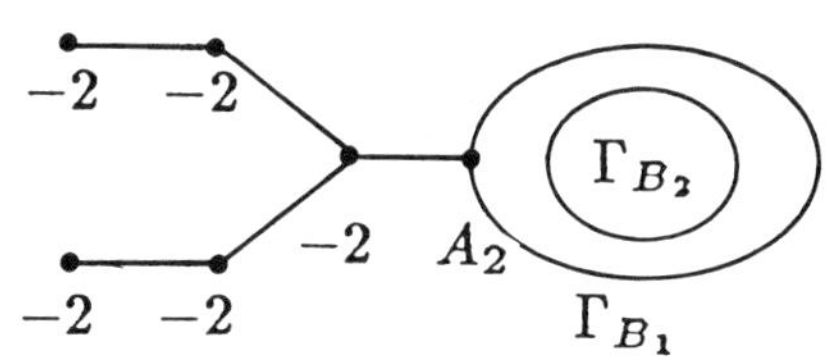

(9) $Z = 1\ 2\ 3\ 4\ 3\ Z_{B_1}$, $z_2 = 1$. $A_2 \cdot Z_{B_1} = -3$. $A_2 \subseteq B_1$. $A_2 \subsetneq B_2$.

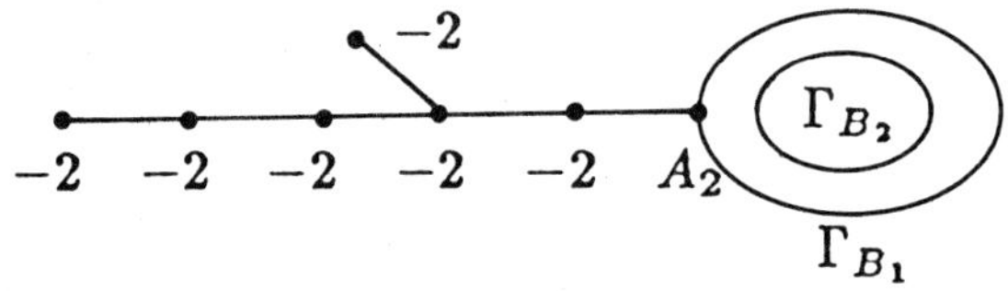

(10) $Z = 2\ 4\ 6\ 5\ 4\ 3\ Z_{B_1}$, $z_2 = 1$. $A_2 \cdot Z_{B_1} = -3$, $A_2 \subseteq B_1$. $A_2 \subsetneq B_2$.

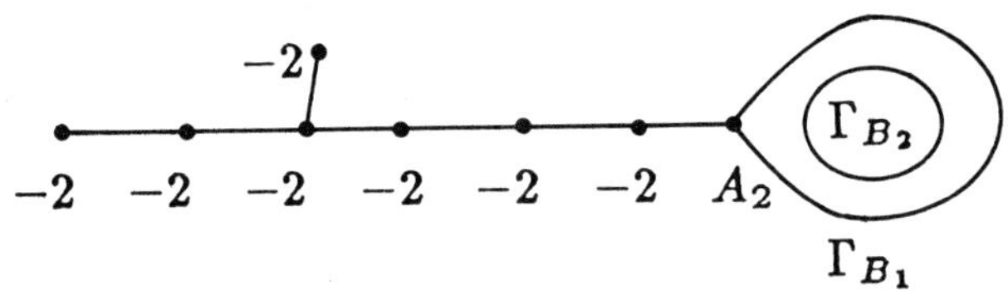

(11) $Z = 1\ 2\ 3\ D$, $D > 0$, $|D| = B_1$, $z_2 = 3$. $A_2 \cdot A_{B_1} = 0$. $A_2 \subseteq B_1$.

$A_2 \subsetneq B_2$.

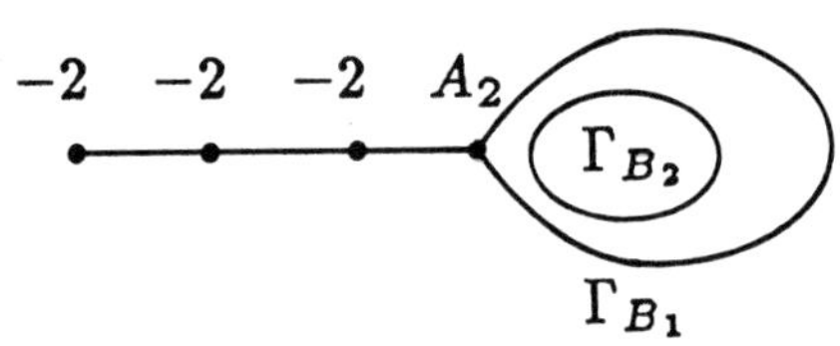

(12) $Z = 3\ D$, $D > 0$. $|D| = B_1$, $z_2 = 5$. $A_2 \cdot Z_{B_1} = 0$. $A_2 \subseteq B_1$, $A_2 \subsetneq B_2$.

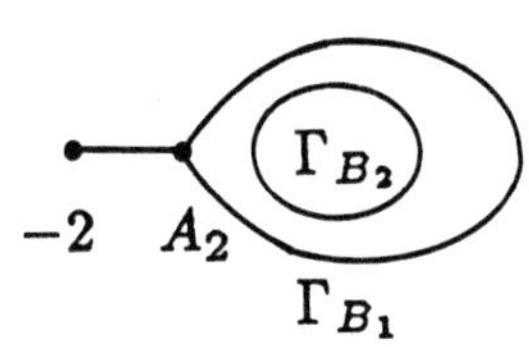

§4. LOWER ESTIMATE OF $K^2 + \chi_T(A)$, $K^2 + n$ FOR WEAKLY ELLIPTIC HYPERSURFACE SINGULARITIES

As we have seen in §2 that $K^2 + \chi_T(A)$ is a numerical invariant of the singularity. By the same argument one sees that $K^2 + n$ is also a numerical invariant of the singularity, where n is the number of irreducible components in the exceptional set A. In this section we shall give sharp lower estimates of these invariants.

THEOREM 4.1. *Let (V,p) be a weakly elliptic hypersurface singularity. Then $K^2 + \chi_T(A) \geqslant -3$ and $K^2 + n \geqslant -2$.*

PROOF. Let m be the multiplicity of (V,p). By (3.18) and (3.19) of [12], $1 = p_a \geqslant \frac{1}{8}(m-1)(m-3)m + 1$, it follows that $m \leqslant 3$. On the other hand, let Y be a maximal ideal cycle with respect to any resolution. By Theorem 2.17 of [10], $m \geqslant -Y \cdot Y$. In particular, we have $Y \cdot Y \geqslant -3$. By the definition of fundamental cycle, Y can be written as $Z + D$ where D is an effective cycle. Since $Y^2 = Z^2 + 2Z \cdot D + D^2$, hence $Y^2 \leqslant Z^2$ because $Z \cdot D \leqslant 0$ and $D^2 \leqslant 0$ by the negative definiteness of the intersection matrix of the exceptional set. We conclude that Z^2 is either -1, -2 or -3. From now on, we shall only consider minimal good resolution. In view of proposition 3.2, the weighted dual graph Γ including genera for the vertices is completely described by proposition 3.3 - 3.8.

By Theorem 3.1, we have

$$K^2 = \sum_{i=0}^{\ell} Z_{B_i}^2 + E^2.$$

Now we list some results as lemmas:

LEMMA 4.2.

$$\chi_T(A) = n + 1 - 2 \sum_{j=1}^{n} g_j - b_1(\Gamma)$$

where n = number of irreducible components in the exceptional set A i.e. $A = \cup_{j=1}^{n} A_j$,

g_j = *the genus of* A_j *and*

b_1 = *first betti number of the graph* Γ .

PROOF. Easy exercise.

LEMMA 4.3. *Let* (V,p) *be a minimal elliptic singularity. Then*

(1) $$\chi_T(A) + K^2 \geqslant -1 \quad if \quad Z \cdot Z = -1$$

(2) $$\chi_T(A) + K^2 \geqslant -2 \quad if \quad Z \cdot Z = -2$$

(3) $$\chi_T(A) + K^2 \geqslant -3 \quad if \quad Z \cdot Z = -3.$$

PROOF. This is a case by case checking by using the table in Laufer [3] and the formula for $\chi_T(A)$.

LEMMA 4.4. *Let* $Z_{B_0}, Z_{B_1}, \ldots, Z_{B_{\ell+1}}$ *be the elliptic sequence with* $Z \cdot Z = -2$, $Z_{B_1} \cdot Z_{B_1} = -1$. *Let* Γ_{B_1} *be the weighted dual graph associated to the minimal good resolution with the exceptional set* B_1, K_1' *be the canonical cycle associated to* Γ_1. *Then*

$$\chi_T(A) + K^2 = \chi_T(B_1) + K_1^2 - 1.$$

PROOF. By proposition 3.4 Γ must be one of the following forms:

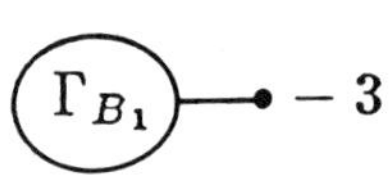

or

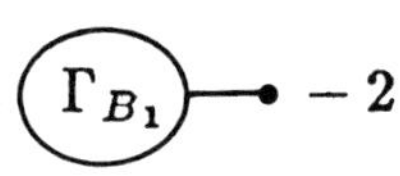

Γ has just one more vertex than Γ_1, so $\chi_T(A) = \chi_T(B_1) + 1$, $K^2 = K^2 - 2$, by using the formulas for K^2 and $\chi_T(A)$. Thus we have

$$\chi_T(A) + K^2 = \chi_T(B_1) + K_1^2 - 1.$$

LEMMA 4.5. *Notation as in Lemma 4.4, but* $Z \cdot Z = -2$, $Z_{B_1} \cdot Z_{B_1} = -2$. *Then*

$$\chi_T(A) + K^2 \geqslant \chi_T(B_1) + K_1^2 - 1.$$

Moreover, if $\chi_T(A) + K^2 = \chi_T(B_1) + K_1^2 - 1$, *then* Γ *is not a subgraph of a weighted dual graph associated to the minimal good resolution of a weakly elliptic hypersurface singularity, in this case we say that* Γ *is a final one.*

PROOF. In this case we have $K^2 = K_1^2 - 2$. By proposition 3.5, if Γ has just one more vertex than Γ_{B_1}, then Γ must be the following one:

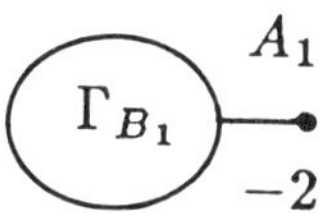

with $z_1 = 2$, $A_1 \cdot Z = -1$.

By checking the cases occuring in proposition 3.7, we see that Γ must be a final one. By Lemma 4.2 we see that $\chi_T(A) + K^2 \geqslant \chi_T(B_1) + 1$.

If Γ has at least two more points than Γ_{B_1}, then $\chi_T(A) \geqslant \chi_T(B_1) + 2$. Thus

$$\chi_T(A) + K^2 \geqslant \chi_T(B_1) + K_1^2.$$

LEMMA 4.6. *Notation as in Lemma 4.4, but* $Z \cdot Z = -3$, $Z_{B_1} \cdot Z_{B_1} = -2$. *Then* $\chi_T(A) + K^2 \geqslant \chi_T(B_1) + K_1^2 - 2$. *Moreover, if* $\chi_T(A) + K^2 = \chi_T(B_1) + K_1^2 - 2$, Γ *must be a final one; if* $\chi_T(A) + K^2 = \chi_T(B_1) + K_1^2 - 1$ *and* Γ *is not a final one, then* Γ *must be one of the following forms*:

(a)

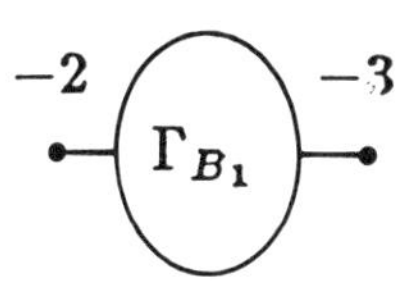

(b) 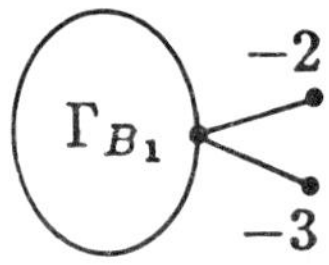

PROOF. In this case $K^2 = K_1^2 - 3$ and $\chi_T(A) = \chi_T(B_1) + k$ where k = number of vertices in Γ − number of vertices in Γ_{B1}, $k \geqslant 1$. By proposition 3.7, if $k < 3$, then Γ must be one of (2), (3), (4), (6), (8) of proposition 3.7. By checking the cases of proposition 3.8, we see that Γ must be a final one if Γ is of the form of (2), (6), (8) in proposition 3.7. If Γ is of the form (3), (4) in proposition 3.7, we see $k = 2$. So we get the result.

LEMMA 4.7. *Notation as in Lemma* 4.4, *but* $Z \cdot Z = -3$, $Z_{B_1} \cdot Z_{B_1} = -1$. *Then* $\chi_T(A) + K^2 \geqslant \chi_T(B_1) + K_1^2 - 2$. *Moreover, if* $\chi_T(A) + K^2 < \chi_T(B_1) + K_1^2$, Γ *must be a final one, or be of the form*

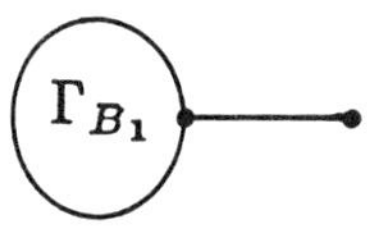

− 4.

PROOF. In this case $K^2 = K_1^2 - 3$, $\chi_T(A) = \chi_T(B_1) + k$ where k = number of vertices in Γ − number of vertices in Γ_{B1}, $k \geqslant 1$. By proposition 3.6 we see that if $k < 3$, Γ must be one of (1), (2), (4) in proposition 3.6. By checking the cases of proposition 3.8, one sees that Γ must be a final one if Γ is one of (2), (4) in proposition 3.6.

LEMMA 4.8. *Notation as in Lemma* 4.4, *but* $Z \cdot Z = -3$, $Z_{B_1} \cdot Z_{B_1} = -3$. *Then* $\chi_T(A) + K^2 \geqslant \chi_T(B_1) + K_1^2 - 2$. *Moreover, if* $\chi_T(A) + K^2 < \chi_T(B_1) + K_1^2$, Γ *must be a final one.*

PROOF. In this case $K^2 = K_1^2 - 3$. By checking the cases in proposition 3.8, one sees that Γ must be a final one if $\chi_T(A) + K^2 < \chi_T(A) + K^2 < \chi_T(B_1) + K_1^2$.

Now we continue to prove $\chi_T(A) + K^2 \geqslant -3$.

CASE (I):

$$\{Z^2_{B_0}, \ldots, Z^2_{B_{\ell+1}}\} \text{ consists of } -1.$$

By Lemma 4.3 (1) and proposition 3.3, we see that $\chi_T(A) + K^2 \geqslant -1$.

CASE (II):

$$\{Z^2_{B_0}, \ldots, Z^2_{B_{\ell+1}}\} \text{ consisits of } -1, -2.$$

Suppose $Z^2_{B_i} = -2$, $Z^2_{B_{i+1}} = -1$. Then by case (i), $\chi_T(B_{i+1}) + K^2_{i+1} \geqslant -1$. By Lemma 4.4, we see that $\chi_T(B_i) + K^2_i \geqslant -2$. By Lemma 4.5 we see that $\chi_T(A) + K^2 \geqslant -3$. Moreover, if $\chi_T(A) + K^2 = -3$, Γ must be a final one.

CASE (III):

$$\{Z^2_{B_0}, \ldots, Z^2_{B_{\ell+1}}\} \text{ consists of } -2.$$

By Lemma 4.3(2), $\chi_T(B_{\ell+1}) + K^2_{\ell+1} \geqslant -2$. By Lemma 4.5, $\chi_T(A) + K^2 \geqslant -3$. Moreover, if $\chi_T(A) + K^2 = -3$, Γ must be a final one.

CASE (IV):

$$\{Z^2_{B_0}, \ldots, Z^2_{B_{\ell+1}}\} \text{ consists of } -1, -3.$$

There are two subcases:

(α) There is a Z_{B_i} in the elliptic sequence such that

$Z^2_{B_i} = -3$, $Z^2_{B_{i+1}} = -1$, and $\Gamma_{B_i} =$ 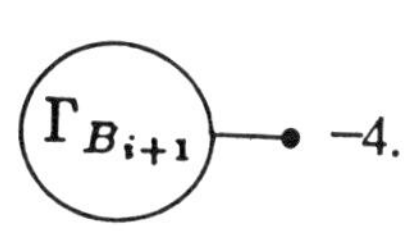

By proposition 3.8, Γ must be one of the following forms.

(a)

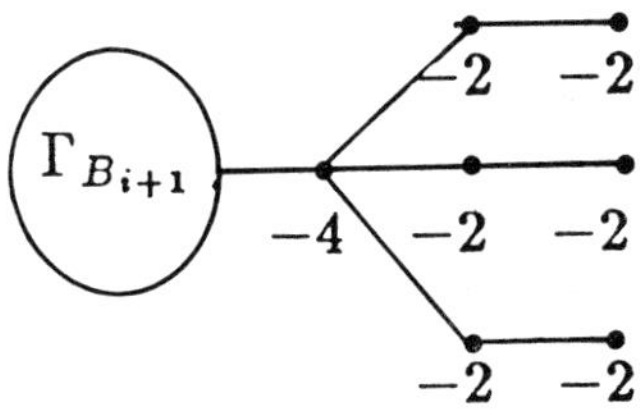

$\chi_T(B_{i+1}) + K_{i+1}^2 \geq -1.$

$\chi_T(A) + K^2 \geq -3.$

(b)

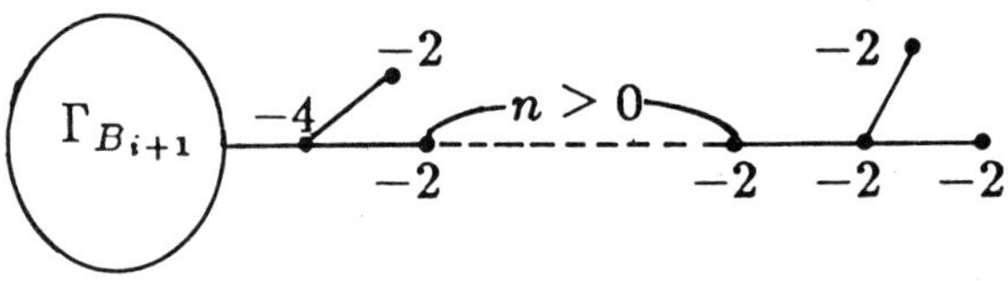

$\chi_T(A) + K^2 \geq -3$

(c)

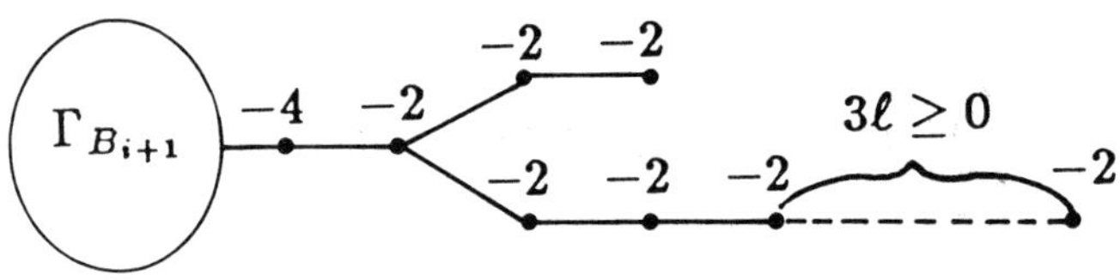

$\chi_T(A) + K^2 \geq -1$

(d)

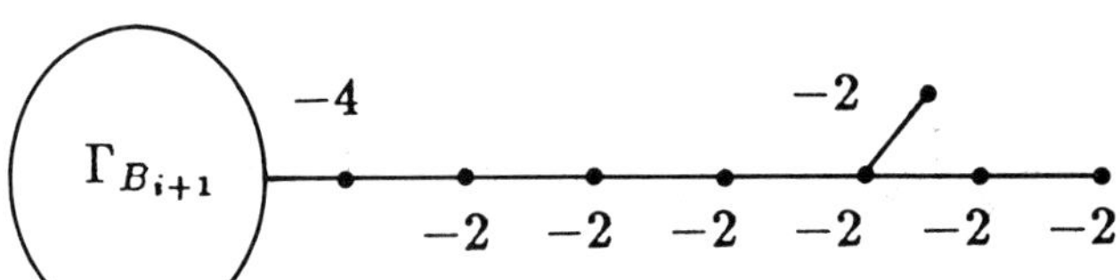

$\chi_T(A) + K^2 \geq +1$

(e)

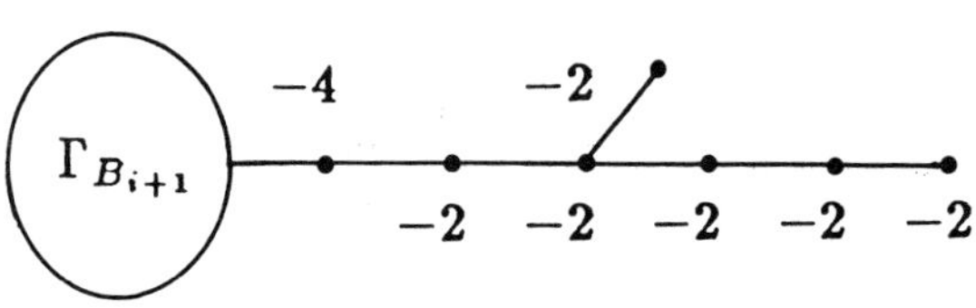

$\chi_T(A) + K^2 \geq 0.$

(f)

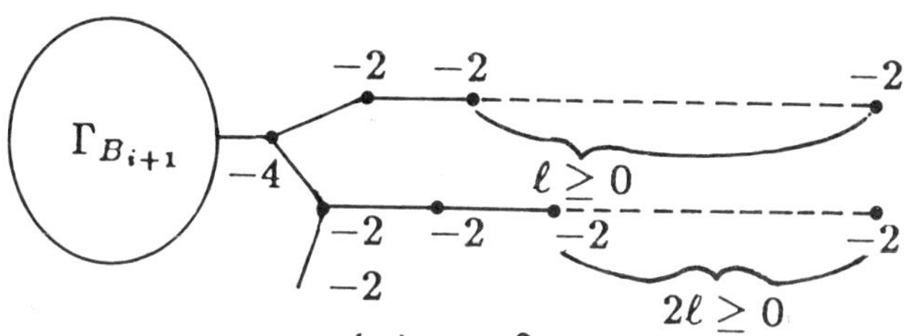

$$\chi_T(A) + K^2 \geq -2$$

(g)

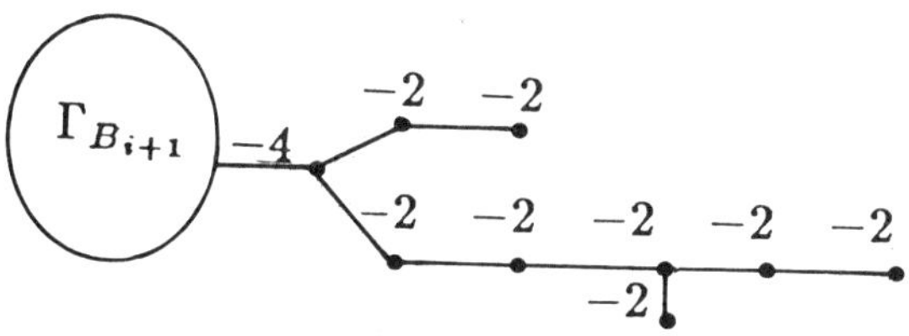

$$\chi_T(A) + K^2 \geq -1.$$

(β) There is no such Z_{B_i} in the elliptic sequence.

Suppose $Z^2_{B_i} = -3$, $Z^2_{B_{i+1}} = -1$, then by case (i) $\chi_T(B_{i+1}) + K^2_{i+1} \geqslant -1$. By Lemma 4.7, $\chi_T(B_i) + K_i^2 \geqslant -1$, if $B_i \neq A$. It follows from Lema 4.8, that $\chi_T(A) + K^2 \geqslant -3$.

So the proof of case (iv) is completed.

CASE (V).

$$\{Z^2_{B_0}, \ldots, Z^2_{B_{\ell+1}}\} \text{ consists of } -2, -3.$$

By checking the table in Laufer [3], we see that the only case that $\chi_T(B_{\ell+1}) + K^2_{\ell+1} = -2$ is $\Gamma_{B_{\ell+1}} = \cdot^{g-1}_{-2}$. By using proposition 3.7, 3.8, we can list all possible Γ with $\Gamma_{B_{\ell+1}} = \cdot^{g=1}_{-2}$. There are 13 such Γ's. Compute $\chi_T(A) + K^2$ for each Γ, we get $\chi_T(A) + K^2 \geqslant -3$. Now suppose $\Gamma_{\ell+1}$ is not $\cdot^{g=1}_{-2}$, then one has $\chi_T(B_{\ell+1}) + K^2_{\ell+1} \geqslant -1$. By Lemma 4.5 and Lemma 4.6 we need only check the cases that there is a Z_{B_i} such that $Z^2_{B_i} = -3$, $Z^2_{B_{I+1}} = -2$ and

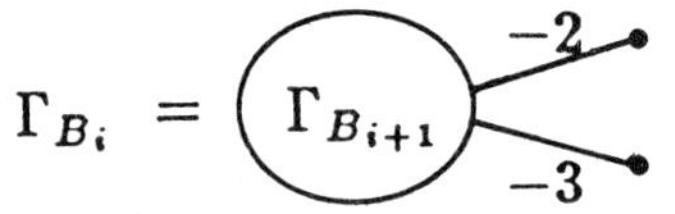

Here −2 vertex and −3 vertex may intersect $\Gamma_{B_{i+1}}$ at different vertices.

By Lemma 4.5, $\chi_T(B_{i+1}) + K_{i+1}^2 \geqslant -1$. So $\chi_T(B_i) + K_i^2 \geqslant -2$. By proposition 3.8, Γ must be one of the following forms:

(a)

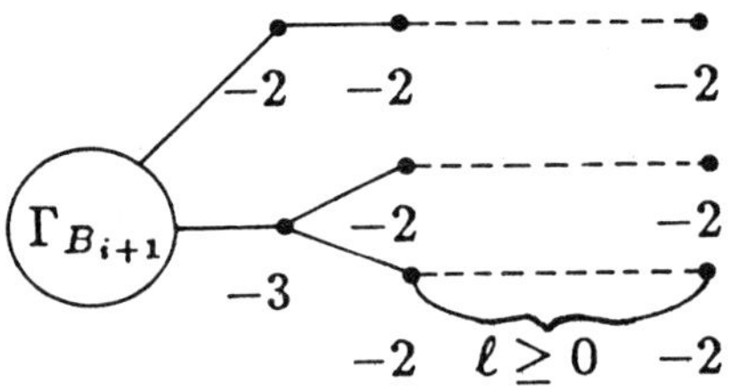

with $\chi_T(A) + K^2 \geqslant -2$

(b)

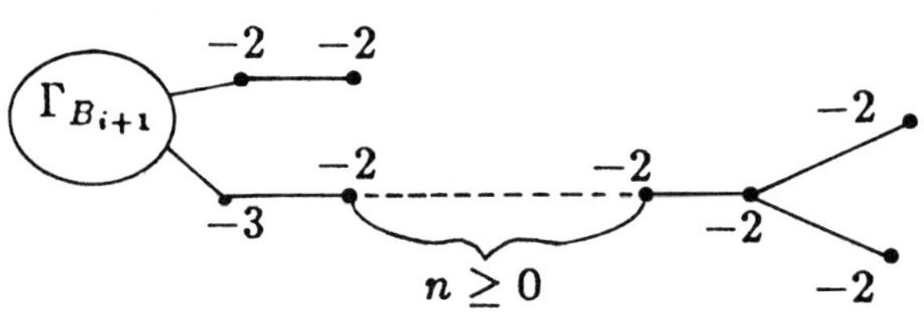

with $\chi_T(A) + K^2 \geqslant n - 1$.

(c)

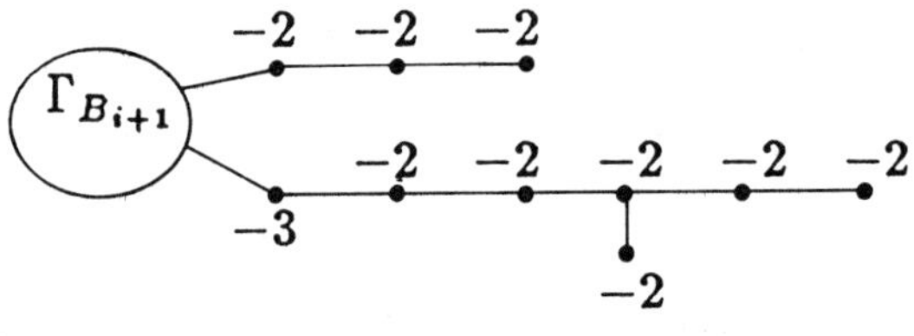

with $\chi_T(A) + K^2 \geqslant 0$.

(d) 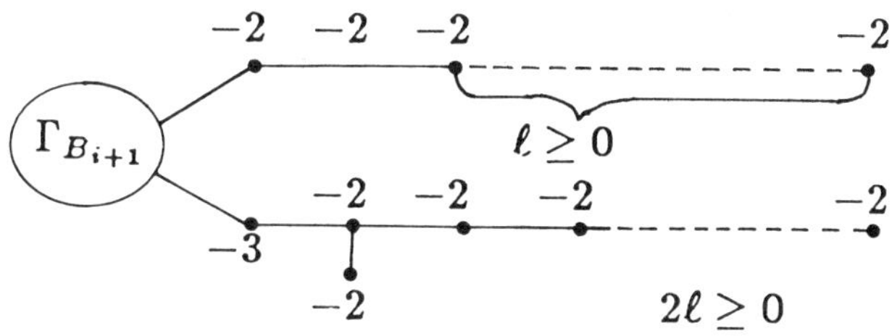

with $\chi_T(A) + K^2 \geqslant -1$.

CASE (VI):

$$\{Z^2_{B_0}, ..., Z^2_{B_{\ell+1}}\} \text{ consists of } -1, -2, -3.$$

Suppose $Z^2_{B_i} = -3$, $Z^2_{B_{i+1}} = -2$, then $\Gamma_{B_{i+1}}$ is not a final one, so $\Gamma_{B_{i+1}}$, must be one of (1), (2), (3), (4) in proposition 3.5, or (1) in proposition 3.4.

If $\Gamma_{B_{i+1}}$ is one of (1), (2) in proposition 3.5, proposition 3.7 tells us that Γ_{B_i} must be

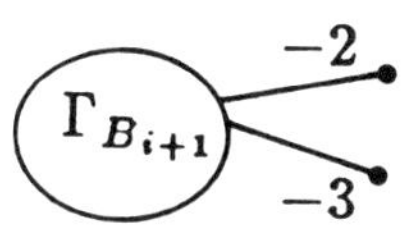

with $\chi_T(B_{i+1}) + K^2_{i+1} \geqslant -2$. As the proof of case (v), we see $\chi_T(A) + K^2 \geqslant -3$.

If $\Gamma_{B_{i+1}}$ is of the form (3) in proposition 3.5, then $Z^2_{B_{i+2}} = -2$, so by Lemma 4.5 and case (ii) $\chi_T(B_{i+2}) + K^2_{i+2} \geqslant -2$. Thus $\chi_T(B_{i+1}) + K^2_{i+1} \geqslant -1$. By proposition 3.7, Γ_{B_i} must be one of the following forms:

(a) 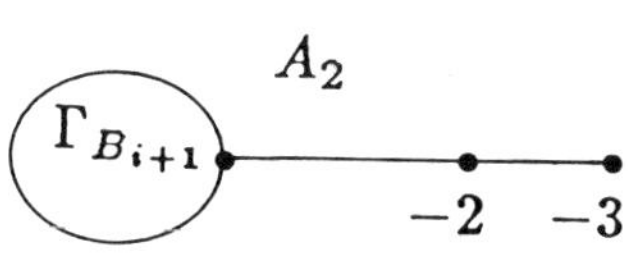 as (6) in 3.7

(b) 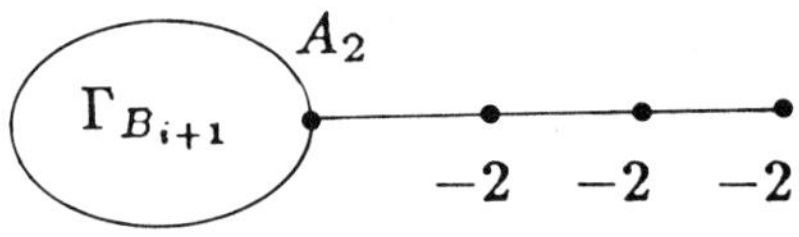 as (7) in 3.7

(c) 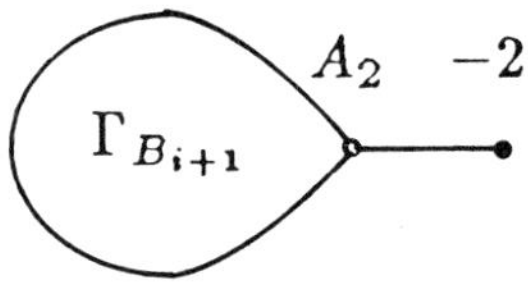 as (8) in 3.7.

In cases (a) and (c), $\Gamma = \Gamma_{B_i}$ (i.e. Γ_{B_i} is a final one). We have that $\chi_T(A) + K^2 \geqslant -3$.

In case (b), $\chi_T(B_i) + K_i^2 \geqslant -1$. Then by Lemma 4.8, we see that $\chi_T(A) + K^2 \geqslant -3$.

If $\Gamma_{B_{i+1}}$ is of the form (4) in proposition 3.5, the proof is the same as above, except $\Gamma_{B_{\ell+1}} = \cdot^{g=1}_{-1}$, since $\Gamma_{B_{\ell+1}} = \cdot^{g=1}_{-1}$ is the only case that $\chi_T(B_{\ell+1}) + K^2_{\ell+1} = -1$. (In all other cases we have $\chi_T(B_{\ell+1}) + K^2_{\ell+1} \geqslant 0$.) In this case $\chi_T(B_{i+1}) + K^2_{i+1}$ could be -2. But by using proposition 3.3, 3.4, 3.5 we see the only possible $\Gamma_{B_{i+1}}$ (which is of form (4) in 3.5) is

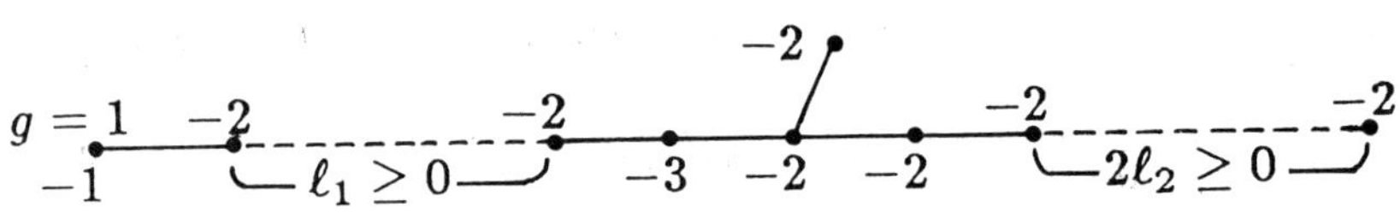

with $\chi_T(B_{i+1}) + K^2_{i+1} = -1$.

Finally, if $\Gamma_{B_{i+1}}$ is of the form (1) in proposition 3.4, by proposition 3.7, Γ_{B_i} must be one of the following forms:

(a) 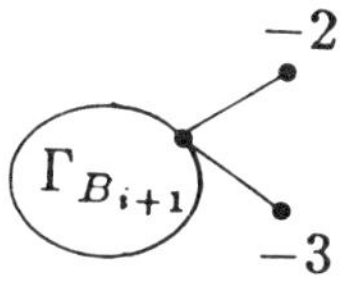
(b) 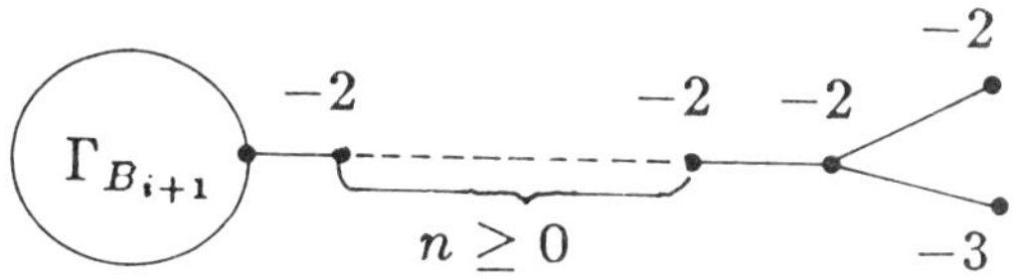

In case (a) we have $\chi_T(B_i) + K_i^2 \geqslant -3$. As the proof of case (v), we see that $\chi_T(A) + K^2 \geqslant \chi_T(B_i) + K_i^2 \geqslant -3$.

In case (b), we have $\chi_T(B_i) + K_i^2 \geqslant -2$. Applying proposition 3.8, one sees that if $\Gamma_{B_i} \neq \Gamma$, Γ must be

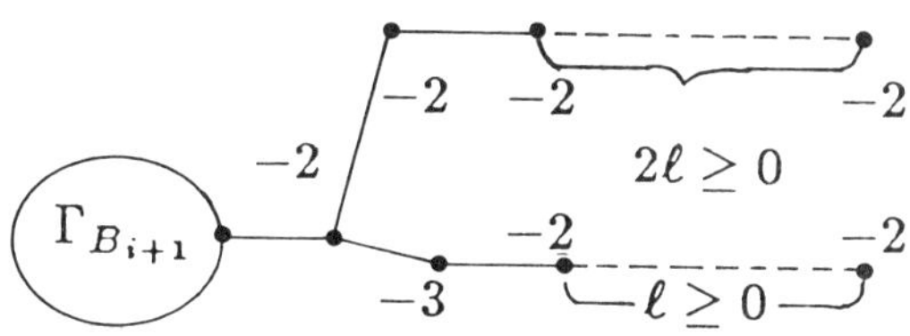

with $\chi_T(A) + K^2 \geqslant -2$.

The proof of case (vi) is completed.

CASE (VII):

$$\{Z^2_{B_0}, \ldots, Z^2_{B_{\ell+1}}\} \text{ consists of } -3.$$

By Lemma 4.7, we need only check the cases that $\chi_T(B_{\ell+1}) + K^2_{\ell+1} \leqslant -1$.

There are two such cases:

(a) 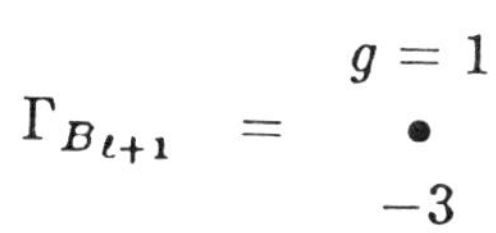

(b) 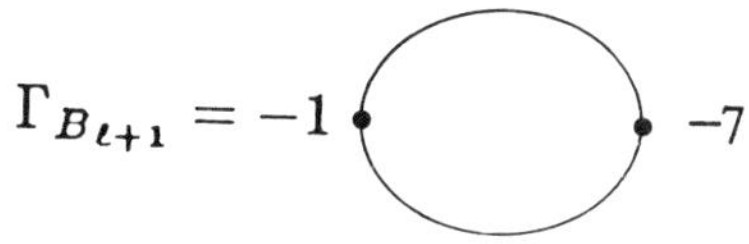

In both cases, there is a $A_1 \subset B_{\ell+1}$ such that $A_1 \cdot Z_{B_{\ell+1}} = -3$, so we can use the proof of case (iv)(α). One sees that $\chi_T(A) + K^2 \geqslant \chi_T(B_{\ell+1}) + K^2_{\ell+1} \geqslant -3$. Since $Z^2 = Z^2_{B_0} \geqslant -3$ as we have seen in the beginning of the proof, the proof of the inequality $\chi_T(A) + K^2 \geqslant -3$ is completed.

Now we are going to prove $K^2 + n \geqslant -2$. From proposition 3.3 - 3.8, we see that an A_i with $g_i > 0$ only occurs in $B_{\ell+1}$ and that the loop in Γ only occurs in $\Gamma_{B_{\ell+1}}$.

Moreover, by checking the table in Laufer [2], one sees that $\Gamma_{B_{\ell+1}}$ cannot have a loop if there is an $A_i \subset B_{\ell+1}$ with $g_i > 0$ and $\Gamma_{B_{\ell+1}}$ cannot have an A_i with $g_i > 0$ if it has a loop. Moreover, $\Gamma_{B_{\ell+1}}$ has at most one loop.

By this observation and Lemma 4.2, we conclude:

(I) $K^2 + n \geqslant \chi_T(A) + K^2 + 1$ if $\Gamma_{B_{\ell+1}}$ has a A_i with $g_i > 0$.

In particular we have $K^2 + n \geqslant -2$.

(II) $K^2 + n = \chi_T(A) + K^2$ if $\Gamma_{B_{\ell+1}}$ has a loop.

(III) $K^2 + n = \chi_T(A) + K^2 - 1$ if $\Gamma_{B_{\ell+1}}$ has neither a loop nor A_i with $g_i > 0$.

If Γ is in case (II), we have $\chi_T(B_{\ell+1}) + K^2_{\ell+1} \geqslant 0$ by checking a few such cases.

If Γ is in case (III), then $\Gamma_{B_{\ell+1}}$ has at least 4 vertices. By a case by case checking, we see that

$$K^2_{\ell+1} + \text{ number of vertices of } \Gamma_{B_{\ell+1}} \geqslant 0.$$

By a similar argument as in the proof of $K^2 + \chi_T(A) \geqslant -3$, we can show that $K^2 + n \geqslant -2$.

EXAMPLES:

(a) Let Γ be $\overset{g=1}{\underset{-3}{\cdot}}$. $K^2 = -3$, $\chi_T(A) = 2 - 2 = 0$, then

$$\chi_T(A) + K^2 = -3.$$

(b) Let Γ be $\overset{g=1}{\underset{-1}{\bullet}} \text{———} \underset{-4}{\bullet}$, $K^2 = -1 - 3 = -4$, $n = 2$, then

$$K^2 + n = -2.$$

These two examples show that our estimate is sharp. Now Theorem 4.1 has been proved.

THE PROOF OF THE THEOREM AND THE COROLLARY IN §1.

Recall, Laufer [1], $\mu = 12p_g - 1 + K^2 + \chi_T(A)$. By Theorem 4.1, $K^2 + \chi_T(A) \geqslant -3$, thus we have $\mu - 12p_g \geqslant -4$.

Recall that Yau [8], Theorem 3.2, $\sigma = -K^2 - n - 8p_g$. By Theorem 4.1, $K^2 + n^2 \geqslant -2$, thus we have

$$\sigma \leqslant 2 - 8p_g .$$

REFERENCES

0. M. Artin, *On isolated rational singularities of surfaces*, Amer. J. Math. **88** (1966), 129-136.

1. A. H. Durfee, *The signature of smoothings of complex surface singularities*, Math. Ann. **232** (1978), 85-98.

2. H. B. Laufer, *On μ for surface singularities*, Proceedings of Symposia in Pure Mathematics **30** (1977), 45-49.

3. H. B. Laufer, *On minimally elliptic singularities*, Amer. J. Math. **99** (1977), 1257-1295.

4. J. Milnor, *Singular points of complex hypersurfaces*, Ann. Math. Studies **91**, in "Princeton University Press 1968".

5. R. Randell, *On some numerical invariants of singularities*, (preprint).

6. M. Tomari, *A geometric characterization of normal two dimensional singularities of multiplicity two with* $p_a \leqslant 1$, Publ. RIMS, Kyoto University **20** (1984), 1-20.

7. M. Tomari, *The inequality* $8p_g < \mu$ *for hypersurface two dimensional isolated double points*, (preprint).

8. Stephen S.-T. Yau, *The signature of Milnor fibres and duality theorem for strongly pseudoconvex manifolds*, Inventiones Math. **46** (1978), 81-97.

9. Stephen S.-T. Yau, *Two theorems on higher dimensional singularities*, Math. Ann. **231** (1977), 55-59.

10. Stephen S.-T. Yau, *On maximally elliptic singularity*, Trans. A.M.S. **257**, No. 2 (1980), 269-329.

11. Stephen S.-T. Yau, *Hypersurface weighted dual graphs of normal singularities of surfaces*, Amer. J. Math. **101, No 4** (1979), 761-812.

12. Stephen S.-T. Yau, *The multiplicity of isolated two-dimensional hypersurface singularities: Zariski Problem*, (preprint).

13. P. Wagreich, *Elliptic singularities of surfaces*, Amer. J. Math. **92** (1970), 419-454.

DEPARTMENT OF MATHEMATICS
UNIVERSITY OF ILLINOIS AT CHICAGO
BOX 4348
CHICAGO, ILLINOIS 60680

Contemporary Mathematics
Volume **90**, 1989

MULTIARRANGEMENTS OF HYPERPLANES AND THEIR FREENESS

Günter M. Ziegler

ABSTRACT. We introduce *hyperplane arrangements with multiplicities* (*multiarrangements*) as natural objects arising in different geometric and combinatorial situations.

This leads to a generalization of the theory of *free* hyperplane arrangements as defined by Terao to *free multiarrangements*.

The restriction of a (simple) free arrangement to one of its hyperplanes is always a free multiarrangement in this sense. Multiarrangements arising from unitary reflection groups are discussed.

1. INTRODUCTION

A *hyperplane arrangement* is a finite set X of hyperplanes (subspaces of codimension one) in a finite dimensional vector spaces. The theory of hyperplane arrangements has made considerable progress in recent years. We refer to P. Cartier's talk [2] in the Bourbaki seminar for a survey. In particular, the theory of *freeness* of hyperplane arrangements initiated by H. Terao [16] has produced interesting insight into the algebraic structure of hyperplane arrangements (see especially [16, 17, 5, 6, 14]).

In this note, we propose an extension of this theory to arrangements $\tilde{X}$ in which every hyperplane $H \in \tilde{X}$ has a positive

0271-4132/89 $1.00 + $.25 per page

integer *multiplicity* $k(H)$. We call them *multiarrangements of hyperplanes*.

Section 2 will give basic definitions and a combinatorial interpretation in terms of nonsimple matroids. The interpretation of matroid loops is considered. We show that the restriction of a hyperplane arrangement to a hyperplane has a natural structure of a multiarrangement of hyperplanes.

In Section 3 we interpret multiarrangements as possibly non-reduced linear divisors on $\mathbf{C}^n$. K. Saito's theory of logarithmic vector fields and differential forms [12] can be extended to the corresponding non-reduced hypersurfaces. This yields definitions of logarithmic vector fields (tangent of order $k(H)$ to every hyperplane H in $\tilde{X}$). We call a multiarrangement *free* if the corresponding module of logarithmic vector fields is a free module. Multiarrangements in vector spaces of dimension $n \leqslant 2$ turn out to be always free. A basis criterion as in [12] is given.

In Section 4 we show that the restriction of a free simple arrangement to one of its hyperplanes is always a free multiarrangement. This relates to a conjecture by P. Orlik. As a second application, we discuss the multiarrangement defined by a finite unitary reflection group and its freeness.

The material of this note is part of the author's Ph.D. thesis [22] written under the direction of Anders Björner, which continues this discussion. The author wants to thank Anders Björner, Louis Billera, Peter Orlik and Gian-Carlo Rota for valuable comments and inspiring discussions.

2. DEFINITIONS AND COMBINATORICS

DEFINITION 1. Let V be an n-dimensional vector space over an arbitrary field $\mathbf{k}$. A *multiarrangement of hyperplanes* is a finite multiset $\tilde{X} = \{H_1^{k(H_1)}, \ldots, H_m^{k(H_m)}\}$ of linear hyperplanes in V, (that is, $(n-1)$-dimensional subspaces of V containing the origin). For a hyperplane $H_i \in \tilde{X}$, $k(H_i)$ is a positive integer, called its *multiplicity*. The *order* of $\tilde{X}$ is $|\tilde{X}| = \tilde{m} = k(H_1) + \cdots + k(H_m)$.

The *associated arrangement* to $\tilde{X}$ is the set $X = \{H_1, ..., H_m\}$ of order $|X| = m$. A multiarrangement $\tilde{X}$ is a *simple arrangement* if $\tilde{X} = X$, that is if $k(H) = 1$ for all $H \in \tilde{X}$.

Let $X = \{H_1, ..., H_m\}$ be a simple arrangement of hyperplanes in V. For every $H \in X$ we choose a non-zero linear functional $\ell_H \in V^*$ that defines H, i.e., such that $H = \{y \in V : \ell_H(y) = 0\}$. Now the set $\{\ell_H : H \in X\}$ of vectors in V^* forms a (represented) matroid. This matroid has received considerable attention. Especially the associated geometric lattice (which is isomorphic to the lattice of intersections of the hyperplanes in X, order by reverse inclusion) was extensively studied by Zaslavsky [21] for enumerative questions and by Orlik and Solomon [5] under topological aspects.

For a multiarrangement $\tilde{X}$, we consider the matroid $M(\tilde{X})$ given by the multiset

$$\{\ell_{H_1}^{k(H_1)}, ..., \ell_{H_m}^{k(H_m)}\}$$

and its linear dependencies.

The matroid $M(\tilde{X})$ has multiple points unless $\tilde{X}$ is simple. The associated simple matroid is $M(X)$, the corresponding lattice of flats is again the intersection lattice of X.

A property of a multiarrangement $\tilde{X}$ is *combinatorial* if it only depends on the associated abstract matroid $M(\tilde{X})$. This generalizes a definition by Terao for the simple case. It was shown that *f*-vector [21] and "being simplicial" are combinatorial properties of simple arrangements. H. Terao's famous conjecture [16, 19] asks whether freeness of an arrangement is combinatorial. We extend this question in Section 3.

EXAMPLE 2. Let X be an arrangement in V, H a hyperplane in V. Then the restriction of X to H is the arrangement $X|_H = \{K \cap H : K \in X \setminus \{H\}\}$. This restriction in fact has the natural structure of a multiarrangement $\widetilde{X|_H}$ with

$$k(K') = \#\{K \in X : K \cap H = K'\}.$$

This construction generalizes naturally to restrictions of multiarrangements. In fact it suggests to even allow *loops* for the matroids $M(\tilde{X})$, with V having a certain multiplicity in $\tilde{X}$ – namely, H having multiplicity $k(H)$ in $\widetilde{X|_H}$, and $\widetilde{X|_H}$ being loop free iff $H \notin X$. However, we will *not* follow this suggestion here and exclude ("delete") loops.

Hence, $\widetilde{X|_H}$ will have order $|\widetilde{X|_H}| = |\tilde{X}| - k(H)$, where we agree that $k(H) = 0$ if $H \notin X$.

EXAMPLE 3. C. Greene [3] considers hyperplane arrangements associated to finite simple graphs. In this context, graphs with multiple edges naturally give rise to multiarrangements. The main result of [3], stating that there is a natural bijection between the regions of the arrangement and acyclic orientations of the graph, generalizes to the case of multigraphs and the associated multiarrangements. We omit details.

We note that topological considerations [22] (compare also to the projective situation in [20]) suggest to allow infinite multiplicity of hyperplanes in $\tilde{X}$, too. For the problems considered in this note, this will not be necessary.

3. FREENESS

K. Saito's theory of logarithmic differential forms and logarithmic vector fields [12] can be extended to the case of a non-reduced hypersurface in a complex manifold. To avoid unnecessary generality, we will describe the main points of this development only for the case where the hypersurface is given by a multiarrangement in $\mathbf{C}^n$ (linear hypersurface singularity), which allows us to work in the rational instead of the holomorphic category and also provides independence of the field and its characteristics.

For the following let $\tilde{X}$ be a multiarrangement in $V = \mathbf{k}^n$, $S = \mathbf{k}[x_1, ..., x_n]$ be the ring of polynomial functions on V, and $S' = \mathbf{k}(x_1, ..., x_n)$ the quotient ring of rational functions. Let

$$Der(V) = \left\{\theta = \sum_{i=1}^{n} p_i \frac{\partial}{\partial x_i} : p_i \in S \text{ for } 1 \leqslant i \leqslant n\right\}$$

be the S-module of (algebraic) derivations on V, and

$$\Omega^s_{alg}(V) = \left\{\omega = \sum_{|I|=s} \omega^I dx_I : \omega^I \in S \text{ for } I \subseteq [n]\right\}$$

for $0 \leqslant s \leqslant n$ the S-module of algebraic s-forms on V. Similarly, let $\Omega^s_{rat}(V) = S' \otimes_S \Omega^s_{alg}(V)$ denote the S-module of rational s-forms.

For $H \in X$ let ℓ_H be a linear form in V^* defining H (as in Section 2). If we identify $\{X_1, X_2, ..., X_n\}$ with a basis of V^*, then every ℓ_H is a linear function in the variables X_i. Now let $\tilde{Q} = \Pi_{H\in X}\ell_H^{k(H)}$ be a defining equation for the multiarrangement $\tilde{X}$. $\tilde{Q}$ is a polynomial in S of degree $deg\ \tilde{Q} = |\tilde{X}|$, well-determined up to a constant factor. Similarly, $Q = \Pi_{H\in X}\ell_H$ will denote a defining equation for the associated simple arrangement X, and $deg\ Q = |X|$.

DEFINITION 4. Let $\tilde{X}$ be a multiarrangement in V.

(a) The S-module

$$Der(\tilde{X}) = \{\theta \in Der(V) : \theta(\ell_H) \in \ell_H^{k(H)} S \quad \text{for all} \quad H \in X\}$$

is called the module of *logarithmic vector fields* along $\tilde{X}$.

(b) The S-modules

$$\Omega^s(\tilde{X}) = \left\{\frac{\omega}{\tilde{Q}} : \omega \in \Omega^s_{alg}(V);\ \omega \wedge d\ell_H \in \ell_H^{k(H)}\Omega^{s+1}_{alg}(V) \text{ for all } H \in X\right\}$$

($1 \leqslant s \leqslant n$) are the modules of *logarithmic s-forms* (s-forms with at most logarithmic pole along $\tilde{X}$).

Note that Definition 4 reduces to the definitions of [16] (in the form of [11]) in the case of simple arrangements, which were originally given by [12] for the general case of reduced complex hypersurface singularities, that is,

$$\Omega^s(X) = \left\{ \frac{\omega}{Q} : \omega \in \Omega^s_{\mathrm{alg}}(V);\ \omega \wedge d\ell_H \in \ell_H \Omega^{s+1}_{\mathrm{alg}}(V) \text{ for all } H \in X \right\}$$

$$= \{\omega \in \Omega^s_{\mathrm{rat}}(V) : Q\omega \in \Omega^s_{\mathrm{alg}}(V);\ Qd\omega \in \Omega^{s+1}_{\mathrm{alg}}(V)\}.$$

As in [15], we can see that $Der(\tilde{X})$ is isomorphic to the module

$$P(\tilde{X}) = \{\mathbf{p} \in S^n : \ell_H^{k(H)} \mid \ell_H \circ \mathbf{p} \quad \text{for all} \quad H \in X\}$$

of polynomial functions $V \to V$ that map every hyperplane into itself with sufficient multiplicity.

THEOREM 5. *The pairing of S-modules*

$$\Omega^1(\tilde{X}) \times Der(\tilde{X}) \to S$$
$$\left[\theta; \frac{\omega}{\tilde{Q}}\right] \mapsto \frac{\omega(\theta)}{\tilde{Q}}$$

is well-defined and non-singular. Hence $Der(\tilde{X})$ *and* $\Omega^1(\tilde{X})$ *are dual S-modules, and* $Der(\tilde{X})$ *is free if and only if* $\Omega^1(\tilde{X})$ *is free.*

PROOF. See [12]. Some arguments are simplified by noting $Der(\tilde{X}) = \cap_{H\in X} Der(\{H^{k(H)}\})$, and assuming $H = \{x_1 = 0\}$ for a particular hyperplane after suitable transformation of coordinates. □

DEFINITION 6. A multiarrangement $\tilde{X}$ is called *free* iff $Der(\tilde{X})$ is a free S-module.

COROLLARY 7. *For dimension* $n \leq 2$, *every multiarrangement* $\tilde{X}$ *is free.*

PROOF. Over a ring of homological dimension at most two, reflexive modules are free. □

The following theorem collects basic facts about bases of $Der(\tilde{X})$ and $\Omega^1(\tilde{X})$. Both the results and proofs are straightforward generalizations from the case of simple arrangements, as in [12] and [16].

THEOREM AND DEFINITION 8. *Let $\tilde{X}$ be a multiarrangement in $V = \mathbf{k}^n$*

(a) $Der(\tilde{X})$ is an S-module of rank n. Let

$$\theta_i = \sum_{j=1}^{n} p_{ij} \frac{\partial}{\partial x_j} \in Der(\tilde{X}) \quad \text{for} \quad 1 \leqslant i \leqslant n.$$

Then $\{\theta_1, ..., \theta_n\}$ is a basis of $Der(\tilde{X})$ if and only if $det(p_{ij}) = c \cdot \tilde{Q}$ for some $c \in \mathbf{k}^*$. If $\tilde{X}$ is free, then $Der(\tilde{X})$ has a homogeneous basis, that is, we can choose $\{\theta_1, ..., \theta_n\}$ such that for all i and j, p_{ij} is either 0 or a homogeneous polynomial of degree m_i. The multiset $\{m_1, ..., m_n\}$ of *multiexponents* for a homogeneous basis of $Der(\tilde{X})$ defined this way does not depend on the particular homogeneous basis chosen. The multiexponents satisfy $m_1 + \cdots + m_n = |\tilde{X}|$.

(b) $\Omega^1(\tilde{X})$ is an S-module of rank n. Let

$$\omega_i = \sum_{j=1}^{n} q_{ij} dx_j \in \Omega^1(\tilde{X}) \quad \text{for} \quad 1 \leqslant i \leqslant n.$$

Then $\{\omega_1, ..., \omega_n\}$ is a basis of $\Omega^1(\tilde{X})$ if and only if $det(q_{ij}) = c \cdot \tilde{Q}^{-1}$ for some $c \in \mathbf{k}^*$. $\{\omega_1, ..., \omega_n\}$ is a dual basis to $\{\theta_1, ..., \theta_n\}$ if (q_{ij}) is the transposed inverse matrix of (p_{ij}). Thus, if $\tilde{X}$ is free, then $\Omega^1(\tilde{X})$ has a homogeneous basis: $\{\omega_1, ..., \omega_n\}$ can be chosen such that for all i and j, q_{ij} is either 0 or a homogeneous rational function of degree $-m_i$.

Corollary 7 in fact implies that every multiarrangement in dimension $n = 2$ is "inductively free" in the sense of [16]. Using Theorem 8, this can be verified constructively.

However, it is nontrivial to write down bases for large 2-dimensional multi-arrangements as in the simple case [16, p. 296]. This has several reasons:

One difficulty stems from the fact that 2-dimensional *simple* arrangements are supersolvable [15, 4] and hence in suitable coordinates have a triangular basis for $Der(\tilde{X})$ (that is, $p_{ij} = 0$ for $i > j$, see [22], [23]) and $\Omega^1(\tilde{X})$ – but this is no longer true for multiarrangements, as the following computations show.

EXAMPLE 9. The multiarrangement $\tilde{X}_1$ defined by $\tilde{Q}_1 = X_1^2X_2^2(X_1 - X_2)$ is free with multiexponents {2, 3} and a "triangular basis" of $Der(\tilde{X})$ given by

$$\begin{aligned}\theta_1 &= X_1^2\frac{\partial}{\partial X_1} + X_2^2\frac{\partial}{\partial X_2}\\ \theta_2 &= (X_1 - X_2)X_2^2\frac{\partial}{\partial X_2}.\end{aligned}$$

Adding an extra hyperplane, we get $\tilde{X}_2$ defined by $\tilde{Q}_2 = X_1^2X_2^2(X_1 - X_2)^2$, which is free with multiexponents {3, 3}. $Der(\tilde{X}_2)$ does not have a "triangular basis." A basis is given by

$$\begin{aligned}\theta_1 &= (X_1 - X_2)X_1^2\frac{\partial}{\partial_1} + (X_1 - X_2)X_2^2\frac{\partial}{\partial X_2}\\ \theta_2 &= X_1^2X_1\frac{\partial}{\partial X_1} + X_2^2(2X_1 - X_2)\frac{\partial}{\partial X_2}.\end{aligned}$$

(We note that for general multiarrangements, the radial Euler vector field $\theta_E = \Sigma X_i(\partial/\partial X_1)$ is not in $Der(\tilde{X})$. In fact $\theta_E \in Der(\tilde{X})$ iff $\tilde{X}$ is simple, and 1 is not a multiexponent of $\tilde{X}$ unless $\tilde{X}$ has a nonempty simple arrangement as a summand.)

Even bigger difficulties in the construction of bases have to do with the observation that multiexponents *cannot* be computed combinatorially. The following example at the same time shows that multiarrangements with triangular basis cannot be characterized combinatorially, either.

PROPOSITION 10. *Multiexponents are not combinatorial in general.*

PROOF/EXAMPLE. Let $\tilde{X}$ be a multiarrangement in $\mathbf{R}^2$, given by $\tilde{Q} = X_1^3 X_2^3 (X_1 - X_2)(X_1 - \alpha X_2)$, with $m = 4$ and $\tilde{m} = 8$ ($\alpha \in \mathbf{R}\backslash\{0,1\}$).

CASE 1. For $\alpha = -1$, $\tilde{X}$ is free with basis of $\Omega^1(\tilde{X})$ given by

$$\omega_1 = \frac{1}{(X_1^2 - X_2^2)} \left[\frac{dX_1}{X_1^3} - \frac{dX_2}{X_2^3} \right]$$

$$\omega_2 = \frac{dX_1}{X_1^3}.$$

Hence in the special case $\tilde{X}$ is free with multiexponents {3, 5}; it has a triangular basis.

CASE 2. For $\alpha \in \mathbf{R}\backslash\{0, \pm 1\}$, $\tilde{X}$ is free with a basis for $\Omega^1(\tilde{X})$ given by

$$\omega_1 = \frac{1}{(X_1 - X_2)} \left[\frac{dX_1}{X_1^3} - \frac{dX_2}{X_2^3} \right]$$

$$\omega_2 = \frac{1}{X_1 - \alpha X_2} \left[\frac{dX_1}{X_2^3} - \frac{dX_2}{\alpha^2 X_2^3} \right],$$

hence $\tilde{X}$ is free with multiexponents {4, 4} in this *generic* case. $\Omega^1(\tilde{X})$ does *not* have a triangular basis. □

4. TWO EXAMPLES AND A CONJECTURE

In the following we will briefly discuss two examples of free multiarrangements occuring "in nature."

THEOREM 11. *Let X be a simple hyperplane arrangement that is free with exponents $\{1, e_2, ..., e_n\}$. Let H be a hyperplane in X and let $\widetilde{X|_H}$ be the multiarrangement in H given by the restriction of X to H with the canonical multiplicities. Then $\widetilde{X|_H}$ is free with multiexponents $\{e_2, ..., e_n\}$.*

PROOF. After change of coordinates we can assume that $\ell_H = X_1$, that is, $H = \{X_1 = 0\}$. Let $\theta_1, ..., \theta_n$ be a homogeneous basis of $Der(X)$, with

$$\theta_1 = \sum_{j=1}^{n} X_j \frac{\partial}{\partial X_j} = \theta_E \quad\text{and}\quad \theta_i = \sum_{j=1}^{n} p_{ij} \frac{\partial}{\partial X_j}.$$

Now $H \in X$ implies $X_1|p_{i1}$ for $1 \leqslant i \leqslant n$, and we can assume $p_{i1} = 0$ for $2 \leqslant i \leqslant n$, by transforming the basis $\theta_i \to \theta_i - (p_{i1}/X_1)\theta_1$ for $i = 2, ..., n$. With this we get $\{\theta_2|_{X_1=0}, ..., \theta_n|_{X_1=0}\}$ as a basis of $Der(\tilde{X})$: an easy computation shows that $\theta_i|_{X_1=0} \in Der(\tilde{X})$ for $2 \leqslant i \leqslant n$, and Theorem 8 together with $\tilde{Q}(\widetilde{X|_H}) = Q(X)/X_1|_{X_1=0}$ imply the basis property. □

Theorem 9 appears to be the "first half" of an argument that would verify P. Orlik's conjecture [19, Problem 2; 9, (1.10)]:

CONJECTURE 12. *For every (simple) free arrangement X and every $H \in X$, $X|_H$ is free.*

In fact the "second half" of the argument would be supplied by a proof of the following, stronger conjecture:

CONJECTURE 13. *If $\tilde{X}$ is a free multiarrangement, is X always free?*

There is some evidence for Conjecture 13. First, it is trivial for $n \leqslant 2$, and the case $|\tilde{X}| = |X| + 1$ is also easy to check. If $\tilde{X}$ is a restriction of a simple supersolvable arrangement to one of its hyperplanes, then $\tilde{X}$ is free and X is also free, because it it supersolvable [15, 4]. If X is generic, then $\tilde{X}$ is free if and only if X is free [22], [23].

We note that P. Orlik's Conjecture 12 was motivated by the observation that for arrangements X defined by Coxeter groups, restrictions to hyperplanes in X are free [10] and restrictions to intersections of hyperplanes in X seem to be free, too [6].

This implies that Conjecture 13 seems to hold in the case where $\tilde{X}$ is the restriction of a Coxeter arrangement to one of its hyperplanes.

A natural test case for a converse of Conjecture 13 (see below) is given by the arrangement defined by finite unitary reflection groups [13]. Let G be such a group, acting on $\mathbf{C}^n$, generated by unitary reflections, and let X_G denote the associated arrangement $X_G = \{Fix(g) : g \in G;\ \dim Fix(g) = n - 1\}$. There are canonical multiplicities for the hyperplanes in X given by

$$k(H) = \#\{g \in G : Fix(g) = H\}.$$

This defines a multiarrangement $\tilde{X}_G = \{H^{k(H)} : H \in X\}$ whose order is the number of unitary reflections in G. A defining equation of $\tilde{X}$ is given by the determinant of the Jacobian matrix of a system of basic invariants for G [13, Thm. 5.2; 1, p. 113]. See [8, 9] for detailed discussions.

The finite irreducible groups generated by unitary reflections in $\mathbf{C}^n$ have been classified by Shephard and Todd [13]. Their list includes many groups (Coxeter groups and others) for which $\tilde{X}$ is simple and hence free by Terao's theorem [18]. Also, many 8 of the groups live in $\mathbf{C}^2$, where every multiarrangement is free by Corollary 7. Of the remaining groups, the only infinite class will be discussed in the following Example 15. Only three more exceptional cases remain to check, if we wish to conclude that all multiarrangements defined by unitary reflection groups are free. They correspond to the groups denoted by (25), (26) and (31) in

[13]. However, these arrangements are so large (with $\tilde{m}$ = 24,33,81 and m = 12,21,19, respectively) that we feel unable to do these verifications without massive support by symbolic computation, as successfully used in [9]. Of course a classification free proof would be desirable. One approach would be to use [18], together with a special explanation why in this case $\tilde{X}$ is free when X is free. We observe, however:

EXAMPLE 14. In general, a multiarrangement $\tilde{X}$ need not be free if the underlying simple arrangement X is free:

The smallest example comes from the supersolvable arrangement X_1 defined by $Q_1 = xy(x-y)(x-z)(y-z)$. The simple arrangement X_1 is free with exponents {1,2,2} whereas the multiarrangement $\tilde{X}_1$ defined by Q_1^2 is not free. The degrees of a minimal set of homogeneous generators of $Der(\tilde{X}_1)$ is {3,4,4,4}, with a relation of degree 5.

To verify these data, we used the help of the computer algebra system MACAULAY.

To decide freeness for arrangements like this one without computer assistance, the theory of multiarrangements and their freeness is not yet in a satisfactory state. In particular, we do not know any addition/deletion theorem in the spirit of [16] – such a result would suffice to decide Conjecture 13 and give criteria for freeness in the situation of Example 14.

More comprehensively, we ask for a combinatorial characterization of free multi-arrangements, from which Conjecture 13 can be derived.

EXAMPLE 15. We consider the multiarrangement $\tilde{X} = \tilde{X}_{p,m,n}$ defined by the unitary reflection group $G(p;m,n)$ [13, p. 277] acting $\mathbf{C}^n$, where $p \geqslant 1$, $n \geqslant 2$, $m|p$. A defining equation is given by

$$\tilde{Q} = (X_1 X_2 \cdot \ldots \cdot X_n)^{(p/m)-1} \prod_{1 \leqslant i<j \leqslant n} (X_i^p - X_j^p),$$

hence $\tilde{X}$ has order

$$n\left[\frac{p}{m} - 1\right] + \frac{n(n-1)}{2}p.$$

Note that $\tilde{X}$ is a simple arrangement iff $m = p$ or $m = p/2$. This includes the cases of all infinite families of Coxeter arrangements, with $G(1;1;n) = A_{n-1} \oplus \phi_1$ (where ϕ_1 is the empty arrangement in $\mathbf{C}^1$ and $\oplus$ denotes the obvious direct sum of arrangements), $G(2;1;n) = B_n$, $G(2;2;n) = D_n$ and $G(p;p;2) = I_2(p)$. For $p \geqslant 3$ and $n \geqslant 3$, the arrangements do not have a real representation, i.e., the corresponding groups are not orthogonal reflection groups.

Now for $i \geqslant -1$, let $\theta_i = \Sigma X_i^{k_i}(\partial/\partial X_i)$, where $k_i = pi + 1$. Then $\theta_i \in Der(\tilde{X})$ for $i > 0$, and $\theta_0 \in Der(\tilde{X})$ if $m = p$ or $m = p/2$, that is, if $\tilde{X} = X$. In fact homogeneous bases for $Der(\tilde{X})$ can explicitly be given as

$$\{(X_1X_2 \cdot \ldots \cdot X_n)^{p-1}\theta_{-1}, \theta_0, \theta_1, \ldots, \theta_{n-2}\} \quad \text{for} \quad m = p$$

and

$$\{(X_1X_2 \cdot \ldots \cdot X_n)^{(p/m)-2}\theta_0, \theta_1, \ldots, \theta_{n-1}\} \quad \text{for} \quad m < p.$$

The verification that these vector fields are in $Der(\tilde{X})$ is easy because of the symmetry of $\tilde{Q}$. The determinant of the basis criterion Theorem 8 reduces to a Vandermonde. From the bases of $Der(\tilde{X})$, we read off the multiexponents as

$$\{(n-1)(p-1), 1, p+1, 2p+1, \ldots, (n-2)p+1\} \quad \text{for} \quad m = p.$$

and

$$\left\{n\left[\frac{p}{m} - 2\right] + 1, p+1, 2p+1, \ldots, (n-1)p+1\right\} \quad \text{for} \quad m < p.$$

We compare the results for $\tilde{X}$ with the associated simple arrangement X: for $m = p$, we have $\tilde{X} = X$ and nothing new, for $m < p$, X is free with basis $\{\theta_0, \ldots, \theta_{n-1}\}$ and exponents $\{1, p+1, \ldots, (n-1)p+1\}$.

REFERENCES

[1] N. Bourbaki, *Groupes et elgebres de Lie, chapitres 4, 5 et 6*, Hermann, Paris, 1968.

[2] P. Cartier, *Les arrangements d'hyperplans: un chapitre de geometrie combinatoire*, Seminaire Bourbaki **561**, 33e annee (1980/81).

[3] C. Greene, *Acyclic orientations (Notes)*, Higher Combinatorics (M. Aigner, ed.), Reidel, Bordrecht 1977, 65-68.

[4] M. Jambu and H. Terao, *Free arrangements of hyperplanes and supersolvable lattices*, Advances in Math. 52 (1984), 248-258.

[5] P. Orlik and L. Solomon, *Combinatorics and topology of complements of hyperplanes*, Inventiones Math. **56** (1980), 167-189.

[6] P. Orlik and L. Solomon, *Coxeter arrangements*, Arcata Symposium on Singularities (ed. P. Orlik), Proc. Symp. Pure Math. **40**, Part 2 (1983), 269-291.

[7] P. Orlik and L. Solomon, *Arrangements defined by unitary reflection groups*, Math. Annalen **261** (1982), 339-357.

[8] P. Orlik and L. Solomon, *The Hessian map in the invariant theory of reflection groups*, Nagoya Math. J., to appear.

[9] P. Orlik and L. Solomon, *Discriminants in the invariant theory of reflection groups*, Nagoya Math. J., to appear.

[10] P. Orlik, L. Solomon and H. Terao, *On Coxeter arrangements and the Coxeter number*, preprint.

[11] P. Orlik, L. Solomon and H. Terao, *Arrangements of hyperplanes and differential forms*, Proc. AMS-NSF Conference on Combinatorics and Algebra, Boulder 1983 (ed. C. Greene), Contemporary Mathematics **34** (1984), AMS, Providence R.I., 29-65.

[12] K. Saito, *Theory of logarithmic differential forms and logarithmic vector fields*, J. Fac. Science, University of Tokyo, Sci. IA **27** (1980), 265-291.

[13] G. C. Shephard and J. A. Todd, *Finite unitary reflection groups*, Can J. Math. **6** (1954), 274-304.

[14] L. Solomon and H. Terao, *A formula for the characteristic polynomial of an arrangement*, Advances in Mathematics, **64** (1987), 305-325..

[15] R. P. Stanley, *T-free arrangements of hyperplanes*, in: Progress in Graphy Theory (eds. J. A. Bondy and U. S. R. Murty), Academic Press (1984), 539.

[16] H. Terao, *Arrangements of hyperplanes and their freeness I*, J. Fac. Science, University of Tokyo, Sci. iA **27**(2) (1980), 293-312.

[17] H. Terao, *Generalized exponents of a free arrangement of hyperplanes and Shephard-Todd-Brieskorn formula*, Inventiones Math. **63** (1981), 159-179.

[18] H. Terao, *Free arrangements of hyperplanes and unitary reflection groups*, Proc. Japan Acad., Ser. A **56** (1980), 389-392.

[19] H. Terao, *The exponents of a free hypersurface*, Arcata Symposium on Singularities (ed. P. Orlik), Proc. Symp. Pure Math. **40**, Part 2 (1983), 561-566.

[20] H. Terao, *Forms with logarithmic pole and the filtration by the order of the pole*, Proc. International Symposium on Algebraic Geometry in Kyoto 1977, Kinokuniya, Tokyo, 1978, 673-685.

[21] T. Zaslavsky, *Facing up to arrangements: face-count formulas for partitions of space by hyperplanes*, Memoirs Amer. Math. Soc. **154** (1983).

[22] G. M. Ziegler, *Algebraic combinatorics of hyperplane arrangements*, Ph.D. Thesis, MIT, 1987.

[23] G. M. Ziegler, *Combinatorial construction of logarithmic differential forms*, Advances in Mathematics, to appear.

M.I.T.
CAMBRIDGE, MASSACHUSETTS 02139

Present address:
DEPARTMENT OF MATHEMATICS
AUGSBURG UNIVERSITY
D-8900 AUGSBURG, WEST GERMANY

ABCDEFGHIJ — 89